Universitext

For other titles in this series, go to
http://www.springer.com/series/223

J.M. Aarts

Plane and Solid Geometry

 Springer

J.M. Aarts
Delft University of Technology
Mediamatics
The Netherlands
j.a.aarts@ewi.tudelft.nl

Translator:
Reinie Erné
Leiden, The Netherlands
erne@math.leidenuniv.nl

Editorial board:
Sheldon Axler, San Francisco State University, San Francisco, CA, USA
Vincenzo Capasso, University of Milan, Milan, Italy
Carles Casacuberta, Universitat de Barcelona, Barcelona, Spain
Angus MacIntyre, Queen Mary, University of London, London, UK
Kenneth Ribet, University of California, Berkeley, CA, USA
Claude Sabbah, Ecole Polytechnique, Palaiseau, France
Endre Süli, Oxford University, Oxford, UK
Wojbor Woyczynski, Case Western Reserve University, Cleveland, OH, USA

ISBN: 978-0-387-78240-9   e-ISBN: 978-0-387-78241-6
DOI: 10.1007/978-0-387-78241-6

Library of Congress Control Number: 2008935537

Mathematics Subject Classification (2000): 51-xx

This is a translation of the Dutch, *Meetkunde*, originally published by Epsilon–Uitgaven, 2000.

© 2008 Springer Science+Business Media, LLC
All rights reserved. This work may not be translated or copied in whole or in part without the written permission of the publisher (Springer Science+Business Media, LLC, 233 Spring Street, New York, NY 10013, USA), except for brief excerpts in connection with reviews or scholarly analysis. Use in connection with any form of information storage and retrieval, electronic adaptation, computer software, or by similar or dissimilar methodology now known or hereafter developed is forbidden.
The use in this publication of trade names, trademarks, service marks, and similar terms, even if they are not identified as such, is not to be taken as an expression of opinion as to whether or not they are subject to proprietary rights.

Printed on acid-free paper

springer.com

To Robert J. and Lucie P.

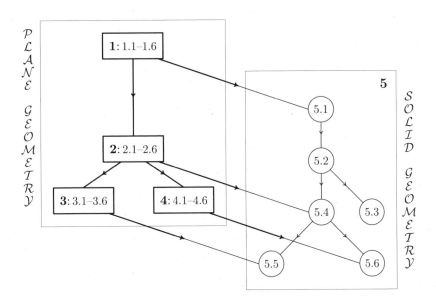

Preface

Nature and the world around us that we ourselves design, furnish, and build contain many geometric patterns and structures. This is one of the reasons that geometry should be studied at school. At first, the study of geometry is experimental. Results are taught and used in numerous examples. Only later do proofs come into play. But are these proofs truly necessary, or can we do without them? A natural answer is that every statement must be provided with a proof, because we want to know whether it is true. However, it is clear that the less experienced student may become frustrated by the presence of too many proofs. Only later will the student understand that proofs not only show the correctness of a statement, but also provide better insight into the relations among various properties of the objects that are being studied. Learning statements without proofs, you risk not being able to see the forest for the trees. For this reason, we will pay much attention to a careful presentation of proofs in this book. In the development of the theory of plane geometry there are, however, many tricky questions, especially at the beginning. The presentation of proofs at that stage is in general more concealing than revealing.

My first objective in writing this book has been to give an accessible exposition of the most common notions and properties of elementary Euclidean geometry in dimensions two and three. These include, in particular, special lines in triangles, congruence criteria, transformations, circles, and conics. All this can be found in the first hundred pages, Chapters 1 and 2. I also briefly discuss fractals and Voronoi diagrams, and give a detailed account of symmetry, cycloids, and notions of solid geometry. Chapters 1, 2, and 4 present a survey of the results of plane geometry at an intermediate level. Chapters 3 and 5 present more advanced topics. The first four chapters deal solely with plane geometry, while the fifth, and final, chapter discusses solid geometry.

Geometry is a useful subject with many applications. But what makes the study of geometry so captivating is the feeling of wonder that comes over you when you ponder questions such as, "Why are those three special lines concurrent?" and "Why do so many special points lie on the same circle?" the feeling that the world is truly a beautiful place! To mention just one example, the nine-point circle contains nine special points of a triangle (Example 2.46), touches both the incircle and the three excircles of the triangle (Theorem 4.30), and is the incircle of the deltoid that is the envelope of the Simson lines of the triangle (Theorem 4.58). In choosing the topics for this book I used the following criteria: is a topic useful, it is surprising, or is it in vogue.

When I started writing this book I aimed to present a complete account of Euclidean geometry in dimensions two and three. But soon I discovered that I could not discuss the foundations of geometry without losing momentum. So I decided to start the discussion at an intermediate stage and to make a special choice of basic assumptions on which to build the theory. Let me briefly mention these assumptions. The first basic assumption is that the plane admits a distance function that assigns a distance to any two points. Using this distance function we can define straight lines. We then make the basic assumption that every straight line is an isometric copy of the real line. On the basis of this assumption we can use various properties of the real numbers to obtain rather simple proofs of results whose traditional proofs are often quite complicated. The next basic assumption is about the existence of parallel lines. Another basic assumption concerns the existence of perpendicular lines; whether two given lines are perpendicular is decided with the help of the inverse Pythagorean theorem. The last basic assumption concerns the measurement of angles. Starting with these basic assumptions, we develop plane geometry. The introduction of local coordinates becomes relatively easy. Needless to say, the use of coordinates simplifies many proofs. Our presentation of the results in this book is not always strictly sequential. Several properties of the nine-point circle, for example, are discussed before the definitions of the circle and circumcircle are given.

With the approach I just mentioned, we can introduce the basic notions and elementary properties of figures in plane geometry in a way that is brief and to the point. We do this in Chapter 1, which also contains a survey of the properties required to read the subsequent chapters.

In Chapter 2, we study distance-preserving maps of the plane, also called isometries. We show that every isometry is a reflection in a line, a translation, a rotation, or a glide reflection. We also introduce the notion of congruence, and give the classic congruence criteria. Next, we give a detailed exposition of the notion of orientation. Finally, we discuss similarities and their role in the theory of fractals.

The subsequent chapters of the book can be read independently of each other. The diagram on page ii shows the relations between various parts of the book. Chapter 3 starts with the study of the symmetry groups of a number of simple plane figures. The main object of the chapter is the study of frieze patterns and periodic tilings, which we classify according to their symmetry groups. To analyze these, we use Voronoi diagrams, which also provide an elegant way of introducing conic sections.

In Chapter 4, we study various curves, namely the circle, conic sections, and cycloids. In the first two sections we give a detailed account of the most important properties of the circle and of the trigonometric functions. This is followed by a discussion of some unexpected applications of the inversion of the plane in a circle. We then turn to the conic sections. In order to obtain a transparent derivation of the equation of a tangent to a conic, we introduce the notion of polarity, for which we need homogeneous coordinates. The last section of the chapter concerns cycloids. We explain the relation between three seemingly unrelated topics: the nine-point circle, the Simson line, and the deltoid.

Chapter 5 presents a bird's-eye view of several topics of solid geometry. We first briefly cover the subjects of the previous chapters, adapted to solid geometry. One of the sections is devoted to a discussion of several ways to make drawings of solid figures. Using reflections in a plane, we can list the types of isometries of Euclidean 3-space: reflections in a plane, translations, rotations, glide reflections, reflections in a point, improper rotations, and screws. In our review of the symmetry of solid figures, we discuss all regular and two semiregular polyhedra. This discussion results in a list of all finite subgroups of the group of direct isometries of 3-space. In the final section of the chapter, we study quadrics. In particular, we consider straight lines on quadrics, the relation between quadrics and conic sections, and the classification of quadrics. The Hessian, which is used in calculus to study stationary points, is mentioned as an application of the latter. We conclude the book with an appendix listing the basic assumptions for both plane and solid geometry.

Math books are not like novels. Reading them becomes fun only when you take a pen and paper and work out the results yourself. For this reason I have included some two hundred exercises in this book. It was not my intention to include brainteasers, but some time and effort will undoubtedly be required to solve the problems. There is no need to become frustrated by this, for devoting time and effort to problemsolving is rewarding. The exercises are related to the material of the section they follow. The solution of an exercise is never tricky, but mostly rather straightforward. Most exercises are just statements, without phrases such as *prove this* or *show that*. The reader is invited to provide a proof of the statement. If an exercise turns out to be too difficult, it should be skipped and reexamined at a later moment. A similar remark applies to the proofs that are presented in this book. Quite often, they look much harder at first reading than they really are. Frequently in mathematics, a problem or proof that was obscure yesterday can be grasped today, and will be trivial tomorrow.

In writing this book, I used the notes from several courses and talks I have given in Delft and in Amsterdam. Parts of Chapters 2 and 3 were developed for a course given to architecture students, and parts of Chapters 4 and 5 were for a course for mechanical-engineering students, both in Delft. Certain subjects, in particular Voronoi diagrams, conics, quadrics, and fractals, were used in courses for secondary-school teachers.

I received help from different directions. Agnes Verweij was coauthor of the material for the course for mechanical-engineering students. She also read through my contributions to the courses for secondary-school teachers; her comments have always been very helpful. K.P. Hart stood by to help on any LaTeX problems I might encounter. He also taught me to work with `mfpic`, the program developed by Tom Leathrum, which I used to make the figures for this book. Eva Coplakova conscientiously worked her way through the whole manuscript, including all the problems; her remarks were very useful. I would like to thank as well everyone else who has helped me.

November 2007

Jan Aarts
Delft

Contents

1 PLANE GEOMETRY 1
   1.1 The Pythagorean Theorem 2
   Exercises 7
   1.2 Metrics and Metric Spaces 8
   Exercises 10
   1.3 Isometries 11
   Exercises 14
   1.4 The Basic Assumptions of Plane Geometry 15
   Exercises 22
   1.5 Local Coordinates and the Equation of a Line 22
   Exercises 30
   1.6 The Inner Product and Determinant 31
   Exercises 36

2 TRANSFORMATIONS 37
   2.1 The Reflection in a Line, Congruence 38
   Exercises 44
   2.2 Angle Measure, Orientation 45
   Exercises 53
   2.3 The Reflection in a Point 53
   Exercises 60
   2.4 Translation and Rotation Groups 61
   Exercises 75
   2.5 Dilation and Similarity 76
   Exercises 84
   2.6 Fractals, Dimension, and Measure 86
   Exercises 93

3 SYMMETRY 95
   3.1 What Is Symmetry? 96
   Exercises 101

xiv   Contents

    3.2   There Are Seven Types of Frieze Patterns 102
    Exercises 108
    3.3   Voronoi Diagrams 109
    Exercises 116
    3.4   Voronoi Cells in Lattices 117
    Exercises 124
    3.5   Periodic Tilings 125
    Exercises 133
    3.6   All Types of Periodic Tilings 133
    Exercises 144

4   CURVES 145
    4.1   Circles, Powers, and Cyclic Quadrilaterals 146
    Exercises 161
    4.2   Trigonometric Functions and Polar Coordinates 163
    Exercises 171
    4.3   Turning the Plane Inside Out 175
    Exercises 183
    4.4   Conics 185
    Exercises 200
    4.5   Homogeneous Coordinates and Polarity 203
    Exercises 222
    4.6   The Parametric Equations of a Curve, Cycloids 225
    Exercises 249

5   SOLID GEOMETRY 255
    5.1   The Basic Assumptions of Solid Geometry 256
    Exercises 265
    5.2   Local Coordinates, the Inner and Outer Products 266
    Exercises 275
    5.3   What Exactly Do I See? 277
    Exercise 284
    5.4   Transformations of Three-Space 284
    Exercises 297
    5.5   Symmetry and Regular Polyhedra 298
    Exercises 313
    5.6   Quadrics and Ruled Surfaces 314
    Exercises 330

A   Basic Assumptions 333

References 337

Index 341

1
PLANE GEOMETRY

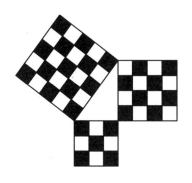

The aim of this book is to tell you something about geometry. How should I begin? Traditionally, geometry was set up in an axiomatic way; it was built up in a strict manner from certain properties of points and lines that are evident to all. For over two thousand years the *Elements* of Euclid [18] was the only example of an axiomatic treatment of Euclidean geometry. A nice and modern discussion of it is presented in [26]. See also [22] and [56]. Most axiomatic presentations of Euclidean geometry start out with the axiom system introduced by Hilbert. This is a rather long list of properties involving incidence, order, congruence, continuity, and parallelism. The order and continuity properties are related to properties of the real line as we learn them in analysis. It is a quite intricate and elaborate path from these axioms to some down-to-earth properties.

For this book I have adopted a middle course between a synthetic and analytic approach. Geometry deals with the measurement and computation of lengths of line segments, areas of figures, and sizes of angles. In the word *geometry*, *geo* refers to earth, and *metry* to measurement: geometry is originally land surveying. See, e.g.,[27]. Nowadays, in the usual, or Euclidean, geometry the plane and three-space are often described as special metric spaces. We will follow this approach. The second section of this chapter deals with metric spaces.

The setup we have chosen for *plane geometry* is somewhat unconventional. One of the starting points of our theory will be that each straight line in the plane or three-space looks like the real line that is encountered in analysis, including its order and continuity properties. A central notion in geometry is congruence. Isometries of metric spaces will take over the role of congruences. In Chap. 2 we will take a closer look at the set of all isometries of the plane.

1 PLANE GEOMETRY

In the current chapter, we will define the notion of *perpendicular lines* using the Pythagorean theorem and its converse. These results may be not evident to all, but almost every student will be familiar with them; in the first section we recall the corresponding theory. A great advantage of this method is that it makes the introduction of coordinates in the plane quite easy. That is done in Sect. 1.5, where the coordinate plane and vectors are defined. Section 1.6 deals with the inner product of vectors. These sections are meant mainly to give a short account of a number of subjects from *vector geometry* that will be used later on in this book; the proofs and deductions are often much simpler using coordinates. We cover much material in the first one and a half chapters of this book, including congruences and many down-to-earth properties of triangles and quadrilaterals.

1.1 The Pythagorean Theorem

The best known theorem in geometry is the Pythagorean theorem: *if a and b are the legs of a right triangle, and c is the hypotenuse, then*

$$\boxed{a^2 + b^2 = c^2 \,.} \tag{1.1}$$

Figure 1.1 shows a triangle ABC with right angle at C. We often denote the

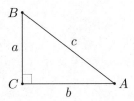

Fig. 1.1. The Pythagorean theorem

side opposite a vertex by the corresponding lowercase letter. Sides a and b are called the *legs* of the triangle, and side c the *hypotenuse*. Mathematicians, however, sometimes use the same symbol for different objects. For example, AB denotes not only the line through A and B, but also the line segment between A and B, and the length of that segment. The letter a denotes the side opposite A and also, for example in (1.1), the length of that side. Although this somewhat disagreeable habit may seem to lead to confusion, in general the meaning is clear from the context.

The present-day reader is no doubt so well versed in working with symbols that he or she sees a^2 (a squared) as a number, but in geometry it is useful to see a^2 as the area of the square with side a. In daily life, too, the word *square* first refers to the geometric object. One of the visual proofs of the Pythagorean theorem is based on this idea; see Fig. 1.2. Within a square with

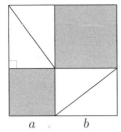 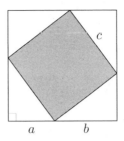

Fig. 1.2. Proof of the Pythagorean theorem

side $a+b$, four congruent right triangles with legs of lengths a and b and hypotenuse of length c are placed in two different configurations. The areas of the shaded parts in the left and right figures are therefore equal; consequently, $a^2 + b^2 = c^2$. We can now also state the Pythagorean theorem as follows: *the area of the square on the hypotenuse of a right triangle is equal to the sum of the areas of the squares on the legs*; see Fig. 1.4, on the left.

The story goes [64, pp. 3–4] that Einstein (1879–1955) found a proof of the Pythagorean theorem at the age of eleven. His proof is based on the following property: if two figures are similar, then their areas are in the same

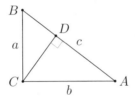

Fig. 1.3. Pythagoras according to Einstein

ratio as the squares of the lengths of corresponding line segments. In the right triangle ABC, Fig. 1.3, CD is the altitude from C onto the hypotenuse AB. The triangles CBD, ACD, and ABC are similar, because the corresponding angles are congruent. Let a, b, and c be the respective hypotenuses of these triangles, and E_a, E_b, and E_c the respective areas. These areas are in the same ratio as the squares of the lengths of the hypotenuses; consequently,

$$E_a = ma^2, \qquad E_b = mb^2, \qquad E_c = mc^2,$$

where m is some positive real number. Since $E_a + E_b = E_c$, it follows that

$$ma^2 + mb^2 = mc^2.$$

Dividing by m gives (1.1).

The proof given by young Einstein shows more than just the Pythagorean theorem. When similar figures are placed on the sides of a right triangle in

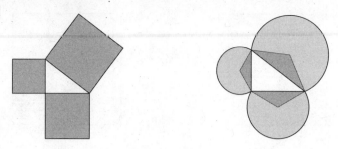

Fig. 1.4. Similar figures on a right triangle

such a way that the sides of the triangle take corresponding positions in these figures, then the sum of the areas of the figures on the legs is equal to the area of the figure on the hypotenuse; see Fig. 1.4.

In the school of Pythagoras (ca. 569–475 BC), triples (a, b, c) of positive *integers* for which (1.1) holds were considered very special. We call them *Pythagorean triples*. Examples are $(3, 4, 5)$, $(5, 12, 13)$, $(7, 24, 25)$, and $(8, 15, 17)$. Of course $(6, 8, 10)$ is also a Pythagorean triple, but it can be reduced to the first triple mentioned here by dividing each of its elements by 2. When no such reduction is possible, the triple is called *primitive*, or *reduced*. Let s and t be positive integers with $s > t$. Then $(2st, s^2 - t^2, s^2 + t^2)$ is a

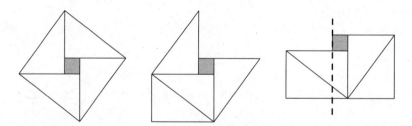

Fig. 1.5. Another proof of the Pythagorean theorem

Pythagorean triple, as one can easily check. It is slightly more difficult to show that every *primitive* Pythagorean triple has this form; see for example [25, Theorem 225]. At some point, one of the visual representations of the Pythagorean theorem was given the nickname the *bride's chair*. In Fig. 1.4, on the left, you can imagine a person carrying the bride's chair on his back. This figure also symbolizes marriage: when taken together, the two smaller squares equal the larger square. For more details, see [18, Book I, p. 417].

The converse of the Pythagorean theorem also holds: *if the equality $a^2 + b^2 = c^2$ holds in triangle ABC, then C is a right angle*. This theorem is used by carpenters and tilers when setting out right angles. Road workers also use the theorem, calling it the 3-4-5 method. This method for making a right-angle gauge (try square) was already described by Vitruvius in *Ten Books on*

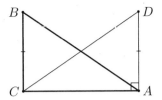

Fig. 1.6. If $a^2 + b^2 = c^2$, then C is a right angle

Architecture [75, Book IX, 6–7], where he tells how Pythagoras used three rods to make a right angle. In Definition 1.22, we will use the 3-4-5 method in an abstract way to define perpendicular lines.

We now give a proof of the converse of the Pythagorean theorem; see Fig. 1.6. We are given a triangle ABC for which $CA^2 + CB^2 = AB^2$. Let us choose a point D such that DA is perpendicular to CA and $AD = CB$. The Pythagorean theorem applied to triangle CAD shows that

$$CD^2 = CA^2 + AD^2 = CA^2 + CB^2 \, .$$

By assumption, the right-hand side is equal to AB^2. Hence $CD = AB$. Consequently, the triangles ACB and CAD have congruent corresponding sides, and are therefore congruent themselves. In particular, $\angle CAD = \angle ACB$, which implies that $\angle C$ is a right angle.

Some of the figures in this section can be used to show the correctness of algebraic formulas. In Fig. 1.7, on the left, we have an illustration from Fig. 1.2 and in the middle one from Fig. 1.5. The figure on the left shows the correctness of the well-known formula

$$(a+b)^2 = a^2 + 2ab + b^2$$

for positive a and b. In the figure in the middle, the side of the shaded square

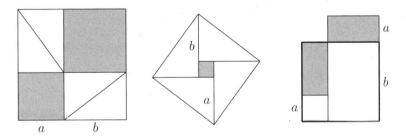

Fig. 1.7. Illustration of a number of formulas

is equal to $b - a$ and the area of the large square is equal to $a^2 + b^2$. This shows that

$$(a-b)^2 = a^2 - 2ab + b^2$$

for positive a and b. The figure on the right shows that
$$b^2 - a^2 = (b-a)(b+a)$$
for positive a and b.

Various properties of triangles can be shown through computations using the Pythagorean theorem. Let us give an example: consider Fig. 1.8. In triangle ABC, AD is the altitude from A onto side BC; the point D lies between B and C. Likewise, BE is the altitude from B, and E lies between A and C. In this situation, we have
$$AB^2 = AC \times AE + BC \times BD .$$

Let us deduce this formula. By applying the Pythagorean theorem to triangles

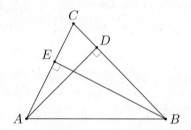

Fig. 1.8. $AB^2 = AC \times AE + BC \times BD$

BEC and ADC, we obtain
$$BC^2 = (AC - AE)^2 + BE^2 = AC^2 + AE^2 - 2AC \times AE + BE^2$$
and
$$AC^2 = (BC - BD)^2 + AD^2 = BC^2 + BD^2 - 2BC \times BD + AD^2 .$$
These formulas give
$$2AC \times AE + 2BC \times BD = AE^2 + BE^2 + BD^2 + AD^2 .$$

Applying the Pythagorean theorem to triangles AEB and ADB leads to the formula mentioned above.

In this section we have shown a number of properties of triangles. Each time, we started out from properties that were either already proved or assumed known, and we deduced the desired property following certain logical principles. Since we can't keep deducing properties from one another without beginning somewhere, it makes sense to fix a number of basic assumptions. In Sect. 1.4 we will state the basic assumptions for plane geometry. The Pythagorean theorem and its converse play a special role in this: the statements of these theorems are used to fix the notion of perpendicular.

Exercises

1.1. Figure 1.5 illustrates another proof of the Pythagorean theorem. Write the corresponding text.

1.2. Let ABC be a triangle with sides $a = 13$, $b = 14$, $c = 15$. This triangle appeared already in Leonardo of Pisa's *De Practica Geometrie* [32]. At first glance, there is nothing special about it, but a closer look shows that it has very simple numerical properties [20, p. 235]. Show that the length of the altitude BD is 12 and the radius of the inscribed circle, or incircle, is 4.
Hint for the computation of the radius of the incircle: the area of the triangle is equal to the sum of the areas of the three triangles obtained by joining the center of the incircle to the vertices of the triangle.

1.3. The figure on the right comes from the book *Ideën [Ideas]* by Multatuli (1820–1887); on p. 529 we read, "Recently, I discovered a new proof of the Pythagorean theorem. Here it is. By constructing six triangles as in the example on the right, each equal to the given right triangle, we obtain two equal squares, A B and C D."
Complete the proof.

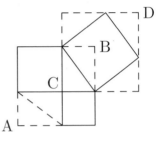

1.4. Consider a circle $\odot(N_1, r_1)$, with center N_1 and radius r_1, and a circle $\odot(N_2, r_2)$. We say that the circles are *orthogonal* if the circles intersect and the tangent lines at each intersection point are perpendicular to each other. Show that the circles $\odot(N_1, r_1)$ and $\odot(N_2, r_2)$ are orthogonal if and only if

$$r_1^2 + r_2^2 = N_1 N_2^2 .$$

Hint: The tangent to $\odot(N, r)$ at a point S is perpendicular to the radius NS.

1.5. The three circles in the figure on the right are tangent to each other and to a straight line. Let r_1 be the radius of the circle on the left, and r_2 that of the circle on the right. The radius of the shaded circle in the middle will be called r_3. We then have

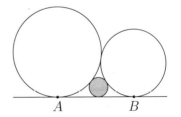

$$\frac{1}{\sqrt{r_3}} = \frac{1}{\sqrt{r_1}} + \frac{1}{\sqrt{r_2}} .$$

Hint: $AB = 2\sqrt{r_1 r_2}$.

1.2 Metrics and Metric Spaces

In Sect. 1.4 we will begin listing the basic assumptions of plane geometry. In later sections we will deduce the remaining properties of plane geometry from these. The first basic assumption is that the Euclidean plane is a metric space. By this we mean the following. For every pair of points X and Y, the distance from one to the other is given; we are not interested in how it is computed or in what units it is given. We denote the distance between two points X and Y by $d(X,Y)$, or simply by XY, and call d the *metric*.

Properties 1.1. *The metric d has the following properties: for every X, Y, Z in the plane, we have*

1. $d(X,Y) \geq 0$;
2. $d(X,Y) = 0$ if and only if $X = Y$;
3. $d(X,Y) = d(Y,X)$;
4. $d(X,Z) \leq d(X,Y) + d(Y,Z)$.

The last property in this list is called the *triangle inequality*. This name comes from the following theorem in plane geometry: Let X, Y, and Z be the vertices of a triangle. Then the length of any side of the triangle, for example XZ, is less than or equal to the sum of the lengths of the other two sides. We have equality if Y lies on the line segment XZ, in which case we call the triangle degenerate. The third property states that the metric is *symmetric*.

Let us give a general definition.

Definition 1.2. *A metric space is a set M with a metric d. We call the elements of M the points of the metric space. The metric is a distance function satisfying the four Properties 1.1, for every X, Y, and Z in M.*

We will denote the points of a metric space by capital letters.

By assumption, the Euclidean plane is a metric space. Let us consider two other examples that we will use frequently.

The Real Line

We denote the set of real numbers by $\mathbb{R}$. Let us fix the following notation: if an element of $\mathbb{R}$ is seen as a point, we use a capital letter; if it is seen as a number, we use a lowercase letter. Let us immediately add that it is impossible to follow this rule consistently. We already deviate from it in the following definition.

We define a metric on $\mathbb{R}$ by setting the distance from a point X to a point Y equal to the absolute value of the difference of X and Y:

$$d(X,Y) = |X - Y|.$$

By definition,

$$|X-Y| = \begin{cases} X - Y \text{ if } X - Y \geq 0, \\ Y - X \text{ if } X - Y < 0. \end{cases}$$

Is d a metric? That is, does d satisfy the four Properties 1.1? It is immediately clear that it satisfies the first two properties. The third property follows from the fact that every real number a satisfies $|a| = |-a|$. To verify the last property, we first note that all real numbers a and b satisfy

$$-|a| \leq a \leq |a| \quad \text{and} \quad -|b| \leq b \leq |b|.$$

Adding these equations together gives

$$-(|a| + |b|) \leq a + b \leq |a| + |b| .$$

By the definition of the absolute value, this last formula implies that

$$|a + b| \leq |a| + |b| .$$

Setting $a = X - Y$ and $b = Y - Z$, we obtain property 4. The metric defined this way is called the *Euclidean metric* on $\mathbb{R}$.

In addition to the triangle inequality, we also have another one that comes up regularly, and is often mentioned at the same time. In plane geometry it reads: the length of any side of a triangle is greater than or equal to the difference of the lengths of the other two sides. In general metric spaces it becomes the following inequality.

Theorem 1.3. *For all points X, Y, and Z of a metric space M, we have*

$$|d(X,Y) - d(Y,Z)| \leq d(X,Z) .$$

Proof. Let X, Y, and Z be three points. By the triangle inequality, we have

$$d(X,Y) \leq d(X,Z) + d(Z,Y) \quad \text{and} \quad d(Y,Z) \leq d(Y,X) + d(X,Z) .$$

By the symmetry of the distance function, this implies that

$$d(X,Y) - d(Y,Z) \leq d(X,Z) \quad \text{and} \quad d(Y,Z) - d(X,Y) \leq d(X,Z) .$$

The desired formula now follows from the definition of the absolute value.

The Discrete Metric

In general we can define more than one metric on any given set. The following example, applied to $\mathbb{R}$, illustrates this.

Example 1.4. Let M be an arbitrary nonempty set, for example a subset of $\mathbb{R}$ or a subset of the Euclidean plane. For X and Y in M, let

$$\rho(X,Y) = \begin{cases} 1 & \text{if } X \neq Y, \\ 0 & \text{if } X = Y. \end{cases} \tag{1.2}$$

We will show that ρ is a metric; consequently, M and ρ form a metric space. It is clear that properties 1, 2, and 3 of Properties 1.1 hold for ρ. We will prove the triangle inequality by contradiction. Assume that the triangle inequality does not hold. Then there exist points X, Y, and Z such that

$$\rho(X,Z) > \rho(X,Y) + \rho(Y,Z). \tag{1.3}$$

This inequality can hold only if $\rho(X,Z) = 1$, $\rho(X,Y) = 0$, and $\rho(Y,Z) = 0$. The last two equalities imply that $X = Y$ and $Y = Z$, hence also $X = Z$. But in that case, $\rho(X,Z) = 0$, which contradicts $\rho(X,Z) = 1$. Since the contradiction follows from the assumption that the triangle inequality does not hold for ρ, this assumption is incorrect, and the triangle inequality holds.

The metric defined by (1.2) is called the *discrete* metric on M. The adjective *discrete* refers to the fact that unlike the Euclidean metric on $\mathbb{R}$, the discrete metric does not allow distinct points with "small distance" between them. Indeed, distinct points of M cannot "lie close to each other," since the distance between two points is either 0 or 1.

Unless stated otherwise, we will always assume that $\mathbb{R}$ is endowed with the Euclidean metric.

Exercises

1.6. For any X and Y in $M = (0, \infty)$, let $\rho(X,Y) = |Y^2 - X^2|$. Show that ρ is a metric.

1.7. The *knight's distance* d_S between two squares on a chess board is defined to be the smallest number of jumps needed for a knight to go from the one square to the other.

(a) Show that d_S is a metric.
(b) Determine $d_S(\text{a1}, \text{a2})$, $d_S(\text{a2}, \text{a3})$, $d_S(\text{a2}, \text{b2})$, $d_S(\text{a2}, \text{c2})$, $d_S(\text{a1}, \text{b2})$, $d_S(\text{b2}, \text{c3})$.
(c) Show that $d_S(\text{a1}, \text{b2}) = d_S(\text{a1}, \text{g7})$.
(d) Can we define the *bishop's distance* in a similar manner?

1.8. Let A, B, C, and D be points in a metric space M with metric d. The following quadrilateral inequality holds:

$$d(A,D) \leq d(A,B) + d(B,C) + d(C,D).$$

1.9. Let X, Y, P, and Q be points in a metric space M with metric d. We have

$$|d(X,P) - d(Y,Q)| \leq d(X,Y) + d(P,Q).$$

1.3 Isometries

Congruence plays an important role in geometry: two figures are said to be *congruent* if there exists an *isometry*, that is, a bijective distance-preserving map, that transforms one figure into the other. Congruence preserves almost all properties. We will study what it means for plane geometry in the next chapter.

Definition 1.5. *Let M be a metric space with metric d. A map $\mathcal{F}$ from M to M is called an* isometry *(of M) if*

1. *$\mathcal{F}$ is surjective, that is, for every Y in M, there exists an X in M such that $\mathcal{F}(X) = Y$;*
2. *for every X and Y in M, we have $d\left(\mathcal{F}(X), \mathcal{F}(Y)\right) = d(X, Y)$.*

Property 2 says that an isometry is *distance-preserving*: the distance from X to Y is equal to that from $\mathcal{F}(X)$ to $\mathcal{F}(Y)$. The prefix *iso* in isometry means equal. Before describing isometries for the examples in the last section, let us first note the following.

Theorem 1.6. *Every isometry of a metric space M is bijective.*

Proof. Let $\mathcal{F}$ be an isometry of M. A map from M to M is called *bijective* if it is injective and surjective. By definition, the isometry $\mathcal{F}$ is surjective. A map $\mathcal{G}$ from M to M is called *injective* if for every X and Y in M, $X \neq Y$ implies $\mathcal{G}(X) \neq \mathcal{G}(Y)$. Now, if $X \neq Y$, then $d(X, Y) > 0$. Since the isometry $\mathcal{F}$ preserves distances, we also have $d\left(\mathcal{F}(X), \mathcal{F}(Y)\right) > 0$, and therefore $\mathcal{F}(X) \neq \mathcal{F}(Y)$. Consequently $\mathcal{F}$ is injective.

Let us first consider the second example of the last section, a set M with the discrete metric. We claim that every bijection $\mathcal{F}$ from M to M is an isometry. Indeed, as a bijection, $\mathcal{F}$ is surjective and injective. Only the second property from the definition of isometry needs further checking. Since $\mathcal{F}$ is injective, every X and Y in M satisfy $X \neq Y$ if and only if $\mathcal{F}(X) \neq \mathcal{F}(Y)$. Consequently $d(X, Y) = 1$ if and only if $d\left(\mathcal{F}(X), \mathcal{F}(Y)\right) = 1$. Since the metric is discrete, we also have $d(X, Y) = 0$ if and only if $d\left(\mathcal{F}(X), \mathcal{F}(Y)\right) = 0$. It follows that $\mathcal{F}$ preserves distances. Conversely, the last theorem shows that every isometry is bijective. This completes the proof of the following statement.

Theorem 1.7. *In a metric space M with the discrete metric, the isometries of M correspond to the bijections.*

Let us describe all isometries of the metric space $\mathbb{R}$ (with the Euclidean metric). What are the isometries of this space? It is immediately clear that not every bijection of $\mathbb{R}$ is an isometry. The function defined by $\mathcal{F}(X) = 2X$ is a bijection, but not an isometry. On the other hand, it is clear that the *identity map* $\mathrm{id}_\mathbb{R}$, given by $\mathrm{id}_\mathbb{R}(X) = X$ for all X in $\mathbb{R}$, is an isometry. Let us consider translations. For a real number a, the *translation over a* is given by

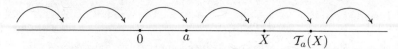

Fig. 1.9. Translation over a

$\mathcal{T}_a(X) = X + a$ for all X in $\mathbb{R}$. The definition of the translation uses the fact that the point X is a real number that can be used in computations. Note that
$$d\left(\mathcal{T}_a(X), \mathcal{T}_a(Y)\right) = |\mathcal{T}_a(X) - \mathcal{T}_a(Y)| = |X - Y| = d(X, Y)$$
for all X and Y in $\mathbb{R}$. A translation is therefore an isometry.

If $\mathcal{T}_b$ is the translation over b, then the *composition* of $\mathcal{T}_a$ and $\mathcal{T}_b$ is the map $\mathcal{T}_a \circ \mathcal{T}_b$ that arises by applying the translations one after the other, with $\mathcal{T}_a$ following $\mathcal{T}_b$: $\mathcal{T}_a \circ \mathcal{T}_b(X) = \mathcal{T}_a\left(\mathcal{T}_b(X)\right)$. We have
$$\mathcal{T}_a \circ \mathcal{T}_b(X) = \mathcal{T}_a\left(\mathcal{T}_b(X)\right) = \mathcal{T}_a(X + b) = (X + b) + a = X + (a + b) \,.$$

The composition of the translations over a and b is also a translation, over $a+b$; it is therefore also an isometry. *Note that the identity map, which maps every point of $\mathbb{R}$ to itself, is a translation*: $\mathrm{id}_\mathbb{R} = \mathcal{T}_0$.

We have not found all isometries of $\mathbb{R}$ yet. There are also the reflections in a point. Let P be a point of $\mathbb{R}$. The *reflection in* P is given by $\mathcal{S}_P(X) = 2P - X$ for all X in $\mathbb{R}$. Note that $\mathcal{S}_P(P) = P$. The formula for $\mathcal{S}_P$ implies that for every X in $\mathbb{R}$,
$$\mathcal{S}_P(X) - P = P - X \,.$$
Consequently, $d\left(\mathcal{S}_P(X), P\right) = d(P, X)$; the points $\mathcal{S}_P(X)$ and X are equidistant from P, but lie on different sides of it. This is why the map $\mathcal{S}_P$ is called the reflection in P. We have

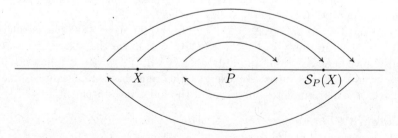

Fig. 1.10. Reflection in P

$$\begin{aligned} d\left(\mathcal{S}_P(X), \mathcal{S}_P(Y)\right) &= |\mathcal{S}_P(X) - \mathcal{S}_P(Y)| \\ &= |(2P - X) - (2P - Y)| = |Y - X| = d(X, Y) \end{aligned}$$

for all X and Y in $\mathbb{R}$. Consequently, a reflection in a point is an isometry. The *composition* $\mathcal{S}_P \circ \mathcal{S}_Q$ is the map that arises by applying the reflections one after the other, with $\mathcal{S}_P$ following $\mathcal{S}_Q$:

$$\mathcal{S}_P \circ \mathcal{S}_Q(X) = \mathcal{S}_P\left(\mathcal{S}_Q(X)\right)$$
$$= \mathcal{S}_P(2Q - X) = 2P - (2Q - X) = X + 2(P - Q).$$

To our surprise, the composition $\mathcal{S}_P \circ \mathcal{S}_Q$ of the reflections in P and Q is a translation over $2(P - Q)$. This translation is in the direction from Q to P: if P lies to the right of Q, that is, $Q < P$, then the translation is toward the right; if P lies to the left of Q, that is, $P < Q$, then the translation is toward the left.

Do we have all isometries of $\mathbb{R}$? This is indeed the case, as we will now prove. The main ingredient of the proof is the fact that an isometry of $\mathbb{R}$ is completely determined by its action on any two distinct points. We state this more carefully in the following lemma.

Lemma 1.8. *Let X, Y, P, and Q be points of $\mathbb{R}$ such that $d(X,Y) = d(P,Q) > 0$. Then there is at most one isometry $\mathcal{F}$ with $\mathcal{F}(X) = P$ and $\mathcal{F}(Y) = Q$.*

Proof. We must show that if $\mathcal{F}$ and $\mathcal{G}$ are isometries with $\mathcal{F}(X) = P = \mathcal{G}(X)$ and $\mathcal{F}(Y) = Q = \mathcal{G}(Y)$, then $\mathcal{F} = \mathcal{G}$. What exactly does $\mathcal{F} = \mathcal{G}$ mean? It means that $\mathcal{F}(Z) = \mathcal{G}(Z)$ for every Z in $\mathbb{R}$. We will prove the equality $\mathcal{F} = \mathcal{G}$ by contradiction. Let us assume that $\mathcal{F} \neq \mathcal{G}$. There then exists a Z in $\mathbb{R}$ with $\mathcal{F}(Z) \neq \mathcal{G}(Z)$. Since $\mathcal{F}$ and $\mathcal{G}$ are isometries, we have

$$d\left(P, \mathcal{F}(Z)\right) = d\left(\mathcal{F}(X), \mathcal{F}(Z)\right) = d(X, Z),$$
$$d\left(P, \mathcal{G}(Z)\right) = d\left(\mathcal{G}(X), \mathcal{G}(Z)\right) = d(X, Z),$$

and we conclude that $d\left(P, \mathcal{F}(Z)\right) = d\left(P, \mathcal{G}(Z)\right)$. Consequently, P is the midpoint of the line segment with endpoints $\mathcal{F}(Z)$ and $\mathcal{G}(Z)$: $P = (\mathcal{F}(Z) + \mathcal{G}(Z))/2$. An analogous computation shows that $Q = (\mathcal{F}(Z) + \mathcal{G}(Z))/2$. But then $P = Q$, which contradicts the assumption that $d(P, Q) > 0$. We conclude that $\mathcal{F} = \mathcal{G}$.

Let us now show that every isometry of $\mathbb{R}$ is either a translation or a reflection; see Fig. 1.11. Let $\mathcal{F}$ be an isometry. Let X and Y be distinct points

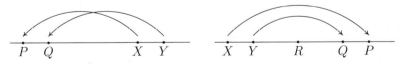

Fig. 1.11. Every isometry of $\mathbb{R}$ is a translation or a reflection

with $X < Y$, and let $P = \mathcal{F}(X)$ and $Q = \mathcal{F}(Y)$. Then $d(X,Y) = d(P,Q)$. We distinguish two cases:

1. $P < Q$ (on the left in the figure): let $a = P - X$. The translation $\mathcal{T}_a$ maps X and Y onto, respectively, P and Q. By the lemma we therefore have $\mathcal{F} = \mathcal{T}_a$.
2. $P > Q$ (on the right in the figure): let $R = (X + P)/2$. We then also have $R = (Y + Q)/2$. We can easily check that the reflection $\mathcal{S}_R$ maps X and Y onto, respectively, P and Q. By the lemma above we therefore have $\mathcal{F} = \mathcal{S}_R$.

The only argument used in this proof is that $d(X, Y) = d(P, Q)$. Consequently, we have proved the following theorem.

Theorem 1.9. *Let X, Y, P, and Q be points of $\mathbb{R}$ such that $d(X,Y) = d(P,Q) > 0$. There is exactly one isometry $\mathcal{F}$ for which $\mathcal{F}(X) = P$ and $\mathcal{F}(Y) = Q$. This isometry is either a translation or a reflection.*

We conclude this section with a remark for later use, concerning the composition of more than two maps.

Remark 1.10. The *composition* of two maps $\mathcal{F}$ and $\mathcal{G}$ from M to M is given by $\mathcal{F} \circ \mathcal{G}(X) = \mathcal{F}(\mathcal{G}(X))$. We first apply the map $\mathcal{G}$ (because $\mathcal{G}$ is the closest to the argument X) and then $\mathcal{F}$. The map $\mathcal{F} \circ \mathcal{G}$ is called $\mathcal{F}$ *of* $\mathcal{G}$, $\mathcal{F}$ *after* $\mathcal{G}$, or $\mathcal{F}$ *composed with* $\mathcal{G}$. In the exercises we will see that the order of the maps is important; in general, a different order gives a different map. When we have three maps $\mathcal{F}$, $\mathcal{G}$, and $\mathcal{H}$ from M to M, the *composition* of these maps is given by

$$\mathcal{F} \circ \mathcal{G} \circ \mathcal{H}(X) = \mathcal{F}(\mathcal{G}(\mathcal{H}(X))) \ . \tag{1.4}$$

The composition of maps is *associative*:

$$(\mathcal{F} \circ \mathcal{G}) \circ \mathcal{H} = \mathcal{F} \circ (\mathcal{G} \circ \mathcal{H}) = \mathcal{F} \circ \mathcal{G} \circ \mathcal{H} \ .$$

In words: the result of the composition does not depend on where the parentheses are placed, that is, on the manner in which the maps are associated. This can be proved by applying the maps on the left and in the middle to an arbitrary point X and verifying that both give the image of X under the right-hand side of (1.4).

Exercises

1.10. The composition of two isometries is again an isometry.

1.11. Every translation of $\mathbb{R}$ is the composition of two reflections of $\mathbb{R}$.

1.12. For any point P of $\mathbb{R}$ there are exactly two isometries of $\mathbb{R}$ that map P onto itself.

1.13. For any a and b in $\mathbb{R}$ we have $\mathcal{T}_a \circ \mathcal{T}_b = \mathcal{T}_b \circ \mathcal{T}_a$.

1.14. For any P and Q in $\mathbb{R}$ we have
$$\mathcal{S}_P \circ \mathcal{S}_Q(X) = 2(P - Q) + X \ .$$
If $P \neq Q$, then $\mathcal{S}_P \circ \mathcal{S}_Q \neq \mathcal{S}_Q \circ \mathcal{S}_P$.

1.15. Consider the set $\{1, 2, 3\}$ endowed with the discrete metric. Let isometries $\mathcal{F}$ and $\mathcal{G}$ be defined by $\mathcal{F}(1) = 2$, $\mathcal{F}(2) = 1$, $\mathcal{F}(3) = 3$, and $\mathcal{G}(1) = 1$, $\mathcal{G}(2) = 3$, $\mathcal{G}(3) = 2$. Then $\mathcal{F} \circ \mathcal{G} \neq \mathcal{G} \circ \mathcal{F}$.
Hint: What does this inequality mean?

1.4 The Basic Assumptions of Plane Geometry

In this section we begin studying plane geometry and fix the basic assumptions for later deductions. The previous sections should be seen as background information and illustrations of these assumptions; they are not part of the theory that we are going to develop. Although it was not something we aimed at specifically, basic assumptions, theorems, and proofs will follow each other in rapid succession. This is almost inevitable; many important details need to be fixed carefully for later reference.

Digression 1. Euclid's *Elements* [18] is a typical example of the axiomatic construction of a theory. For centuries this monumental work of Euclid (ca. 325–ca. 265 BC) was a model for the axiomatic approach to constructing mathematical theories. Roughly speaking, the axiomatic approach goes as follows. First we fix a number of *primitive notions*, notions that require no further explanation. In the classical approach these are often *point* and *line*. All other notions are defined using the primitive notions. The *axioms* give the links between the different notions and describe properties that are not or cannot be deduced. The remaining statements in the theory, the *theorems*, are deduced from the axioms following generally accepted logical principles and proof structures.

The way this book was set up, a number of notions from set theory are assumed known, among others *element, set, function*. We moreover assume that the reader is familiar with the real line. We will define, among others, the notions *point, distance, line segment, line,* and *angle*. We fix the properties that we do not prove in *basic assumptions*. We use the words *basic assumption* instead of *axiom* because the basic assumptions differ from the axioms that are usually seen as the starting point of geometry in the traditional construction. In particular, the definition of perpendicular is rather unconventional.

Basic Assumption 1.14, which we will state shortly, simply states that every line has the structure of a real line. Consequently, the points on a line are ordered, and we can easily define the ratio of line segments. In [48, Chap. 20], ordering and ratio are defined using Euclid's postulates, and the two methods we mention here are compared.

In the statement of the first basic assumption, we refer to Definition 1.2.

Basic Assumption 1.11. *The Euclidean plane is a metric space: a set V with a metric d. The set V has more than one point.*

This basic assumption also fixes the notation for the rest of this book: whenever we speak of "the plane" or "V," we mean the Euclidean plane with its metric d.

Before defining straight lines, let us first say what a line segment is in an arbitrary metric space.

Definition 1.12. *For any pair of points A and B in a metric space M with metric d, the* line segment *$[AB]$ is given by*

$$[AB] = \{\, X \in M : d(A, X) + d(X, B) = d(A, B) \,\} \,.$$

In words: the line segment $[AB]$ is the set of all points X of M such that the sum of the distances from X to A and from X to B is equal to the distance from A to B.

This definition corresponds to what is commonly used for the metric space $\mathbb{R}$: for real numbers A and B, $[AB]$ is the line segment with endpoints A and B. The definition does not exclude the case $A = B$: the line segment is then the set consisting of a single point. Note that both A and B are elements of $[AB]$. It is also immediately clear that $[AB] = [BA]$. Moreover, for any point Y *outside* $[AB]$, that is, $Y \notin [AB]$, we have

$$d(A, B) < d(A, Y) + d(Y, B) \,.$$

Indeed, by the triangle inequality we always have $d(A, B) \leq d(A, Y) + d(Y, B)$. Equality holds if and only if $Y \in [AB]$. One could say that the line segment $[AB]$ is the shortest path from A to B.

Definition 1.13. *For any pair of* distinct *points A and B of V, the* straight line *AB is given by*

$$AB = \{\, X : X \in [AB] \text{ or } B \in [AX] \text{ or } A \in [XB] \,\} \,.$$

See Fig. 1.12: the points X, Y, and Z are elements of the line AB; the caption of the figure shows why. Lines are often denoted by lowercase letters l, m, n. If the point X is an element of line l, we also say that X *lies on l*, that l

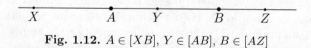

Fig. 1.12. $A \in [XB]$, $Y \in [AB]$, $B \in [AZ]$

passes through X, or that each of X, l *is incident* to the other; we abbreviate this as $X \in l$. Note that A and B both lie on line AB.

1.4 The Basic Assumptions of Plane Geometry

How many lines pass through two given points? For any pair of distinct points A and B, we would like there to be just one line passing through both. We can deduce this property using the following basic assumption, which fixes the structure of lines in the plane.

Basic Assumption 1.14. *Every line l in the plane admits an isometric surjection φ from l to $\mathbb{R}$.*

An *isometric surjection* is a distance-preserving surjection. First of all, φ is surjective: for every c in $\mathbb{R}$ there exists a P on l such that $\varphi(P) = c$. Secondly, φ preserves distances:

$$|\varphi(X) - \varphi(Y)| = d(X, Y) \text{ for all } X \text{ and } Y \text{ in } l.$$

In the basic assumption, the map φ is also injective (see the proof of Theorem 1.6), hence it is bijective. Through this basic assumption, the real line $\mathbb{R}$ has become the prototype of the lines in the plane. The following theorem gives an additional property of isometric surjections.

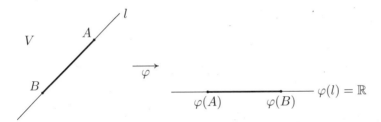

Fig. 1.13. The image of a line segment under φ

Theorem 1.15. *Let φ be an isometric surjection from a line l in V to $\mathbb{R}$. Then for any two points points A and B of l, we have $\varphi([AB]) = [\varphi(A)\varphi(B)]$.*

In words: φ maps the line segment $[AB]$ onto the line segment $[\varphi(A)\varphi(B)]$; see Fig. 1.13.

Proof. Let X be a point on the line AB. Then $X \in [AB]$ if and only if

$$d(A, X) + d(X, B) = d(A, B).$$

Since φ preserves distances, this holds if and only if

$$|\varphi(A) - \varphi(X)| + |\varphi(X) - \varphi(B)| = |\varphi(A) - \varphi(B)|.$$

This, in turn, is true if and only if $\varphi(X)$ lies between $\varphi(A)$ and $\varphi(B)$, that is, $\varphi(X) \in [\varphi(A)\varphi(B)]$. The theorem follows.

Theorem 1.16. *For every pair of distinct points A and B of the plane, there is exactly one line m passing through A and B.*

The (unique) line through two distinct points is called the *join* of these points.

Proof. Let us first note that the line $n = AB$ passes through A and B. It remains to show that if A and B lie on the line $m = CD$, then $n = m$. Once this is proved, we know that every line through A and B coincides with n. By Basic Assumption 1.14, there is an isometric surjection φ from m to $\mathbb{R}$. A

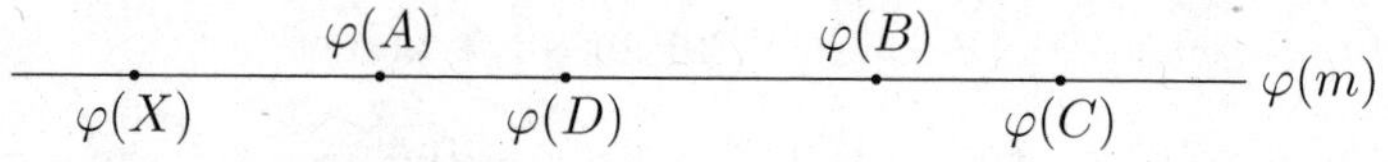

Fig. 1.14. The isometric image of $m = CD$

possible image of m under the map φ is given in Fig. 1.14. Other dispositions of $\varphi(A)$, $\varphi(B)$, $\varphi(C)$, and $\varphi(D)$ are of course also possible; those situations can be dealt with analogously. It is important to note that we always have $\varphi(A) \neq \varphi(B)$ and $\varphi(C) \neq \varphi(D)$, since φ is injective and $A \neq B$ and $C \neq D$.

Let X be an arbitrary point of m. We distinguish three cases: $X \in [CD]$, $D \in [CX]$, and $C \in [XD]$. Let us assume that we are in the second case; see Fig. 1.14. The other cases can be dealt with analogously. The points $\varphi(X)$, $\varphi(A)$, and $\varphi(B)$ lie on the real line. In the situation sketched in the figure, we have
$$|\varphi(X) - \varphi(A)| + |\varphi(A) - \varphi(B)| = |\varphi(X) - \varphi(B)| \ .$$
Since φ is an isometry, this implies that $d(X, A) + d(A, B) = d(X, B)$ and therefore $A \in [XB]$, whence $X \in AB$. In general, every point X of CD lies on AB. Hence $m \subseteq n$: the line m is a subset of the line n.

What have we proved? We have shown that if the points A and B of line n lie on line $m = CD$, then $m \subseteq n$. In particular, $C \in n$ and $D \in n$. Thus, the points C and D of line m lie on $n = AB$. By the same token, $n \subseteq m$. We conclude that $n = m$.

Let us now look at the position of lines with respect to each other. First we need the assertion that there is more than one line in the plane. This is given by the following basic assumption.

Basic Assumption 1.17. *The Euclidean plane contains three noncollinear points.*

This basic assumption implies that there are at least three distinct lines.

Definition 1.18. *We call a point P an* intersection point *of the lines l and m if P lies on both l and m. In this case we also say that the lines l and m are* concurrent *or that they* concur at *P* We say that the lines *m* and *n* are parallel if $m = n$ or $m \cap n = \emptyset$; we denote this by $m \,//\, n$. If *m* and *n* are not parallel, we say that *m* and *n* are intersecting *lines*.

The following theorem gives an important property of intersecting lines.

Theorem 1.19. *Two intersecting lines have exactly one intersection point.*

Proof. Let l and m be two intersecting lines. Since they are not parallel, the lines l and m have an intersection point and $l \neq m$. Let P be a common point of l and m. If it is not unique, there exists another point, say Q, that lies on both l and m. According to Theorem 1.16 we then have $l = m$, which gives a contradiction. Therefore the intersection point is unique.

Because of this last theorem, we can speak of *the* intersection point of two intersecting lines. The following basic assumption and Basic Assumption 1.17 guarantee the existence of parallel, noncoincident lines.

Basic Assumption 1.20. *For every point P and every line m of the plane, there exists a unique line passing through P and parallel to m.*

It immediately follows from the definition of parallel that *every line l is parallel to itself*, $l \,//\, l$, and that if $l \,//\, m$, then $m \,//\, l$. The first property expresses the *reflexivity* of the parallel relation, the second its *symmetry*. The following theorem asserts its *transitivity*.

Theorem 1.21. *Given lines l, m, and n, if $l \,//\, m$ and $m \,//\, n$, then $l \,//\, n$.*

Proof. We give a proof by contradiction. Suppose l is not parallel to n; then l and n have an intersection point P. By assumption, both l and n are parallel to m. According to Basic Assumption 1.20 there is exactly one line through P that is parallel to m. Consequently $l = n$. This contradicts the assumption that l is not parallel to n, which concludes the proof of the theorem.

The proofs given in Sect. 1.1 use congruence and similarity. In this phase of the setup of plane geometry we do not yet have these notions at our disposal, since they will be studied only in the next chapter. Many of the properties needed at that point follow from the Pythagorean theorem and its converse. Those properties are implicitly present in the following definition of perpendicular and the basic assumption concerning the existence of mutually perpendicular lines.

Definition 1.22. *Let l and m be intersecting lines with common point C. We say that l is perpendicular to m if for every point A of l and every point B of m we have*

$$d(C, A)^2 + d(C, B)^2 = d(A, B)^2 \; ; \tag{1.5}$$

we denote this by $l \perp m$.

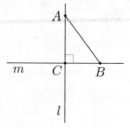

Fig. 1.15. $l \perp m$: $d(C,A)^2 + d(C,B)^2 = d(A,B)^2$

Consider Fig. 1.15. It is immediately clear that the statements $l \perp m$ and $m \perp l$ are equivalent. If $l \perp m$, l is said to be a *perpendicular* on m; we also say that l and m are *perpendicular to each other*. In a figure, we sometimes show that $l \perp m$ by placing a small square at their intersection point C. Definition 1.22 clearly states that (1.5) must hold for all A on l and all B on m. Consequently, if we know that lines l and m are perpendicular to each other, we can use this formula, which is in fact the Pythagorean theorem. This will simplify the introduction of coordinates; see Sect. 1.5. The existence of perpendicular lines is asserted in the following basic assumption.

Basic Assumption 1.23. *For every point P and every line m of the plane, there exists a line l through P that is perpendicular to m: $P \in l$ and $l \perp m$.*

This basic assumption states the existence, for every point P and every line m, of a perpendicular on m passing through P. The following theorem implies the uniqueness of such a perpendicular for given P and m. We call the intersection point of the perpendicular on m and the line m the *foot* of the perpendicular.

Theorem 1.24. *If lines l and m are both perpendicular to a line n, then $l // m$. Through every point there is a unique perpendicular onto n.*

Proof. See Fig. 1.16. Let P be the intersection point of l and n, and Q that of m and n. We distinguish two cases: $P \neq Q$ and $P = Q$. Consider the first case: $P \neq Q$. Suppose that l and m meet at R. Then by Definition 1.22,

$$d(P,R)^2 + d(P,Q)^2 = d(Q,R)^2 \quad \text{and} \quad d(Q,R)^2 + d(Q,P)^2 = d(P,R)^2 \ .$$

This implies that $2d(P,Q)^2 = 0$ and therefore $P = Q$, which contradicts the fact that $P \neq Q$. The assumption that l and m intersect each other is therefore incorrect. We conclude that $l // m$. Now consider the second case: $P = Q$. Let S be another point on n and let $n_\perp$ be the perpendicular on n at S. By what we have just proved, $l // n_\perp$ and $m // n_\perp$. By Theorem 1.21, this implies that $l // m$.

Finally, let us prove the second statement of the theorem. Let l and m be perpendiculars onto n passing through R. Then by the first statement

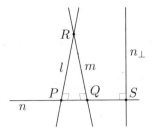

Fig. 1.16. Perpendiculars onto n are parallel

of the theorem, $l \,//\, m$. Since R lies on both l and m, it follows from Basic Assumption 1.20 that $l = m$.

For the proof of the following theorem we refer to Exercise 1.17.

Theorem 1.25. *If $l \,//\, m$ and $l \perp n$, then $m \perp n$.*

We conclude this section with a number of definitions. A *quadrilateral* $ABCD$ is a set consisting of four points A, B, C, and D, the *vertices* of the quadrilateral, no three of which are collinear. The four line segments $[AB]$, $[BC]$, $[CD]$, and $[DA]$ are the *sides* of the quadrilateral. The lines AB, BC,

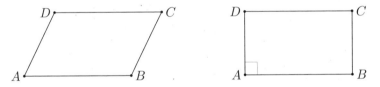

Fig. 1.17. (a) A parallelogram; (b) a rectangle

CD, and DA are called the *sidelines* of the quadrilateral. The quadrilateral $ABCD$ is called a *parallelogram* if $AB \,//\, DC$ and $AD \,//\, BC$; if, moreover, two intersecting sidelines of a parallelogram, for example AD and AB, are perpendicular to each other, we call the quadrilateral $ABCD$ a *rectangle*. The following theorem gives a number of properties of the rectangle; see Exercise 1.18 for a proof.

Theorem 1.26. *The following properties hold in a rectangle $ABCD$:*

1. *Any two intersecting sidelines are perpendicular to each other.*
2. *$d(A,B) = d(D,C)$ and $d(B,C) = d(A,D)$.*

The second property of the theorem also holds for a parallelogram; see Theorem 1.29.

Exercises

1.16. Let M be a set consisting of more than one point, endowed with the discrete metric. For any two distinct points A and B, we have $[AB] = \{A, B\}$, that is, the line segment $[AB]$ consists of the two points A and B.

1.17. The proof of Theorem 1.25 can be divided up into the following steps. We use the notation of the theorem.

(a) m and n are intersecting lines. Call the intersection point Q.
(b) The perpendicular $n_\perp$ on n at Q is parallel to l.
(c) $n_\perp$ coincides with m; hence $m \perp n$.

1.18. In this exercise we consider the proof of Theorem 1.26. The first property can be proved using Theorem 1.25. For the second property, we can first use Definition 1.22 to prove that

$$\text{if } d(A,B) > d(D,C), \text{ then } d(D,C) > d(A,B).$$

This implies that $d(A,B) = d(D,C)$.
Hint: First consider the triangles ACB and DBC, then ADB and DAC.

1.19. For every line AB there is a unique isometry φ from AB to $\mathbb{R}$ such that $\varphi(A) = 0$ and $\varphi(B) = d(A,B)$.

1.20. Use Basic Assumption 1.14 to show that the line segment $[AB]$ contains a unique point E such that $d(A,E) = d(E,B)$; the point E is called the *midpoint* of $[AB]$. The perpendicular l onto AB in E is called the *perpendicular bisector* of $[AB]$. We have:

(a) A point X lies on l if and only if $d(X,A) = d(X,B)$.
   Hint: If $d(X,A) = d(X,B)$, then $d(A,D)^2 = d(B,D)^2$, where D is the foot of the perpendicular from X onto the line AB.
(b) In triangle ABC, the perpendicular bisectors of the sides AB, BC, CA meet in one point.

1.21. Consider a set F consisting of four points A, B, C, and D, endowed with the discrete metric. Show that this metric space satisfies Basic Assumptions 1.11, 1.17, and 1.20, but not the other basic assumptions.

1.22. Let A, B, C, and D be four distinct points such that $AB \mathbin{/\!/} DC$. If A, B, and C are collinear, then D is also collinear with the other three.

1.5 Local Coordinates and the Equation of a Line

Coordinates are an important, if not crucial, tool for studying geometry. The coordinates used in plane geometry are closely related to the *vector space* $\mathbb{R}^2$.

1.5 Local Coordinates and the Equation of a Line

The elements of $\mathbb{R}^2$ are called vectors. Such an element is an ordered pair (x_1, x_2) of real numbers; we often denote it by the corresponding boldface letter: $\mathbf{x} = (x_1, x_2)$. The numbers x_1 and x_2 are called the *components* of the vector $\mathbf{x}$. *By convention* $\mathbf{x} = \mathbf{y}$ *if and only if* $x_1 = y_1$ *and* $x_2 = y_2$. By definition, the components of a vector form an ordered pair: this implies that the vector (x_1, x_2) is unequal to (x_2, x_1) unless $x_1 = x_2$. Using this notation, we write
$$\mathbb{R}^2 = \{\, \mathbf{x} = (x_1, x_2) : x_1 \in \mathbb{R},\ x_2 \in \mathbb{R} \,\}.$$
For any vector $\mathbf{x}$,
$$\|\mathbf{x}\| = \sqrt{x_1^2 + x_2^2}$$
is the *norm* of $\mathbf{x}$. We draw the vector space $\mathbb{R}^2$ in the usual manner; see Fig. 1.18. Using the three-four-five method or simply a right triangle, we first draw a Cartesian coordinate system $x_1 O x_2$. We mark off the real numbers on

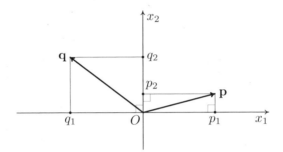

Fig. 1.18. Vectors in $\mathbb{R}^2$

both axes, using the same scale and choosing O at the intersection point of the axes. This point is called the *origin* of the coordinate system. We often denote the vector $\mathbf{p} = (p_1, p_2)$ by an arrow that begins at the origin and ends at the intersection point of the perpendicular on the x_1-axis at p_1 and the perpendicular on the x_2-axis at p_2. The horizontal axis represents the set $\{\,(x_1, 0) : x_1 \in \mathbb{R}\,\}$, which we often call the x_1-axis. Likewise, we call the set $\{\,(0, x_2) : x_2 \in \mathbb{R}\,\}$ the x_2-axis.

Algebraic operations with vectors are carried out on the components. In particular, for any vectors $\mathbf{a}$, $\mathbf{b}$ and any real number λ,
$$\mathbf{a} + \mathbf{b} = (a_1, a_2) + (b_1, b_2) = (a_1 + b_1, a_2 + b_2),$$
$$\mathbf{a} - \mathbf{b} = (a_1, a_2) - (b_1, b_2) = (a_1 - b_1, a_2 - b_2),$$
$$\lambda \mathbf{a} = \lambda(a_1, a_2) = (\lambda a_1, \lambda a_2).$$

We call the vector $\mathbf{a} + \mathbf{b}$ the *sum* of the vectors $\mathbf{a}$ and $\mathbf{b}$, and the vector $\mathbf{a} - \mathbf{b}$ the *difference*. The vector $\lambda \mathbf{a}$, where λ is any real number, is called a *scalar multiple* of $\mathbf{a}$. The vector $\mathbf{o} = (0, 0)$ is the *zero vector*. In our representation of $\mathbb{R}^2$, the zero vector is at the origin: $O = \mathbf{o}$. Using the definitions, we can deduce a number of rules for vector computation. For example, $\mathbf{a} + \mathbf{b} = \mathbf{b} + \mathbf{a}$ for every $\mathbf{a}$ and $\mathbf{b}$. Indeed,

$$\mathbf{a} + \mathbf{b} = (a_1, a_2) + (b_1, b_2) = (a_1 + b_1, a_2 + b_2) = (b_1 + a_1, b_2 + a_2) = \mathbf{b} + \mathbf{a}.$$

Other rules are that for any $\mathbf{a}$, $\mathbf{b}$, and $\mathbf{c}$ we have

$$(\mathbf{a} + \mathbf{b}) + \mathbf{c} = \mathbf{a} + (\mathbf{b} + \mathbf{c}),$$
$$\mathbf{a} + \mathbf{o} = \mathbf{a}.$$

We also see that for any λ in $\mathbb{R}$ and any $\mathbf{a}$ in $\mathbb{R}^2$,

$$\|\lambda \mathbf{a}\| = \sqrt{\lambda^2 a_1^2 + \lambda^2 a_2^2} = |\lambda| \, \|\mathbf{a}\|.$$

Later on in this section, we will consider the geometric interpretation of these algebraic operations. For vectors $\mathbf{p}$ and $\mathbf{q}$, let

$$\rho(\mathbf{p}, \mathbf{q}) = \|\mathbf{q} - \mathbf{p}\|.$$

For example, for $\mathbf{p} = (3, 1)$ and $\mathbf{q} = (-2, 3)$ we have $\mathbf{q} - \mathbf{p} = (-5, 2)$ and $\rho(\mathbf{p}, \mathbf{q}) = \|\mathbf{q} - \mathbf{p}\| = \sqrt{29}$. We will prove below that ρ is a metric and that $\mathbb{R}^2$ with this metric is an isometric copy of the Euclidean plane. Because of this we will call $\mathbb{R}^2$ with metric ρ the *coordinate plane*. The algebraic structure of $\mathbb{R}^2$ is a big advantage for us: being able to compute with points in the plane simplifies many proofs. In Sect. 1.4 we followed the convention of denoting points in the plane by capital letters. However, the points in the coordinate plane are vectors in $\mathbb{R}^2$; it is therefore sometimes better to denote points in the coordinate plane by lowercase letters, as we do with vectors.

There is much freedom in the choice of the origin and the axes of the coordinate plane, as is asserted in the following theorem.

Theorem 1.27. *The function ρ is a metric. For every point C of the Euclidean plane and every pair (l_1, l_2) of perpendicular lines through C, there is an isometric surjection Φ from the Euclidean plane to the coordinate plane $\mathbb{R}^2$ such that $\Phi(C) = \mathbf{o}$, and Φ maps l_1 onto the x_1-axis and l_2 onto the x_2-axis. For any points P and Q in the Euclidean plane we have $d(P, Q) = \rho(\Phi(P), \Phi(Q))$.*

Proof. By Basic Assumption 1.14, there is an isometric surjection φ_1 from the line l_1 onto the x_1-axis. By composing such an isometric surjection, if necessary, with a suitable translation, we can assume that $\varphi_1(C) = 0$. There are in fact two isometric surjections satisfying this condition. If φ_1 is one of them, we obtain the other one by composing the reflection of $\mathbb{R}$ in 0 with φ_1, where the reflection is applied after φ_1. The choice of φ_1 made here is indicated by an arrow on l_1; see Fig. 1.19. Likewise, there is an isometric surjection φ_2

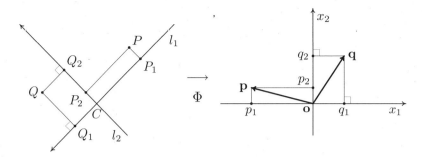

Fig. 1.19. Affixing a coordinate plane

from the line l_2 onto the x_2-axis such that $\varphi_2(C) = 0$. Here too we indicate the choice made between the two possibilities by an arrow, this time on line l_2.

Let X be an arbitrary point in the plane, and let X_i be the foot of the perpendicular from X onto l_i, for $i = 1, 2$. Let

$$\Phi(X) = (\varphi_1(X_1), \varphi_2(X_2)) = (x_1, x_2) = \mathbf{x}.$$

In Fig. 1.19 we have drawn $\mathbf{p} = \Phi(P)$ and $\mathbf{q} = \Phi(Q)$; note that $q_1 = d(Q_1, C)$, $p_1 = -d(P_1, C)$, and so on. From the left-hand side of Fig. 1.19, and, among others, Definition 1.22 and Theorem 1.26, we deduce that for every P and Q,

$$d(P, Q) = \sqrt{d(P_1, Q_1)^2 + d(P_2, Q_2)^2}.$$

It now follows from the definitions of Φ and ρ that $d(P, Q) = \rho(\mathbf{p}, \mathbf{q})$. Since d is a metric and Φ is a bijection, we immediately conclude that ρ is also a metric and that the map Φ is an isometric surjection from the Euclidean plane onto the coordinate plane.

We call $(\mathbb{R}^2, \rho)$ as in the theorem above a *coordinate system at the point C*. The components of the vectors are called the *local coordinates* at C. Note that we have just proved that at any point of $\mathbb{R}^2$, a coordinate system can be fixed with given perpendicular directions for the axes. We will illustrate this shortly with examples. Let us first consider the algebraic structure of the coordinate plane.

The Parametric Equation of a Straight Line

Let $\mathbf{p}$ and $\mathbf{q}$ be distinct points of the coordinate plane. The *parametric equation* of the line $\mathbf{pq}$ is

$$\boxed{\mathbf{x} = (1 - \lambda)\mathbf{p} + \lambda \mathbf{q}, \quad \lambda \in \mathbb{R}.} \tag{1.6}$$

This means that

1. for every λ the point $(1 - \lambda)\mathbf{p} + \lambda \mathbf{q}$ lies on the line $\mathbf{pq}$;

2. for every point **x** on **pq** there is a unique λ for which (1.6) holds.

Equation (1.6) describes the position of the point **x** as a function of the parameter λ; as λ moves along the real line, **x** moves along the line **pq**. Showing this equation represents the line **pq**, that is, that it satisfies both conditions that were just mentioned, is not that easy; this is due to the peculiar definition of a straight line introduced in the last section. We first define a map ψ from $\mathbb{R}$ to $\mathbb{R}^2$ by setting

$$\psi(y) = \mathbf{p} + \frac{y}{\|\mathbf{q}-\mathbf{p}\|}(\mathbf{q}-\mathbf{p}) .$$

Let $c = \rho(\mathbf{p},\mathbf{q}) = \|\mathbf{q}-\mathbf{p}\|$. We immediately see that $\psi(0) = \mathbf{p}$ and $\psi(c) = \mathbf{q}$.

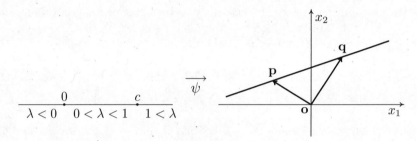

Fig. 1.20. Deduction of the parametric equation of the line **pq**

The function ψ preserves distances. Indeed, for any y and z in $\mathbb{R}$ we have

$$\rho(\psi(y),\psi(z)) = \|\psi(y)-\psi(z)\| = \left\|\frac{y-z}{c}(\mathbf{q}-\mathbf{p})\right\| = |y-z| .$$

Where does $\psi(y)$ end up in $\mathbb{R}^2$? We claim that $\psi(y)$ lies on the line **pq**. To prove this we distinguish three cases; see Fig. 1.20. First suppose that y lies between 0 and c, that is, $0 \leq y \leq c$. There then exists a real number λ satisfying $0 \leq \lambda \leq 1$ such that $y = \lambda c$, $d(0,y) = \lambda c$, and $d(y,c) = (1-\lambda)c$. Since ψ preserves distances, we have $\rho(\mathbf{p},\psi(y)) = \lambda c$ and $\rho(\psi(y),\mathbf{q}) = (1-\lambda)c$. It follows that

$$\rho(\mathbf{p},\psi(y)) + \rho(\psi(y),\mathbf{q}) = c = \rho(\mathbf{p},\mathbf{q}) .$$

Consequently, $\psi(y)$ lies on the line segment [**pq**]. For later use we also note that we have shown *that $\psi(\lambda c)$ splits the line segment* [**pq**] *into sections that are in the same ratio as the sections into which λc splits* [0c], *namely* $\lambda : (1-\lambda)$; see Fig. 1.21.

Does every point of [**pq**] lie in the image of ψ? According to Basic Assumption 1.14, there is an isometric surjection φ from **pq** to $\mathbb{R}$. By Theorem 1.9, there is an isometry of $\mathbb{R}$ that maps $\varphi(\mathbf{p})$ and $\varphi(\mathbf{q})$, respectively, onto 0 and c.

1.5 Local Coordinates and the Equation of a Line

To keep the notation uniform, we assume that $\varphi(\mathbf{p}) = 0$ and $\varphi(\mathbf{q}) = c$. In particular, φ maps $[\mathbf{pq}]$ bijectively onto the line segment $[0c]$ of $\mathbb{R}$, preserving distances; this follows from Theorem 1.15. This implies that for every point $\mathbf{y}$ of $[\mathbf{pq}]$ there is a unique λ with $0 \leq \lambda \leq 1$ such that $\rho(\mathbf{p}, \mathbf{y}) = \lambda c$ and $\rho(\mathbf{y}, \mathbf{q}) = (1 - \lambda)c$. Consequently, $\mathbf{y} = \psi(\lambda c)$. We conclude that ψ is an isometric surjection from $[0c]$ to $[\mathbf{pq}]$.

Fig. 1.21. Partitioning of the line segments $[0c]$ and $[\mathbf{pq}]$

The other cases that need to be considered to prove that $\psi(y)$ always lies on $\mathbf{pq}$ are that in which 0 lies between $y = \lambda c$ and c ($\lambda < 0$), and that in which c lies between 0 and y ($\lambda > 1$). The proofs for these cases are similar to the one given above. We conclude by scaling ψ, defining a map ψ^* from $\mathbb{R}$ to $\mathbb{R}^2$ by

$$\psi^*(\lambda) = \psi(\lambda c) = \psi(\lambda \|\mathbf{q} - \mathbf{p}\|) = \mathbf{p} + \lambda(\mathbf{q} - \mathbf{p}) = (1 - \lambda)\mathbf{p} + \lambda \mathbf{q}.$$

This map satisfies both conditions mentioned after (1.6).

Example 1.28 (Centroid). We apply what we have learned to prove that the medians of a triangle are concurrent; the point where they meet is called the *centroid* of the triangle. In triangle ABC, let D be the midpoint of BC, E

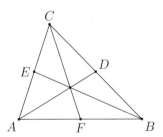

Fig. 1.22. The medians of a triangle are concurrent

the midpoint of AC, and F the midpoint of AB. We choose an arbitrary coordinate system, and indicate the vectors associated to points with the corresponding bold lowercase letters. The parametric equation of the line CB

is $\mathbf{x} = (1 - \lambda)\mathbf{b} + \lambda\mathbf{c}$. Consequently, $\mathbf{d} = (1/2)(\mathbf{b} + \mathbf{c})$. The *median AD* has equation
$$\mathbf{x} = (1 - \lambda)\mathbf{a} + \lambda\left(\tfrac{1}{2}\mathbf{b} + \tfrac{1}{2}\mathbf{c}\right).$$
Substituting $\lambda = 2/3$ gives the point $\mathbf{z} = (\mathbf{a} + \mathbf{b} + \mathbf{c})/3$, which lies on the median AD. Since the formula for $\mathbf{z}$ is symmetric in $\mathbf{a}$, $\mathbf{b}$, and $\mathbf{c}$, $\mathbf{z}$ also lies on the other medians; $\mathbf{z}$ is therefore the centroid. The value of λ implies that the centroid splits each of the medians into segments in the ratio $2 : 1$.

Addition and Scalar Multiplication

If in (1.6) we substitute $\mathbf{o}$ and $\mathbf{a}$, respectively, for $\mathbf{p}$ and $\mathbf{q}$, we obtain $\mathbf{x} = \lambda\mathbf{a}$. We see that the scalar multiples $\lambda\mathbf{a}$ of $\mathbf{a}$ lie along the same line l through $\mathbf{o}$ as the vector $\mathbf{a}$; see Fig. 1.23, on the left. If $\mathbf{a} \neq \mathbf{o}$, the line is unique. The line l is then called the *span* of $\mathbf{a}$. We can easily verify that $\lambda\mathbf{a} + \mu\mathbf{a} = (\lambda + \mu)\mathbf{a}$. Let us now consider addition: if $\mathbf{a}$ and $\mathbf{b}$ have distinct spans, then $\mathbf{o}$, $\mathbf{b}$, $\mathbf{a} + \mathbf{b}$,

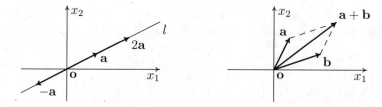

Fig. 1.23. (a) Scalar multiplication; (b) addition

and $\mathbf{a}$ are, in this order, the *vertices of a parallelogram*; we say that this parallelogram is *spanned* by the vectors $\mathbf{a}$ and $\mathbf{b}$. To prove this it suffices to show that the line l through $\mathbf{o}$ and $\mathbf{b}$ is parallel to the line m through $\mathbf{a}$ and $\mathbf{a} + \mathbf{b}$. Indeed, if we prove this, interchanging $\mathbf{a}$ and $\mathbf{b}$ shows that the line through $\mathbf{o}$ and $\mathbf{a}$ is parallel to the line through $\mathbf{b}$ and $\mathbf{a} + \mathbf{b}$.

The line l has parametric equation
$$\mathbf{x} = \lambda\mathbf{b}, \text{ that is, } \begin{cases} x_1 = \lambda b_1, \\ x_2 = \lambda b_2. \end{cases}$$
Eliminating λ gives $x_1 b_2 - x_2 b_1 = 0$. This is the linear equation of l. Substituting $\mathbf{a}$ and $\mathbf{a} + \mathbf{b}$ for $\mathbf{p}$ and $\mathbf{q}$, respectively, in (1.6) gives the following equation for m:
$$\mathbf{x} = \mathbf{a} + \lambda\mathbf{b}, \text{ that is, } \begin{cases} x_1 = a_1 + \lambda b_1, \\ x_2 = a_2 + \lambda b_2. \end{cases}$$
Eliminating λ gives $x_1 b_2 - x_2 b_1 = a_1 b_2 - a_2 b_1$, the linear equation of m. The intersection point of l and m is a solution of the system

1.5 Local Coordinates and the Equation of a Line

$$\begin{cases} x_1 b_2 - x_2 b_1 = 0, \\ x_1 b_2 - x_2 b_1 = a_1 b_2 - a_2 b_1. \end{cases}$$

Unless $a_1 b_2 - a_2 b_1 = 0$, this system is inconsistent, in which case $l \mathbin{/\mkern-4mu/} m$. If $a_1 b_2 - a_2 b_1 = 0$, **a** and **b** have the same span, which contradicts our assumption.

Theorem 1.29. *In a parallelogram $ABCD$ we have $d(A, B) = d(D, C)$ and $d(B, C) = d(A, D)$.*

Proof. See Fig. 1.24 (a). We choose the origin of the coordinate system at A; $A = \mathbf{o}$. We use the associated lowercase bold letters to indicate the vectors corresponding to the other vertices of the parallelogram. We have just seen that the vectors $\mathbf{o}, \mathbf{b}, \mathbf{d}+\mathbf{b}, \mathbf{d}$, in that order, are the vertices of a parallelogram. Since $AB \mathbin{/\mkern-4mu/} DC$ and $AD \mathbin{/\mkern-4mu/} BC$, it follows from Basic Assumption 1.20 that $\mathbf{c} = \mathbf{d}+\mathbf{b}$. Hence $\|\mathbf{c}-\mathbf{d}\| = \|\mathbf{b}\|$, whence $d(D, C) = d(A, B)$ and $\|\mathbf{c}-\mathbf{b}\| = \|\mathbf{d}\|$; consequently, $d(B, C) = d(A, D)$.

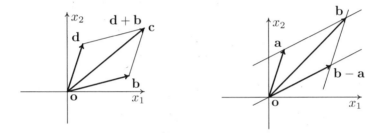

Fig. 1.24. Parallelograms: (a) sum of vectors; (b) difference of vectors

In Fig. 1.24 (b), we have shown how the difference $\mathbf{b} - \mathbf{a}$ of the vectors $\mathbf{b}$ and $\mathbf{a}$ is drawn when the vectors have distinct spans. Draw a line through $\mathbf{o}$ parallel to line $\mathbf{ab}$; draw a line through $\mathbf{b}$ parallel to $\mathbf{oa}$. The vector $\mathbf{b} - \mathbf{a}$ corresponds to the intersection point. Indeed, the solution of the equation $\mathbf{a} + \mathbf{x} = \mathbf{b}$ is $\mathbf{x} = \mathbf{b} - \mathbf{a}$, as one can check. It follows that $\mathbf{o}, \mathbf{b} - \mathbf{a}, \mathbf{b}, \mathbf{a}$, in this order, are the vertices of a parallelogram.

We sometimes also use another parametric equation for the line through $\mathbf{p}$ and $\mathbf{q}$ than that of (1.6). It is deduced from the first by substituting $\mathbf{q} = \mathbf{p} + \mathbf{a}$ (note that $\mathbf{a} \neq \mathbf{o}$):

$$\boxed{\mathbf{x} = \mathbf{p} + \lambda \mathbf{a}, \quad \lambda \in \mathbb{R}.} \tag{1.7}$$

If $\mathbf{p}$ and $\mathbf{a}$ have different spans, then $\mathbf{o}, \mathbf{a}, \mathbf{p} + \mathbf{a}$, and $\mathbf{p}$ are the vertices of a parallelogram. We see that (1.7) is the parametric equation of the line through $\mathbf{p}$ parallel to the span of $\mathbf{a}$.

Exercises

1.23. Let $\mathbf{x} = \mathbf{p} + \lambda\mathbf{a}$ and $\mathbf{x} = \mathbf{q} + \lambda\mathbf{b}$ be two lines. These lines are parallel if and only if $\mathbf{a}$ and $\mathbf{b}$ have the same span.
Hint: The first line is parallel to the span of $\mathbf{a}$.

1.24. In the coordinate plane, distinct points $\mathbf{a}$, $\mathbf{b}$, $\mathbf{c}$, $\mathbf{d}$, in this order, are the vertices of a parallelogram if and only if $\mathbf{a} + \mathbf{c} = \mathbf{b} + \mathbf{d}$ and $\mathbf{a}$, $\mathbf{b}$, $\mathbf{c}$, $\mathbf{d}$ are noncollinear.
Hint: Use Exercises 1.23 and 1.22 of Sect. 1.4.

1.25. The *diagonals* of a quadrilateral $ABCD$ are the line segments $[AC]$ and $[BD]$. The diagonals of a parallelogram bisect each other.

1.26. Given a quadrilateral $ABCD$, let E, F, G, and H be the respective midpoints of the sides AB, BC, CD, and DA.
The points E, F, G, and H either are collinear or are, in this order, the vertices of a parallelogram.

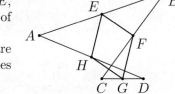

1.27. In triangle ABC, let D, E, F be the midpoints of the sides BC, AC, AB, respectively. Then $AFDE$ is a parallelogram. We have $ED \parallel AB$ and $d(E, D) = d(A, B)/2$. The line DE is called a *mid-parallel* of ABC.

1.28. In triangle ABC, the perpendicular from C onto AB is called an *altitude*. We are going to prove that the three altitudes of a triangle are concurrent; their intersection point is called the *orthocenter* of the triangle. Given a triangle ABC, draw the line l through A parallel to BC, the line m through B parallel to CA, and the line n through C parallel to AB. Let D be the intersection point of m and n, E that of n and l, and F that of l and m.

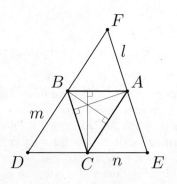

Show that the perpendicular bisectors of triangle DEF coincide with the altitudes of triangle ABC.

1.29. Given a triangle ABC, define a function f from the plane to $\mathbb{R}$ by setting

$$f(X) = d(X, A)^2 + d(X, B)^2 + d(X, C)^2, \quad X \in V.$$

The function f has a minimum at the centroid of the triangle.
Hint: Choose the coordinate system such that A and B lie on the x_1-axis, while C lies on the x_2-axis.

1.6 The Inner Product and Determinant

As we saw in the last section, vectors can be very useful in studying geometric problems. Can coordinates also be used to check whether lines are perpendicular to each other? They can, through the inner product.

Definition 1.30. *The* inner product *of vectors* $\mathbf{x}$ *and* $\mathbf{y}$ *of* $\mathbb{R}^2$ *is given by*

$$\langle \mathbf{x}, \mathbf{y} \rangle = x_1 y_1 + x_2 y_2 \; .$$

The inner product is a real number and *not* a vector. Let us begin by stating a number of rules for computing the inner product; these rules can easily be checked on the components. For any vectors $\mathbf{a}$, $\mathbf{b}$, and $\mathbf{c}$ and any real number λ,

$$\langle \mathbf{a}, \mathbf{a} \rangle = \|\mathbf{a}\|^2 \;,\; \langle \mathbf{a}, \mathbf{b} \rangle \langle \mathbf{b}, \mathbf{a} \rangle \;,$$
$$\langle (\lambda \mathbf{a}), \mathbf{b} \rangle = \lambda \langle \mathbf{a}, \mathbf{b} \rangle \;,\; \langle \mathbf{a}, (\mathbf{b} + \mathbf{c}) \rangle = \langle \mathbf{a}, \mathbf{b} \rangle + \langle \mathbf{a}, \mathbf{c} \rangle \;,$$
$$\langle \mathbf{a} + \mathbf{b}, \mathbf{a} + \mathbf{b} \rangle = \|\mathbf{a}\|^2 + 2 \langle \mathbf{a}, \mathbf{b} \rangle + \|\mathbf{b}\|^2 \;.$$

The usefulness of the inner product becomes clear in the following theorem.

Theorem 1.31. *Let l and m be lines in the coordinate plane, given by the parametric equations $\mathbf{x} = \mathbf{p} + \lambda \mathbf{a}$ and $\mathbf{x} = \mathbf{q} + \mu \mathbf{b}$, respectively. The lines l and m are perpendicular to each other if and only if $\langle \mathbf{a}, \mathbf{b} \rangle = 0$.*

Proof. First note that if $l \;//\; m$, then $\langle \mathbf{a}, \mathbf{b} \rangle \neq 0$. Indeed, if $l \;//\; m$, $\mathbf{a}$ and $\mathbf{b}$ have the same span. Since $\mathbf{a} \neq \mathbf{o} \neq \mathbf{b}$, we then have $\mathbf{b} = \nu \mathbf{a}$ for some $\nu \neq 0$, and therefore $\langle \mathbf{a}, \mathbf{b} \rangle = \nu \|\mathbf{a}\|^2$, which is nonzero. To prove the theorem, we first assume that l and m are perpendicular to each other. In that case the lines intersect, say at the point $\mathbf{r}$; see Fig. 1.25, on the right (the coordinate axes are not drawn in this figure). The equations of l and m can be rewritten in the

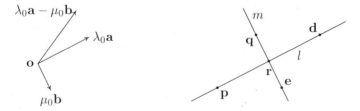

Fig. 1.25. Lines l and m perpendicular to each other

form $\mathbf{x} = \mathbf{r} + \lambda' \mathbf{a}$ respectively $\mathbf{x} = \mathbf{r} + \mu' \mathbf{b}$; indeed, l passes through $\mathbf{r}$ and is parallel to the span of $\mathbf{a}$, while m passes through $\mathbf{r}$ and is parallel to the span of $\mathbf{b}$. What does it mean for these equations if l and m are perpendicular to each other? Let us choose points $\mathbf{d}$ on l and $\mathbf{e}$ on m, both distinct from $\mathbf{r}$; then $\mathbf{d} = \mathbf{r} + \lambda_0 \mathbf{a}$ and $\mathbf{e} = \mathbf{r} + \mu_0 \mathbf{b}$ for suitable nonzero λ_0 and μ_0. Computations show that

$$\rho(\mathbf{r},\mathbf{d})^2 = \|\mathbf{d}-\mathbf{r}\|^2 = \|\lambda_0\mathbf{a}\|^2 = \lambda_0^2\|\mathbf{a}\|^2 \, ,$$
$$\rho(\mathbf{r},\mathbf{e})^2 = \|\mathbf{e}-\mathbf{r}\|^2 = \|\mu_0\mathbf{b}\|^2 = \mu_0^2\|\mathbf{b}\|^2 \, ,$$
$$\rho(\mathbf{d},\mathbf{e})^2 = \|\mathbf{e}-\mathbf{d}\|^2 = \|\mu_0\mathbf{b}-\lambda_0\mathbf{a}\|^2$$
$$= \langle \mu_0\mathbf{b}-\lambda_0\mathbf{a}, \mu_0\mathbf{b}-\lambda_0\mathbf{a}\rangle$$
$$= \mu_0^2\|\mathbf{b}\|^2 - 2\lambda_0\mu_0\langle\mathbf{a},\mathbf{b}\rangle + \lambda_0^2\|\mathbf{a}\|^2 \, ;$$

see Fig. 1.25 on the left. If $l \perp m$, we must have $\rho(\mathbf{d},\mathbf{e})^2 = \rho(\mathbf{r},\mathbf{d})^2 + \rho(\mathbf{r},\mathbf{e})^2$. Consequently, $\langle\mathbf{a},\mathbf{b}\rangle = 0$. If, conversely, we know that $\langle\mathbf{a},\mathbf{b}\rangle = 0$, then by the remark made at the beginning of the proof, the lines l and m intersect each other. It then follows from the computation given above that $l \perp m$.

In Exercise 1.28 of the last section we already mentioned that the altitudes of a triangle are concurrent, at a point called the *orthocenter* of the triangle. We will now prove this using the inner product.

Example 1.32 (Orthocenter). Given a triangle ABC, we draw the altitudes AD and BE. These altitudes intersect at H; if AD were parallel to BE, then by Theorems 1.24 and 1.25, AC would be parallel to BC, which is impossible. We choose the origin of the coordinate system at H; see Fig. 1.26. The axes are not drawn in the figure on the right. As usual, we denote the vector associated

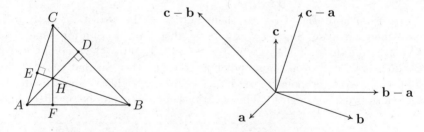

Fig. 1.26. The altitudes of a triangle are concurrent

to A by $\mathbf{a}$, and so on. We obtain the following parametric equations:

$$HA: \quad \mathbf{x} = \lambda_1\mathbf{a}, \qquad BC: \quad \mathbf{x} = \mathbf{b} + \mu_1(\mathbf{c}-\mathbf{b}),$$
$$HB: \quad \mathbf{x} = \lambda_2\mathbf{b}, \qquad AC: \quad \mathbf{x} = \mathbf{a} + \mu_2(\mathbf{c}-\mathbf{a}).$$

Since $HA \perp BC$ and $HB \perp AC$, we have $\langle\mathbf{a},\mathbf{c}-\mathbf{b}\rangle = 0$ and $\langle\mathbf{b},\mathbf{c}-\mathbf{a}\rangle = 0$. Moreover,
$$\langle\mathbf{a},\mathbf{c}-\mathbf{b}\rangle - \langle\mathbf{b},\mathbf{c}-\mathbf{a}\rangle = \langle\mathbf{a}-\mathbf{b},\mathbf{c}\rangle \, ,$$
so $\langle\mathbf{a}-\mathbf{b},\mathbf{c}\rangle = 0$. This implies that HC is perpendicular to BA, that is, that HC is the altitude on AB. The three altitudes therefore all pass through H.

Many other properties can easily be proved using vectors. The parallelogram law is a nice example of this.

1.6 The Inner Product and Determinant 33

Example 1.33 (Parallelogram Law). If $ABCD$ is a parallelogram, then

$$d(A,C)^2 + d(B,D)^2 = 2\left(d(A,B)^2 + d(B,C)^2\right) .$$

In words: the sum of the squares of the two diagonals is equal to the sum of the squares of the four sides. This is called the *parallelogram law*. To prove that

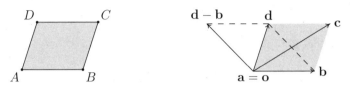

Fig. 1.27. The parallelogram law

the formula is correct, we choose the origin of the coordinate system at A. We see that $d(A,B) = \|\mathbf{b}\|$ and $d(A,D) = \|\mathbf{d}\|$. Since $ABCD$ is a parallelogram, we have $d(B,C) = d(A,D)$ and $\mathbf{c} = \mathbf{b} + \mathbf{d}$. Moreover, $d(B,D) = \|\mathbf{d} - \mathbf{b}\|$. Consequently,

$$\begin{aligned}\|\mathbf{d}+\mathbf{b}\|^2 + \|\mathbf{d}-\mathbf{b}\|^2 &= \langle \mathbf{d}+\mathbf{b}, \mathbf{d}+\mathbf{b}\rangle + \langle \mathbf{d}-\mathbf{b}, \mathbf{d}-\mathbf{b}\rangle \\ &= \|\mathbf{d}\|^2 + 2\langle \mathbf{b},\mathbf{d}\rangle + \|\mathbf{b}\|^2 + \|\mathbf{d}\|^2 - 2\langle \mathbf{b},\mathbf{d}\rangle + \|\mathbf{b}\|^2 \\ &= 2\left(\|\mathbf{b}\|^2 + \|\mathbf{d}\|^2\right) .\end{aligned}$$

This proves the parallelogram law.

The following theorem states two important inequalities that are closely related to the triangle inequality. The first is again called the *triangle inequality*, the second is known as the *Cauchy–Bunyakovskiĭ–Schwarz inequality*.

Theorem 1.34. *The following inequalities hold for every $\mathbf{x}$ and $\mathbf{y}$ in $\mathbb{R}^2$:*

1. $\|\mathbf{x}+\mathbf{y}\| \leq \|\mathbf{x}\| + \|\mathbf{y}\|$.
2. $|\langle \mathbf{x},\mathbf{y}\rangle| \leq \|\mathbf{x}\|\,\|\mathbf{y}\|$.

Proof. Consider the points $\mathbf{o}$, $\mathbf{x}$, and $\mathbf{x}+\mathbf{y}$. The triangle inequality says that

$$\rho\left(\mathbf{o},(\mathbf{x}+\mathbf{y})\right) \leq \rho(\mathbf{o},\mathbf{x}) + \rho\left(\mathbf{x},(\mathbf{x}+\mathbf{y})\right), \text{ that is, } \|\mathbf{x}+\mathbf{y}\| \leq \|\mathbf{x}\| + \|\mathbf{y}\| .$$

This proves the first inequality. By substituting $-\mathbf{y}$ for $\mathbf{y}$ we obtain $\|\mathbf{x}-\mathbf{y}\| \leq \|\mathbf{x}\| + \|\mathbf{y}\|$. Since $\|\mathbf{x}-\mathbf{y}\|^2 = \langle \mathbf{x}-\mathbf{y}, \mathbf{x}-\mathbf{y}\rangle$, squaring this last inequality gives

$$\|\mathbf{x}\|^2 - 2\langle \mathbf{x},\mathbf{y}\rangle + \|\mathbf{y}\|^2 \leq \|\mathbf{x}\|^2 + 2\|\mathbf{x}\|\,\|\mathbf{y}\| + \|\mathbf{y}\|^2 ,$$

and therefore $-\langle \mathbf{x},\mathbf{y}\rangle \leq \|\mathbf{x}\|\cdot\|\mathbf{y}\|$. Analogously we obtain $\langle \mathbf{x},\mathbf{y}\rangle \leq \|\mathbf{x}\|\,\|\mathbf{y}\|$. The Cauchy–Bunyakovskiĭ–Schwarz inequality follows from this.

Digression 2. This seems like the right place to mention a common method for setting up plane geometry. First one introduces the vector space $\mathbb{R}^2$. Then the norm and inner product are introduced in the same way as in the last section. Next one proves the Cauchy–Bunyakovskiĭ–Schwarz inequality, for example as is done in Exercise 1.34, and the resulting triangle inequality (compare this to Exercise 1.33). After this, lines are introduced, for example using parametric equations. As we have seen, the notions of parallel and perpendicular can be introduced in terms of vectors. Finally, one can show that the basic assumptions stated in Sect. 1.4 hold. Therefore, $\mathbb{R}^2$ has the structure needed to develop plane geometry.

The Distance from a Point to a Line

We conclude this section with the computation of the distance from a point to a line. To determine the distance from a point P to a line l, we drop the perpendicular m from P onto l. Let Q be the foot of m, that is, the intersection

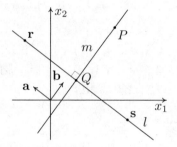

Fig. 1.28. Determining the distance from P to l

point of l and m. Then, by definition, $d(P,Q)$ is the *distance from P to Q*. By the definition of perpendicular, we easily see that $d(P,Q) < d(P,X)$ for every X on l other than Q. To determine the distance from P to l, we first rewrite the equation of l in a special form. We start with the equation (1.7) of l in some coordinate system:

$$\mathbf{x} = \mathbf{q} + \lambda \mathbf{a}, \text{ that is, } \begin{cases} x_1 = q_1 + \lambda a_1, \\ x_2 = q_2 + \lambda a_2. \end{cases}$$

The line l is parallel to the span of $\mathbf{a}$. By eliminating λ we obtain the following equation of l:

$$a_2 x_1 - a_1 x_2 = a_2 q_1 - a_1 q_2 \,.$$

Setting $\mathbf{b} = (b_1, b_2) = (a_2, -a_1)$ and $c = a_2 q_1 - a_1 q_2$, we can rewrite this equation in the form

$$\boxed{b_1 x_1 + b_2 x_2 = c\,,} \quad \text{that is,} \quad \boxed{\langle \mathbf{b}, \mathbf{x} \rangle - c = 0\,.} \tag{1.8}$$

1.6 The Inner Product and Determinant

The line l is therefore determined by an equation of degree one. We see that $\langle \mathbf{b}, \mathbf{a} \rangle = 0$. Consequently, the span of $\mathbf{b}$ is perpendicular to the span of $\mathbf{a}$, hence also perpendicular to l. This proves the first statement of the following theorem.

Theorem 1.35. *If the line l has equation $\langle \mathbf{b}, \mathbf{x} \rangle - c = 0$ with $c = \langle \mathbf{b}, \mathbf{q} \rangle$, then l and the span of $\mathbf{b}$ are perpendicular to each other. Moreover, the distance from a point $\mathbf{p}$ to l is equal to $|\langle \mathbf{b}, \mathbf{p} \rangle - c|/\|\mathbf{b}\|$.*

Proof. See Fig. 1.28. Draw the perpendicular m from P onto l and let Q be its foot. Since m is parallel to the span of $\mathbf{b}$, the vector $\mathbf{p} - \mathbf{q}$ is a multiple of $\mathbf{b}$, say $\mathbf{p} - \mathbf{q} = \lambda \mathbf{b}$. Then $\mathbf{p} = \mathbf{q} + \lambda \mathbf{b}$, and the distance from $\mathbf{p}$ to l is equal to the distance from $\mathbf{p}$ to $\mathbf{q}$, which is $|\lambda|\,\|\mathbf{b}\|$. Since $\mathbf{q}$ lies on l, $\langle \mathbf{b}, \mathbf{q} \rangle - c = 0$. We therefore have

$$\frac{|\langle \mathbf{b}, \mathbf{p} \rangle - c|}{\|\mathbf{b}\|} = \frac{|\langle \mathbf{b}, \mathbf{q} + \lambda \mathbf{b} \rangle - c|}{\|\mathbf{b}\|} = \frac{|\langle \mathbf{b}, \mathbf{q} \rangle + \langle \mathbf{b}, \lambda \mathbf{b} \rangle - c|}{\|\mathbf{b}\|} = |\lambda|\,\|\mathbf{b}\|,$$

the distance from $\mathbf{p}$ to l.

Example 1.36 (Determinant). We can use the formula from the last theorem to find a formula for the area of the parallelogram spanned by vectors $\mathbf{a}$ and $\mathbf{b}$. First we take a vector $\mathbf{c}$ such that $\|\mathbf{c}\| = 1$ and $\langle \mathbf{b}, \mathbf{c} \rangle = 0$. This vector must

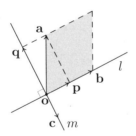

Fig. 1.29. Geometric interpretation of the determinant

equal either $(b_2, -b_1)/\|\mathbf{b}\|$ or $(-b_2, b_1)/\|\mathbf{b}\|$. We choose the first possibility. Let l and m be the respective spans of $\mathbf{b}$ and $\mathbf{c}$; see Fig. 1.29. The equation of l is $\langle \mathbf{c}, \mathbf{x} \rangle = 0$. According to Theorem 1.35, the distance from $\mathbf{a}$ to l, that is, the height of the parallelogram, is equal to $|\langle \mathbf{c}, \mathbf{a} \rangle|$. The area of the parallelogram is therefore $\|\mathbf{b}\|\,|\langle \mathbf{c}, \mathbf{a} \rangle|$. By the definition of $\mathbf{c}$, we have

$$\|\mathbf{b}\|\langle \mathbf{c}, \mathbf{a} \rangle = \langle \|\mathbf{b}\|\mathbf{c}, \mathbf{a} \rangle = \langle (b_2, -b_1), (a_1, a_2) \rangle = a_1 b_2 - a_2 b_1 .$$

Consequently, the area is $|a_1 b_2 - a_2 b_1|$. The number $a_1 b_2 - a_2 b_1$ is called the *determinant* of $\mathbf{a}$ and $\mathbf{b}$ and is written

$$\boxed{\det(\mathbf{ab}) = a_1 b_2 - a_2 b_1 .}$$

The order of the vectors is important. By dropping perpendiculars from the point **a** onto l and m we obtain $\mathbf{a} = \mathbf{p} + \mathbf{q}$, where **p** and **q** are the *projections* of **a** on respectively l and m. It follows from the computations above that

$$\|\mathbf{q}\| = |\langle \mathbf{c}, \mathbf{a} \rangle| = \frac{|\det(\mathbf{ab})|}{\|\mathbf{b}\|}.$$

We can now also compute $\|\mathbf{p}\|$ using the Pythagorean theorem. After some heavy computation, we obtain

$$\|\mathbf{p}\| = \frac{|\langle \mathbf{a}, \mathbf{b} \rangle|}{\|\mathbf{b}\|}, \text{ whence } |\langle \mathbf{a}, \mathbf{b} \rangle| = \|\mathbf{p}\| \|\mathbf{b}\|.$$

Exercises

1.30. $\langle \mathbf{a}, \mathbf{b} \rangle = (1/4) \left(\|\mathbf{a} + \mathbf{b}\|^2 - \|\mathbf{a} - \mathbf{b}\|^2 \right)$.

1.31. A *rhombus* is a parallelogram with four equal sides. A parallelogram is a rhombus if and only if the diagonals are perpendicular to each other.
Hint: Use Exercise 1.25 of Sect. 1.5.

1.32. A parallelogram is a rectangle if and only if the diagonals have the same length.

1.33. Deduce the triangle inequality $\|\mathbf{x} + \mathbf{y}\| \leq \|\mathbf{x}\| + \|\mathbf{y}\|$ from the Cauchy–Bunyakovskiĭ–Schwarz inequality.

1.34. For any **x** in $\mathbb{R}^2$, $\mathbf{y} \neq \mathbf{o}$ in $\mathbb{R}^2$, and λ in $\mathbb{R}$, we have

$$\langle \mathbf{x} + \lambda \mathbf{y}, \mathbf{x} + \lambda \mathbf{y} \rangle = \left(\lambda \|\mathbf{y}\| + \frac{\langle \mathbf{x}, \mathbf{y} \rangle}{\|\mathbf{y}\|} \right)^2 + \|\mathbf{x}\|^2 - \frac{\langle \mathbf{x}, \mathbf{y} \rangle^2}{\|\mathbf{y}\|^2}.$$

For what value of λ is the expression on the right-hand side minimal? Use this formula to prove the Cauchy–Bunyakovskiĭ–Schwarz inequality.

1.35. Let $ax_1 + bx_2 = c$ be the equation of a line l. If $b \neq 0$, we call $\mu_l = -a/b$ the *slope* of l. If line m has slope μ_m, then $l \perp m$ if and only if $\mu_l \mu_m = -1$.

1.36. The equation of the perpendicular bisector of the line segment $[\mathbf{ab}]$ is

$$\langle \mathbf{b} - \mathbf{a}, \mathbf{x} \rangle = \tfrac{1}{2} \left(\|\mathbf{b}\|^2 - \|\mathbf{a}\|^2 \right).$$

1.37. For any **a** and **b**, $\det(\mathbf{ab}) = -\det(\mathbf{ba})$ and $\det(\mathbf{aa}) = 0$. For any **a** and **b** with $\mathbf{a} \neq \mathbf{o} \neq \mathbf{b}$, $\det(\mathbf{ab}) = 0$ if and only if there is a λ such that $\mathbf{a} = \lambda \mathbf{b}$.

1.38. Let $a_1 x_1 + a_2 x_2 + a_3 = 0$ and $b_1 x_1 + b_2 x_2 + b_3 = 0$ be the respective equations of lines l and m.

(a) $l \mathbin{/\mkern-5mu/} m$ if and only if $a_1 b_2 - a_2 b_1 = 0$.
(b) $l \perp m$ if and only if $a_1 b_1 + a_2 b_2 = 0$.

1.39. If $\langle \mathbf{x}, \mathbf{y} \rangle = \|\mathbf{x}\| \|\mathbf{y}\|$ and $\mathbf{x} \neq \mathbf{o} \neq \mathbf{y}$, then $\|\mathbf{x} + \mathbf{y}\| = \|\mathbf{x}\| + \|\mathbf{y}\|$ and **x** and **y** lie on the line segment $[\mathbf{o}\,(\mathbf{x} + \mathbf{y})]$.

2
TRANSFORMATIONS

Much of this chapter concerns isometries of the plane V; these are surjective, distance-preserving maps from V to V. Why are isometries so important? Many of the concepts developed in the previous chapter were defined using metrics and general set-theoretic properties. Because of this, these concepts are invariant under isometric surjections; if a figure has such a property, so does its image under an isometric surjection.

The *reflection in a line* deserves special attention. This is by far the most important isometry, because every isometry of the plane is a composition of suitably chosen reflections. We will use this property to analyze the isometries of the plane. In addition to the reflection in a line, we distinguish the *reflection in a point, translation, rotation*, and *glide reflection*. In Table 2.1 we give a simple overview of the different isometries, including a number of characteristic properties. In passing, we also discuss mirror symmetry and point symmetry.

Similarities are another important type of transformation. We discuss them in Sect. 2.5. In that section, we anticipate the treatment of the nine-point circle in Chap. 4 and begin studying it; we use similarities to prove the property that gives the circle its name. In Sect. 2.6 we discuss fractals, which are figures that are similar to themselves in a very particular way.

Parallel to our analysis of transformations, we also study *congruences* and *similarities*. In this way, we obtain well-known congruence and similarity criteria in a natural manner. The notion of an *angle* is an indispensable element in these considerations; we introduce it in Sect. 2.2. We also introduce the important concept of *angle orientation* and the fact that a reflection reverses the orientation. We can divide the isometries into two classes: the *direct* isometries, which preserve the orientation, and the *indirect* or *opposite* isometries, which reverse the orientation.

2.1 The Reflection in a Line, Congruence

The Euclidean plane V was introduced in the last chapter as a metric space with certain properties, which are stated in basic assumptions. In this chapter we discuss the structure of the isometries of V in more detail. An *isometry* of V is a surjective distance-preserving map from V to V; see Sect. 1.3. It is automatically bijective. All concepts introduced in the last chapter in terms of metrics are invariant under isometries. These include the line segment and the line; see the proofs of Theorem 1.15 and Theorem 1.16. This means that the image of a line segment, respectively a line under an isometry, is a line segment of equal length, respectively a line.

The reflection in a line is possibly the most important isometry of the plane. It is the building block of all other isometries.

Definition 2.1. *Let l be a line in the plane. The* reflection in the line l *is the map S_l from the plane to itself defined as follows:*

1. *For every point X of l, we have $S_l(X) = X$.*
2. *For any point X not on l, $S_l(X)$ is the unique point such that l is the perpendicular bisector of the line segment $[X\,S_l(X)]$.*

We call l the reflection axis *of S_l.*

If X does not lie on the reflection axis l, we can find $S_l(X)$ by drawing the perpendicular on l through X. The point $S_l(X)$ lies on this perpendicular, has the same distance to the foot as X, and lies on the opposite side of the foot from X; see Fig. 2.1. We call the point X a *fixed point* of an isometry $\mathcal{F}$ if $\mathcal{F}(X) = X$. The fixed points of the reflection S_l are exactly the points of l. It follows that any two reflections S_l and S_m are equal if and only if $l = m$.

Let us prove that the reflection S_l is indeed an isometry. We first show that it is surjective, and then that it preserves distances. It follows directly from the definition of a reflection that

$$S_l \circ S_l = \mathrm{id}_V.$$

In other words, applying the same reflection twice gives the identity map, which leaves every point in place. We conclude that a reflection is surjective: every point P is the image of $S_l(P)$ under the reflection S_l. Let X and Y be distinct points. Let us first assume that the line segment $[XY]$ does not meet the reflection axis l; see Fig. 2.1, on the left. Let us assume, moreover, that Y does not lie on the line $XS_l(X)$; if it does lie on that line, the proof needs to be adapted, which we leave to the reader. Let P and Q be the feet

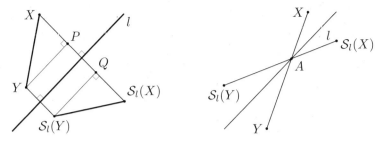

Fig. 2.1. The reflection S_l is an isometry

of the perpendiculars from respectively Y and $S_l(Y)$ to the line joining X and $S_l(X)$.

It is clear that $YS_l(Y)QP$ is a rectangle and that the reflection axis l bisects the sides $[Y\,S_l(Y)]$ and $[PQ]$. It follows that $d(Y,P) = d(S_l(Y),Q)$ and $d(X,P) = d(S_l(X),Q)$. Definition 1.22 (Pythagoras) now implies that $d(X,Y) = d\left(S_l(X), S_l(Y)\right)$. In general, see Fig. 2.1 on the right, we consider the line segments $[XA]$ and $[AY]$ in which the axis l divides $[XY]$ and apply similar reasoning to each of the two segments.

We can also use coordinates to prove that a reflection is an isometry. In a coordinate system whose x_1-axis lies along the reflection axis l, the reflection S_l is given by the formula $S_l\left((x_1, x_2)\right) = (x_1, -x_2)$. The correctness of this formula follows from the fact that the x_1-axis is the perpendicular bisector of the line segment from (x_1, x_2) to $(x_1, -x_2)$. We can now easily verify that the reflection preserves distances.

The name *reflection* is evocative. If we imagine a physical mirror placed perpendicularly to this page, see Fig. 2.2, the eye sees both the figure F and

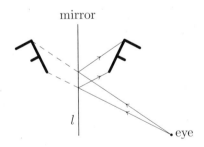

Fig. 2.2. Construction of the reflected rays using the mirror image

its reflection $S_l(F)$. In fact, we use the virtual extension inside the mirror of the light rays through which we see the reflected image to reconstruct the real light rays. The reflection axis is also called an *axis of rotation*. The image that is associated to this name is the plane turning over in space, where the reflection axis is used as rotation axis. The final result of turning around an axis is of course equal to the reflection in that axis. We will say that a figure F in the plane has a *mirror symmetry* if there exists a line l (the reflection axis) for which $\mathcal{S}_l(F) = F$. The *symmetry group* of a figure F is the set of isometries of the plane that map F onto itself. Saying that a figure has a mirror symmetry therefore means that there is a reflection in the symmetry group of the figure. In Fig. 2.3, the figures on the left have mirror symmetries, while those on the

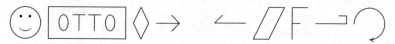

Fig. 2.3. (a) Figures with mirror symmetry; (b) figures without mirror symmetry

right do not.

We will now study isometries and the role of reflections. This discussion has some overlap with that in Sect. 1.3. Later on in this chapter, in Sect. 2.4, we will characterize isometries using the number of fixed points. The following theorem is the first step toward that characterization.

Theorem 2.2. *An isometry $\mathcal{F}$ of the plane V with at least two fixed points is either a reflection or the identity map id_V.*

Proof. Consider Fig. 2.4. Let P and Q be two distinct fixed points of the

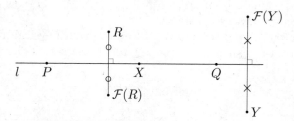

Fig. 2.4. An isometry with two fixed points

isometry $\mathcal{F}$. If $\mathcal{F} = \mathrm{id}_V$, that is, if every point is a fixed point of $\mathcal{F}$, there is nothing left to prove. We may therefore assume that $\mathcal{F} \neq \mathrm{id}_V$. Consequently, there exists a point R such that $\mathcal{F}(R) \neq R$. For any fixed point X of $\mathcal{F}$ we have
$$d(X, R) = d\left(\mathcal{F}(X), \mathcal{F}(R)\right) = d\left(X, \mathcal{F}(R)\right) .$$
It follows that X lies on the perpendicular bisector l of the line segment $[R\,\mathcal{F}(R)]$. This holds in particular for the points P and Q, which implies that $l = PQ$, the line through P and Q. Let Y be a point outside l; this is not a fixed point. Consequently, as above, both P and Q lie on the perpendicular bisector of $[Y\,\mathcal{F}(Y)]$, which is therefore also l. By the definition of $\mathcal{S}_l$, it follows that $\mathcal{F} = \mathcal{S}_l$, which proves that $\mathcal{F}$ is a reflection.

2.1 The Reflection in a Line, Congruence   41

We can deduce the following statement from the proof we have just given.

Theorem 2.3. *The only isometry of the plane with three noncollinear fixed points is the identity map.*

Proof. Let A, B, and C be three noncollinear fixed points of an isometry $\mathcal{F}$. Suppose there exists a point R such that $\mathcal{F}(R) \neq R$. By the proof of the last theorem, every fixed point of $\mathcal{F}$ lies on the perpendicular bisector of the line segment $[R\mathcal{F}(R)]$. This gives a contradiction, since the fixed points A, B, and C are not collinear. Consequently, every point is a fixed point, and $\mathcal{F} = \mathrm{id}_V$.

We noted earlier that the reflections are the building blocks of the isometries. We will now make that remark more precise: every isometry of the plane is the composition of at most three reflections. This statement is worked out in the following theorem.

Theorem 2.4. *Let A, B, and C be three noncollinear points in the plane. Let, moreover, A_3, B_3, and C_3 be points such that $d(A,B) = d(A_3, B_3)$, $d(B,C) = d(B_3, C_3)$, and $d(C,A) = d(C_3, A_3)$. There is exactly one isometry $\mathcal{F}$ of the plane with $\mathcal{F}(A) = A_3$, $\mathcal{F}(B) = B_3$, and $\mathcal{F}(C) = C_3$.*

Every isometry $\mathcal{G}$ of the plane is the composition of at most three reflections.

Proof. Consider Fig. 2.5. Let us first observe that the points A_3, B_3, and C_3

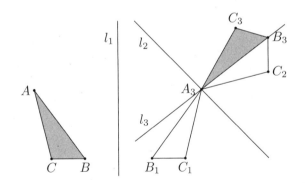

Fig. 2.5. Composition of reflections

are noncollinear. If, for example, C_3 were to lie on the segment $[A_3, B_3]$, we would have $d(A_3, C_3) + d(C_3, B_3) = d(A_3, B_3)$, and therefore $d(A,C) + d(C,B) = d(A,B)$. This would imply that C lies on $[AB]$, which contradicts the hypothesis. We will construct the isometry $\mathcal{F}$ in three steps. First draw the perpendicular bisector l_1 of the line segment $[AA_3]$; if $A = A_3$, this step is left out. The reflection $\mathcal{S}_1$ in l_1 transforms triangle ABC into triangle

$A_3B_1C_1$; it maps point A to the desired position. Next draw the perpendicular bisector l_2 of line segment $[B_1B_3]$; if $B_1 = B_3$, this step is left out. The reflection $\mathcal{S}_2$ in l_2 transforms triangle $A_3B_1C_1$ into triangle $A_3B_3C_2$; it maps the line segment $[AB]$ to the desired position. If $C_2 = C_3$, triangle ABC has been transformed into triangle $A_3B_3C_3$. If this is not the case, we conclude with the following step. Since the reflections $\mathcal{S}_1$ and $\mathcal{S}_2$ are isometries, we have $d(A_3, C_2) = d(A_3, C_3)$ and $d(B_3, C_2) = d(B_3, C_3)$. The points A_3 and B_3 therefore lie on the perpendicular bisector l_3 of $[C_2C_3]$. The reflection $\mathcal{S}_3$ in l_3 therefore maps C_2 onto C_3 and leaves A_3 and B_3 in place. We let $\mathcal{F}$ denote the composition $\mathcal{S}_3 \circ \mathcal{S}_2 \circ \mathcal{S}_1$ of the reflections $\mathcal{S}_1$, $\mathcal{S}_2$, and $\mathcal{S}_3$; $\mathcal{F}$ is the isometry we were looking for.

We still need to prove that $\mathcal{F}$ is unique. For this we use Remark 1.10. Let $\mathcal{G}$ be an isometry with $\mathcal{G}(A) = A_3$, $\mathcal{G}(B) = B_3$, and $\mathcal{G}(C) = C_3$. Let $\mathcal{H} = \mathcal{S}_1 \circ \mathcal{S}_2 \circ \mathcal{S}_3 \circ \mathcal{G}$. Then we can easily check that $\mathcal{H}(A) = A$, $\mathcal{H}(B) = B$, and $\mathcal{H}(C) = C$. It follows from Theorem 2.3 that $\mathcal{H} = \mathrm{id}_V$. Since every reflection $\mathcal{S}$ satisfies $\mathcal{S} \circ \mathcal{S} = \mathrm{id}_V$, we have

$$\begin{aligned} \mathcal{F} &= \mathcal{S}_3 \circ \mathcal{S}_2 \circ \mathcal{S}_1 = \mathcal{S}_3 \circ \mathcal{S}_2 \circ \mathcal{S}_1 \circ \mathcal{H} \\ &= \mathcal{S}_3 \circ \mathcal{S}_2 \circ \mathcal{S}_1 \circ \mathcal{S}_1 \circ \mathcal{S}_2 \circ \mathcal{S}_3 \circ \mathcal{G} \\ &= \mathcal{S}_3 \circ \mathcal{S}_2 \circ \mathcal{S}_2 \circ \mathcal{S}_3 \circ \mathcal{G} = \mathcal{G}. \end{aligned}$$

This proves the uniqueness of $\mathcal{F}$.

Let us now prove the last statement of the theorem. Let $\mathcal{G}$ be an isometry of the plane. Choose three noncollinear points A, B, and C. By what we have just shown, there exists an isometry $\mathcal{F}$ that is the composition of at most three reflections such that $\mathcal{F}(A) = \mathcal{G}(A)$, $\mathcal{F}(B) = \mathcal{G}(B)$, and $\mathcal{F}(C) = \mathcal{G}(C)$. By the uniqueness of $\mathcal{F}$, which we have just proved, $\mathcal{F} = \mathcal{G}$. Hence $\mathcal{G}$ is the composition of at most three reflections.

Digression 3. *The first property stated in the definition of isometry is surjectivity. This condition is often left out, keeping only the condition that an isometry be distance-preserving. In the case of isometries of the plane, the resulting definition is equivalent! This can be explained as follows. As noted above, a reflection $\mathcal{S}$ is surjective, because $\mathcal{S} \circ \mathcal{S} = \mathrm{id}_V$. We have just proved that every isometry of the plane is the composition of at most three reflections. The proof does not use the surjectivity of the map $\mathcal{G}$, just that it preserves distances. Consequently, as a composition of surjective maps, $\mathcal{G}$ is also surjective, and therefore an isometry.*

Congruence

How can we decide whether there exist isometries transforming a given figure into another? This is where congruence criteria come in. The congruence criteria concern triangles; if two triangles have certain properties in common, there is an isometry of the plane that maps one of the triangles onto the other.

2.1 The Reflection in a Line, Congruence

Theorem 2.6 explains why we restrict ourselves to triangles in the congruence criteria. The first congruence criterion follows directly from Theorem 2.4. Let us first explain what a congruence is.

Definition 2.5. *We call two figures F_1 and F_2 in the plane congruent, denoted by $F_1 \cong F_2$, if there exists an isometric surjection $\mathcal{H}$ from F_1 to F_2.*

In short, congruent figures have the same properties. Of course, by properties we mean those that can be stated using the notion of distance. The congruence criteria concern triangles. Let us first recall the agreement to denote the side opposite a vertex by the corresponding lowercase letter; this letter often also denotes the length of the side. The first criterion concerns three equal sides.

Congruence Criterion 1 (SSS). *Two triangles ABC and $A_1B_1C_1$ are congruent if $a = a_1$, $b = b_1$, and $c = c_1$.*

The correctness of this criterion is obvious. Indeed, by the assumptions, the conditions of Theorem 2.4 are fulfilled, that is, $d(A, B) = d(A_1, B_1)$, etc. Consequently, there is an isometry that maps A to A_1, B to B_1, and C to C_1. Note that we obtain more than we asked for: we were looking for an isometric surjection from $\triangle ABC$ to $\triangle A_1B_1C_1$ and found an isometry from V to V that transforms $\triangle ABC$ into $\triangle A_1B_1C_1$. The following theorem shows that this was not a coincidence.

Theorem 2.6. *Let F_1 and F_2 be congruent figures, where F_1 contains at least three noncollinear points. If $\mathcal{H}$ is an isometric surjection from F_1 to F_2, there is exactly one isometry $\mathcal{G}$ from V to V such that $\mathcal{G}$ restricted to F_1 corresponds to $\mathcal{H}$: $\mathcal{G}(X) = \mathcal{H}(X)$ for all X in F_1.*

This theorem is worth pausing to think about. If two triangles ABC and $A_1B_1C_1$ are congruent, then by definition there is an isometric surjection, say $\mathcal{H}$, from triangle ABC to triangle $A_1B_1C_1$. After, if necessary, permuting the names of the vertices of triangle $A_1B_1C_1$, we may assume that $\mathcal{H}(A) = A_1$, $\mathcal{H}(B) = B_1$, and $\mathcal{H}(C) = C_1$. Then $d(A, B) = d(A_1, B_1)$, and so on. By Theorem 2.4, there is exactly one isometry $\mathcal{G}$ from V to V such that $\mathcal{G}(A) = A_1$, $\mathcal{G}(B) = B_1$, and $\mathcal{G}(C) = C_1$. This isometry is defined on the whole plane, and is the only isometry of the plane satisfying $\mathcal{G}(A) = A_1$, $\mathcal{G}(B) = B_1$, and $\mathcal{G}(C) = C_1$. We will now show that the action of $\mathcal{G}$ on F_1 is the same as that of $\mathcal{H}$. As a byproduct we will obtain the uniqueness of the isometric surjection $\mathcal{H}$; this is a direct consequence of the uniqueness of $\mathcal{G}$.

Proof. Choose three noncollinear points in F_1, and call these A, B, and C. Since F_1 and F_2 are congruent, there is an isometric surjection $\mathcal{H}$ from F_1 to F_2. Moreover, since $\mathcal{H}$ is distance-preserving, $d(A, B) = d(\mathcal{H}(A), \mathcal{H}(B))$, and so on. By Theorem 2.4, there is exactly one isometry $\mathcal{G}$ from V to V that maps A, B, C onto respectively $\mathcal{H}(A)$, $\mathcal{H}(B)$, $\mathcal{H}(C)$. Let us now prove, by contradiction, that $\mathcal{G}(X) = \mathcal{H}(X)$ for all points X of F_1. Assume that $\mathcal{H}(Y) \neq \mathcal{G}(Y)$ for some Y in F_1. Since $\mathcal{H}$ and $\mathcal{G}$ preserve distances, we have

$$d(\mathcal{G}(A), \mathcal{G}(Y)) = d(A, Y) = d(\mathcal{H}(A), \mathcal{H}(Y)) .$$

Since $\mathcal{H}(A) = \mathcal{G}(A)$, we see that $\mathcal{H}(A)$ lies on the perpendicular bisector of the line segment $[\mathcal{H}(Y)\mathcal{G}(Y)]$. As this also holds for $\mathcal{H}(B)$ and $\mathcal{H}(C)$, we conclude that $\mathcal{H}(A)$, $\mathcal{H}(B)$, and $\mathcal{H}(C)$ are collinear. Consequently $\mathcal{H}(A) \in [\mathcal{H}(B)\mathcal{H}(C)]$, $\mathcal{H}(B) \in [\mathcal{H}(A)\mathcal{H}(C)]$, or $\mathcal{H}(C) \in [\mathcal{H}(A)\mathcal{H}(B)]$. As $\mathcal{H}$ preserves distances, it follows that $A \in [BC]$, $B \in [AC]$, or $C \in [AB]$. But then A, B, and C are also collinear. This gives a contradiction. We conclude that $\mathcal{H}(X) = \mathcal{G}(X)$, for all X in F_1.

Exercise

2.1. An isometry $\mathcal{F}$ with exactly one fixed point is the composition of two reflections.
Hint: Consider three noncollinear points A, B, and C with $A = \mathcal{F}(A)$.

2.2. The three mid-parallels of a triangle divide the triangle into four congruent triangles.

2.3. If two reflections $\mathcal{S}_1$ and $\mathcal{S}_2$ satisfy $\mathcal{S}_1 \circ \mathcal{S}_2 = \mathrm{id}_V$, then $\mathcal{S}_1 = \mathcal{S}_2$.

2.4. For $i = 1, 2$, let $\mathcal{S}_i$ be the reflection in the line l_i. If $l_1 \perp l_2$, then $\mathcal{S}_1 \circ \mathcal{S}_2 = \mathcal{S}_2 \circ \mathcal{S}_1$.
Hint: Choose a coordinate system with axes along l_1 and l_2.

2.5. Let $\mathcal{S}_1$ and $\mathcal{S}_2$ be reflections satisfying $\mathcal{S}_1 \circ \mathcal{S}_2 = \mathcal{S}_2 \circ \mathcal{S}_1$. Show that $(\mathcal{S}_1 \circ \mathcal{S}_2) \circ (\mathcal{S}_1 \circ \mathcal{S}_2) = \mathrm{id}_V$.

2.6. If P and Q are fixed points of an isometry $\mathcal{F}$ of the plane, every point on the line PQ is a fixed point of $\mathcal{F}$.
Hint: Recall the definition of a straight line.

2.7. Any two lines are congruent. Two line segments $[AB]$ and $[CD]$ are congruent if and only if they have the same length ($d(A, B) = d(C, D)$).

2.8. Let l and m be two lines, let $F_1 \subseteq l$ be a subset containing more than one point, and suppose that $F_2 \subseteq m$ and $F_1 \cong F_2$. Then:

(a) If $\mathcal{H}$ is an isometric surjection from F_1 to F_2, there exists exactly one isometric surjection $\mathcal{G}$ from l to m such that $\mathcal{G}$ restricted to F_1 corresponds to $\mathcal{H}$.
(b) There are exactly two isometries of V whose action on F_1 corresponds to that of $\mathcal{H}$.

2.2 Angle Measure, Orientation

Up to now we have avoided using angles. The reason for this is that the notion of angle is not as simple as it may seem. We begin by explaining that two intersecting lines determine four angles. Let us first consider one line, l. The rest of the plane, $V \setminus l$, consists of two *open half-planes* H_{l1} and H_{l2}; see Fig. 2.6. Two points P and Q belong to the same open half-plane if and

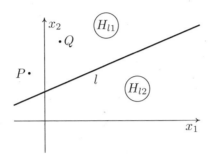

Fig. 2.6. The line l divides the plane into two parts

only if the line segment $[PQ]$ and the line l have no common point, that is, if $[PQ] \cap l = \emptyset$. In this case we also say that P and Q lie on the *same side* of l. If P and Q do not lie on l and $[PQ] \cap l \neq \emptyset$, we say that P and Q lie on *different sides* of l. Even though these statements seem obvious, it is still worth substantiating them with an analytic proof. Let us introduce a coordinate system. The equation of l is $\langle \mathbf{b}, \mathbf{x} \rangle - c = 0$; see (1.8). Define a function f from V to $\mathbb{R}$ by setting

$$f(\mathbf{x}) = \langle \mathbf{b}, \mathbf{x} \rangle - c, \quad \mathbf{x} \in V .$$

For any $\mathbf{x}$ in V, $\mathbf{x} \in l$ if and only if $f(\mathbf{x}) = 0$. Let

$$H_{l1} = \{\, \mathbf{x} : f(\mathbf{x}) > 0 \,\} \quad \text{and} \quad H_{l2} = \{\, \mathbf{x} : f(\mathbf{x}) < 0 \,\} .$$

The open half-planes into which l divides the plane are H_{l1} and H_{l2}. The statements above follow directly from the following lemma.

Lemma 2.7. *For $i = 1, 2$ we have that for all points $\mathbf{p}, \mathbf{q}$ that do not lie on l,*

$$\mathbf{p} \in H_{li} \text{ and } \mathbf{q} \in H_{li} \quad \text{if and only if} \quad [\mathbf{pq}] \cap l = \emptyset.$$

Proof. Let $\mathbf{y} \in [\mathbf{pq}]$. Then $\mathbf{y} = \lambda \mathbf{p} + (1 - \lambda)\mathbf{q}$ for some λ with $0 \leq \lambda \leq 1$. Let us determine $f(\mathbf{y})$:

$$\begin{aligned} f(\mathbf{y}) = \langle \mathbf{b}, \mathbf{y} \rangle - c &= \langle \mathbf{b}, \lambda \mathbf{p} + (1 - \lambda)\mathbf{q} \rangle - c \\ &= \lambda \langle \mathbf{b}, \mathbf{p} \rangle + (1 - \lambda)\langle \mathbf{b}, \mathbf{q} \rangle - c \\ &= \lambda(\langle \mathbf{b}, \mathbf{p} \rangle - c) + (1 - \lambda)(\langle \mathbf{b}, \mathbf{q} \rangle - c) \\ &= \lambda f(\mathbf{p}) + (1 - \lambda)f(\mathbf{q}) . \end{aligned}$$

If, for example, $\mathbf{p} \in H_{l1}$ and $\mathbf{q} \in H_{l1}$, then $f(\mathbf{p}) > 0$ and $f(\mathbf{q}) > 0$. It then follows from the computation that $f(\mathbf{y}) > 0$, whence $\mathbf{y} \in H_{l1}$. If $\mathbf{p} \in H_{l1}$ and $\mathbf{q} \in H_{l2}$, then $f(\mathbf{p}) > 0$ and $f(\mathbf{q}) < 0$. We can easily check that in this case

$$\lambda_0 \mathbf{p} + (1-\lambda_0)\mathbf{q} \in l \text{ for } \lambda_0 = -f(\mathbf{q})/(f(\mathbf{p}) - f(\mathbf{q})) \,.$$

Note that $0 < \lambda_0 < 1$.

Digression 4. *In 1880, Pasch (1843–1930) proved the following theorem, which was later named after him. This theorem is a typical example of a property that, when drawn, leaves no doubt as to its correctness. It was used tacitly in Euclid's Elements. Until the theorem was proved by Pasch, people were not aware of the fact that anything required proving. Hilbert (1862–1943) included the property as an axiom for plane geometry.*

Theorem 2.8 (Pasch's Axiom). *Let A, B, C be three noncollinear points, and let l be a line passing through none of them. If l meets the line segment $[AB]$, l also meets either the line segment $[BC]$ or the line segment $[AC]$, but not both.*

Proof. The proof follows without much trouble from Lemma 2.7; see Fig. 2.7. Let H_1 and H_2 be the half-planes into which l divides the plane. Since the

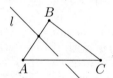

Fig. 2.7. Pasch's axiom

line l meets the line segment $[AB]$, the lemma implies that A and B lie in different half-planes; we may assume that $A \in H_1$ and $B \in H_2$. The point C does not lie on l, hence either $C \in H_1$ or $C \in H_2$. If $C \in H_1$, C and B lie in different half-planes, and $[BC]$ and l intersect. In the other case $[AC]$ and l intersect.

For a line l, we define the *closed half-planes* G_{li} as $G_{li} = H_{li} \cup l$, $i = 1, 2$.

Definition 2.9. *Let l and n be intersecting lines meeting at P. These lines define four angles, namely*

$$P_1 = G_{l1} \cap G_{n1}, \ P_2 = G_{l2} \cap G_{n1}, \ P_3 = G_{l2} \cap G_{n2}, \ P_4 = G_{l1} \cap G_{n2}.$$

See Fig. 2.8; P is the intersection point of l and n. The angles P_1 and P_2, which together form a closed half-plane, are called *supplementary angles*. The angles P_1 and P_4 are also supplementary. We say that P_1 and P_3 are *vertical*

2.2 Angle Measure, Orientation   47

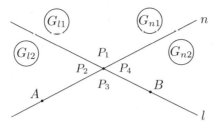

Fig. 2.8. Two intersecting lines form four angles

angles. The angles P_2 and P_4 are also vertical. We often denote an angle using three of its points: the intersection point P of the two lines that form the angle and one point other than P on each of the lines. For example, in Fig. 2.8, angle P_3 is the same as APB; the intersection point P is placed in the middle. For clarity, we note the following. The two lines that define angle P_3 form four open half-planes. The point A lies in exactly one of these open half-planes, namely H_{l2}. The point B also lies in exactly one of these half-planes, namely H_{n2}. The angle P_3 is the intersection of the corresponding closed half-planes: $P_3 = G_{l2} \cap G_{n2}$. For the moment we do not make a distinction between the angles APB and BPA. By the angle P we mean one of the angles P_i.

Definition 2.10. *Let P be the intersection point of lines l and m, and Q the intersection point of lines r and s. We say that angle P is congruent to angle Q, denoted by $P \cong Q$, if there is an isometric surjection from angle P to angle Q.*

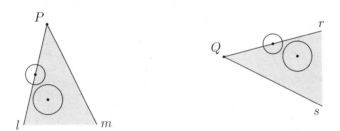

Fig. 2.9. The angles P and Q are congruent

In the definition we speak of the congruence of angles and not of equality; the latter deals with the measure of an angle, and will be considered later. It follows from Theorem 2.6 that if an angle P is congruent to an angle Q, there is an isometry $\mathcal{F}$ of the (whole) plane that transforms angle P into angle Q. We then have $\mathcal{F}(P) = Q$. This seems evident, but on closer inspection it requires an explanation. We assume that $P = G_{l1} \cap G_{m2}$, where we use the notation of the last two definitions. We call the intersection of the corresponding open half-planes $H_{l1} \cap H_{m2}$ the *interior* of angle P; it has been shaded in Fig. 2.9.

The parts of l and m that belong to the angle are called the *legs* of angle P; they form the boundary of the angle. We can characterize the interior of angle P as follows: for every point N of the interior there is an $r > 0$ such that the *circle* $\odot(N, r) = \{ X : d(X, N) = r \}$ lies entirely in angle P. This characterization uses only metric properties. It follows that $\mathcal{F}$ maps the interior of angle P onto the interior of angle Q. For every point R on the legs other than the intersection point, there is an $r > 0$ such that half the circle $\odot(R, r)$ lies inside angle P. It then follows that $\mathcal{F}$ maps the legs of angle P onto the legs of angle Q. Consequently $\mathcal{F}(P) = Q$.

Theorem 2.11. *If perpendicular lines l and m meet at the point P, the resulting angles $P_1, \ldots, P_4$ satisfy*

$$\text{angle } P_1 \cong \text{ angle } P_2 \cong \text{ angle } P_3 \cong \text{ angle } P_4.$$

We call the angle P_i, $i = 1, \ldots, 4$, a right angle.

The theorem follows by considering the effect of the reflections $\mathcal{S}_l$ and $\mathcal{S}_m$; see Fig. 2.10, on the left. If in addition to the lines l and m we also consider

Fig. 2.10. All right angles are congruent

perpendicular lines l_1 and m_1 that meet at Q, we can find an isometry $\mathcal{F}$ such that $\mathcal{F}(l) = l_1$ and $\mathcal{F}(P) = Q$. Since the perpendicular position is invariant under isometry, it follows from Theorem 1.24 that $\mathcal{F}(m) = m_1$. By applying $\mathcal{S}_l$, $\mathcal{S}_m$, respectively $\mathcal{S}_m \circ \mathcal{S}_l$, we can map angle P_1 to angle P_2, P_4, respectively P_3. The isometries $\mathcal{F}$, $\mathcal{F} \circ \mathcal{S}_l$, $\mathcal{F} \circ \mathcal{S}_m$, and $\mathcal{F} \circ \mathcal{S}_m \circ \mathcal{S}_l$ each map angle P_1 onto a different angle Q_j. The same holds for angles P_2, P_3, and P_4. Consequently, angle P_i is congruent to angle Q_j for all i and j. *Right angles are therefore congruent to each other.*

Sometimes it is also useful to consider the closed half-plane as an angle; we then call the half-plane a *straight angle*. The proof that *all straight angles are congruent* is even easier than that for right angles given above.

Now that we know what congruence means in terms of angles, we can state the second congruence criterion. We use the same notation as in the first criterion, **SSS**. This time we have two equal sides with congruent included angle.

Congruence Criterion 2 (SAS). *Two triangles ABC and $A_1 B_1 C_1$ are congruent if $a = a_1$, $c = c_1$, and angle B is congruent to angle B_1.*

Proof. Since angle B is congruent to angle B_1, there is an isometry Φ of V that maps angle B onto angle B_1. From the remarks made after Definition 2.10, we deduce that $\Phi(B) = B_1$ and that $\Phi(A)$ and $\Phi(C)$ lie on the legs of angle B_1. To simplify the notation, we place ourselves in the situation after the application of Φ and write A instead of $\Phi(A)$, and so on. If the points A and A_1 lie on the same leg of angle B_1, they coincide. In that case C and C_1 also coincide and there is nothing left to prove. Let us consider the other situation, where A and A_1 lie on different legs of B_1; see Fig. 2.11. Draw the perpendicular

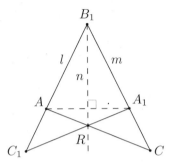

Fig. 2.11. The congruence criterion SAS

bisector n of the line segment $[AA_1]$. The point B_1 lies on n, since $d(A, B) = d(A_1, B_1)$ and $B = B_1$. Next apply the reflection $\mathcal{S}_n$ to $\triangle ABA_1$: $\mathcal{S}_n(B) = B_1$, $\mathcal{S}_n(A) = A_1$, and $\mathcal{S}_n(A_1) = A$. We see that $\mathcal{S}_n$ interchanges the lines l and m. Consequently, $\mathcal{S}_n(C) = C_1$, which concludes the proof. Let us just remark that the intersection point R of n and AC lies on $[A_1C_1]$, because $\mathcal{S}_n(R) = R$. The three lines n, AC, and A_1C_1 therefore concur.

Angle Measure

We will now introduce the measure of an angle. We consider angle P, shaded in Fig. 2.12, on the left, and draw a circle $\odot(P, 1)$ with center P and radius 1. The intersection of the circle and angle P is an arc, drawn thicker in the figure. We will not deduce this property, but rather state it with other properties in

Fig. 2.12. Angle measure and its additivity

a basic assumption. We use the length of the intercepted arc as the measure of the size of angle P. We denote this measure by $\angle P$. The unit of $\angle P$ is the *radian*; if the length of the arc is equal to 1, then $\angle P = 1$ rad. In the following basic assumption we state the properties of *angle measure*.

Basic Assumption 2.12. *The intersection of any angle P and the circle $\odot(P,1)$ is an arc; we denote the length of that arc by $\angle P$. The following properties hold:*

1. $0 < \angle P < \pi$ *for every angle P.*
2. *If angle P and angle Q are congruent, then $\angle P = \angle Q$.*
3. *See Fig. 2.12, on the right. For every real number c satisfying $0 < c < \angle APB$, there exists a point Q in the interior of APB such that $c = \angle APQ$.*
4. *See Fig. 2.12, on the right. If a point Q lies in the interior of angle APB, then*

$$\angle APB = \angle APQ + \angle QPB.$$

For other definitions of angle measure we refer to [48] or [56]. The first book includes a detailed account of the definition of arc length.

Property 4 above is called the *additivity* of the angle measure: we say that $\angle APB$ is the *sum* of $\angle APQ$ and $\angle QPB$. This property is closely related to the additivity of the arc length. Since the circumference of a circle with radius 1 is equal to 2π and all right angles are congruent, the additivity of the arc length implies that the measure of a right angle is equal to $\pi/2$ rad. At this point one could object that the basic assumptions do not include the additivity of the arc length, and that the deduction we have just given has no solid basis. But the deduction does give sufficient arguments for the normalization of the angle measure of a right angle at $\pi/2$. If we consider the half-plane as a straight angle, the straight angle is assigned the angle measure π rad. Part 2 of Basic Assumption 2.12 states that congruent angles have the same angle measure. The converse is also true.

Theorem 2.13. *If $\angle P = \angle Q$, the angles P and Q are congruent.*

Proof. Let A and B be distinct points on the legs of angle P such that $d(P,A) = 1$ and $d(P,B) = 1$. Let C and D be distinct points on the legs of angle Q such that $d(Q,C) = 1$ and $d(Q,D) = 1$. By Sect. 2.1, Exercise 2.7, there is an isometry $\mathcal{F}$ with $\mathcal{F}(P) = Q$ and $\mathcal{F}(A) = C$. If, as in Fig. 2.13, the points D and $\mathcal{F}(B)$ lie in different half-planes of the line QC, we first apply a reflection in QC.

We call the resulting map $\mathcal{G}$; we have $\mathcal{G}(P) = Q$, $\mathcal{G}(A) = C$. Moreover, D and $\mathcal{G}(B)$ lie on the same side of the line QC. The equality of the angles denoted by small circles in the figure implies that $\mathcal{G}(B) = D$; otherwise, there would be a contradiction with Basic Assumption 2.12, property 4.

Thanks to this theorem we can rephrase the congruence criterion **SAS** as follows.

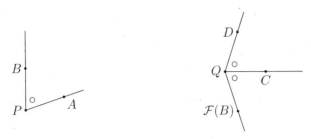

Fig. 2.13. If $\angle P = \angle Q$, then the angles are congruent

Congruence Criterion 3 (SAS; 2nd version). *Two triangles ABC and $A_1B_1C_1$ are congruent if $a = a_1$, $c = c_1$, and $\angle B = \angle B_1$.*

Because of Basic Assumption 2.12, property 2, and Theorem 2.13, we will not always be too strict in differentiating between congruent angles and angles of equal measure. If we say that two angles *are equal to each other*, we mean that they have the same measure; the angles are then also congruent. With this terminology we can say that all right angles are equal to one another and that all straight angles are equal to one another.

Let us give the definitions of two terms that are used often. But first this: if we say that an angle is equal to the sum of two angles, we mean that the measure of that angle is equal to the sum of the measures of the two angles in question. We call two angles *complementary* if their sum equals $\pi/2$. Two angles are said to be *supplementary* if their sum equals π. Two supplementary angles can be positioned next to each other in such a way that they form a straight angle. This definition is consistent with our use of the notion of *supplementary angles* right after Definition 2.9.

Angle Orientation

In certain situations we need to distinguish the angles APB and BPA. For this we use the determinant. Let us choose a coordinate system; see Fig. 2.14. The points A, P, and B are denoted in the figure by the corresponding bold lowercase letters. The *standard basis vectors* in a coordinate plane are $\mathbf{e}_1 = (1, 0)$ and $\mathbf{e}_2 = (0, 1)$. We say that angle APB has *positive orientation* if we have $\det((\mathbf{a} - \mathbf{p})(\mathbf{b} - \mathbf{p})) > 0$, and that it has *negative orientation* if we have $\det((\mathbf{a} - \mathbf{p})(\mathbf{b} - \mathbf{p})) < 0$. For example, the angle $\mathbf{e}_1\mathbf{o}\mathbf{e}_2$ has positive orientation, while the angle $\mathbf{e}_2\mathbf{o}\mathbf{e}_1$ has negative orientation. The notation we use here may seem somewhat unusual: the bold lowercase letters denote points in the coordinate plane, and angle $\mathbf{e}_1\mathbf{o}\mathbf{e}_2$ is one of the angles formed by the lines $\mathbf{o}\mathbf{e}_1$ and $\mathbf{o}\mathbf{e}_2$.

More generally, the orientation of angle APB is the opposite of that of angle BPA. The orientation of an angle depends on the choice of a coordinate system. We are going to prove that a reflection reverses the orientation of an angle in any coordinate system.

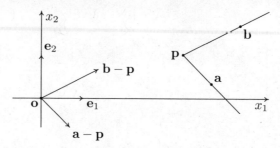

Fig. 2.14. Determining the orientation of angle **apb**

Theorem and Definition 2.14. *Every reflection $\mathcal{S}$ reverses orientation: for any angle APB, the orientation of $\mathcal{S}(A)\mathcal{S}(P)\mathcal{S}(B)$ is the opposite of that of APB.*

Proof. Consider Fig. 2.15. Let l be the reflection axis of $\mathcal{S}$. We introduce a coordinate system in which l has equation $\langle \mathbf{c}, \mathbf{x}\rangle - d = 0$, where we may assume that $\|\mathbf{c}\| = 1$ (see Theorem 1.35). Let us compare the orientation of angle **apb**

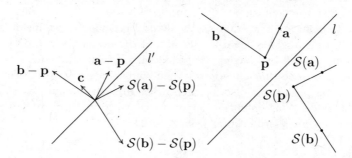

Fig. 2.15. A reflection reverses orientations

with that of angle $\mathcal{S}(\mathbf{a})\mathcal{S}(\mathbf{p})\mathcal{S}(\mathbf{b})$. We want to show that

$$\det\left((\mathbf{a}-\mathbf{p})(\mathbf{b}-\mathbf{p})\right) = -\det\left((\mathcal{S}(\mathbf{a})-\mathcal{S}(\mathbf{p}))(\mathcal{S}(\mathbf{b})-\mathcal{S}(\mathbf{p}))\right).$$

Note that $\mathcal{S}(\mathbf{a}) - \mathcal{S}(\mathbf{p})$ and $\mathcal{S}(\mathbf{b}) - \mathcal{S}(\mathbf{p})$ arise from respectively $\mathbf{a}-\mathbf{p}$ and $\mathbf{b}-\mathbf{p}$ through the reflection in the line l' with equation $\langle \mathbf{c}, \mathbf{x}\rangle = 0$. We denote this reflection by $\mathcal{S}'$, and write $\mathbf{a}' = \mathbf{a} - \mathbf{p}$ and $\mathbf{b}' = \mathbf{b} - \mathbf{p}$. Then $\mathcal{S}'(\mathbf{a}') = \mathcal{S}(\mathbf{a}) - \mathcal{S}(\mathbf{p})$ and $\mathcal{S}'(\mathbf{b}') = \mathcal{S}(\mathbf{b}) - \mathcal{S}(\mathbf{p})$. By Theorem 1.35, we obtain

$$\mathcal{S}'(\mathbf{a}') = \mathbf{a}' - 2\langle \mathbf{a}', \mathbf{c}\rangle \mathbf{c} \quad \text{and} \quad \mathcal{S}'(\mathbf{b}') = \mathbf{b}' - 2\langle \mathbf{b}', \mathbf{c}\rangle \mathbf{c}.$$

To deduce these formulas, note that $\langle \mathbf{a}', \mathbf{c}\rangle$ is the distance from $\mathbf{a}'$ to the line l'. The following computations imply the formula with the determinants given above.

$$\det\left((\mathbf{a}' - 2\langle \mathbf{a}', \mathbf{c}\rangle \mathbf{c})(\mathbf{b}' - 2\langle \mathbf{b}', \mathbf{c}\rangle \mathbf{c})\right)$$
$$= (a'_1 - 2\langle \mathbf{a}', \mathbf{c}\rangle c_1)(b'_2 - 2\langle \mathbf{b}', \mathbf{c}\rangle c_2) - (a'_2 - 2\langle \mathbf{a}', \mathbf{c}\rangle c_2)(b'_1 - 2\langle \mathbf{b}', \mathbf{c}\rangle c_1)$$
$$= a'_1 b'_2 - a'_2 b'_1 + 2\langle \mathbf{a}', \mathbf{c}\rangle(b'_1 c_2 - b'_2 c_1) + 2\langle \mathbf{b}', \mathbf{c}\rangle(a'_2 c_1 - a'_1 c_2)$$
$$= a'_1 b'_2 - a'_2 b'_1 + 2c_1^2(-a'_1 b'_2 + b'_1 a'_2) + 2c_2^2(-a'_1 b'_2 + b'_1 a'_2)$$
$$= -\det(\mathbf{a}' \mathbf{b}') .$$

Exercises

2.9. Any point Q in the interior of an angle APB satisfies $\angle QPB < \angle APB$.

2.10. An isometry $\mathcal{F}$ that is the composition of an odd number of reflections reverses orientation.

2.11. We call a triangle *isosceles* if it has two equal sides. In an isosceles triangle, the angles opposite the equal sides are themselves equal.

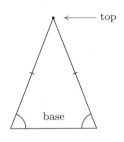

We usually call the equal sides of an isosceles triangle the *legs* and the third side the *base*; when drawing the triangle, we set it on its base. The angle opposite the base is called the *top angle* of the triangle, and the vertex opposite the base is called the *top*. With these conventions, this exercise reads as follows: the base angles of an isosceles triangle are equal.

2.12. Let K, L, and M be three distinct points on a line l, and let P be a point outside of l. Show by contradiction that the distances $d(P, K)$, $d(P, L)$, and $d(P, M)$ cannot all three be equal.

Conclusion: a circle and a line have at most two intersection points.

2.13. In an isosceles triangle, the median from the top is also the altitude.

2.14. If in a triangle the median from a vertex is also the altitude, the triangle is isosceles.

2.15. The figure sketches the course of a ball in a game of billiards when it hits one cushion (dashed line) or two cushions (continuous line) before hitting another ball. Try to describe this using reflections.

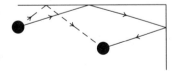

2.3 The Reflection in a Point

In this section we study the reflection in a point. Many important theorems from plane geometry can be proved using the properties of this transformation.

Definition 2.15. *Let P be a point in the plane. The reflection in the point P is the map $\dot{S}_P$ from the plane to itself defined as follows:*

1. $\dot{S}_P(P) = P$.
2. *For any point X of the plane other than P, $\dot{S}_P(X)$ is the unique point such that P is the midpoint of the line segment $[X\,\dot{S}_P(X)]$.*

If $X \neq P$, $\dot{S}_P(X)$ lies on the line XP, on the other side of P from X; the distance from $\dot{S}_P(X)$ to P is equal to that from X to P. See Fig. 2.16, on the left, where both the figure F and its image under $\dot{S}_P$ are drawn. It follows immediately from the definition that

$$\dot{S}_P \circ \dot{S}_P = \mathrm{id}_V\,.$$

In other words, applying the same reflection twice gives the identity map. Note that the reflection $\dot{S}_P$ maps every line through P onto itself. Using local

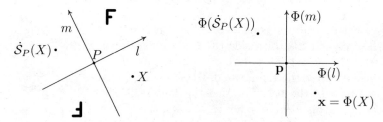

Fig. 2.16. The reflection $\dot{S}_P$ in P

coordinates at P, we immediately see that the reflection in the point P is the composition of two line reflections, with reflection axes meeting perpendicularly at P. Indeed, if l and m are lines through P that are perpendicular to each other, then l and m can be seen as the axes of a coordinate system at P. Let Φ be the map from the plane to the resulting coordinate plane. In local coordinates, the action of the reflection S_l in the line l followed by the reflection S_m in the line m is given by $(x_1, x_2) \to (-x_1, -x_2)$, which is exactly what the reflection in the point P does. This shows that the reflection in a point is the composition of two line reflections. The only condition imposed on the two lines l and m is that they meet perpendicularly at P. But then it follows that the product of two reflections with perpendicular axes is a reflection in a point. Theorem 2.14 now implies the following result.

Theorem and Definition 2.16. *The reflection in the point P is the composition of two line reflections whose axes meet perpendicularly at P. Conversely, the product of two reflections with perpendicular axes is the reflection in the intersection point of the axes.*

Every reflection in a point $\dot{S}_P$ preserves orientation, that is, the orientation of any angle AQB is equal to that of angle $\dot{S}_P(A)\dot{S}_P(Q)\dot{S}_P(B)$.

We call a figure F in the plane *point-symmetric* if there is a point P such that $\dot{\mathcal{S}}_P(F) = F$. Figure 2.17 shows figures with point symmetry on the left,

Fig. 2.17. (a) Figures with point symmetry; (b) without ("Ahoj" is Czech for both "hello" and "goodbye")

and without point symmetry on the right.

Let l and n be two lines intersecting at P. By Definition 2.9, this intersection determines four angles; see Fig. 2.8. The reflection $\dot{\mathcal{S}}_P$ maps both l and n onto themselves and interchanges vertical angles. Consequently, we have the following theorem:

Theorem 2.17. *Vertical angles are equal.*

By our interpretation of *equal*, this means that vertical angles have the same measure and also that vertical angles are congruent. We have now arrived at a very important property of the point reflection.

Theorem 2.18. *Consider a point P. Every line l is parallel to its image under the reflection $\dot{\mathcal{S}}_P$ in P: $l \mathbin{/\!/} \dot{\mathcal{S}}_P(l)$.*

Proof. Consider Fig. 2.18. Let l be a line. The reflection $\dot{\mathcal{S}}_P$ can be seen as the

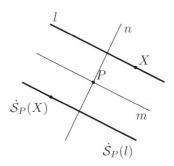

Fig. 2.18. The image of a line under a reflection in a point

composition of two reflections with axes m and n that meet perpendicularly at P. We still have a degree of freedom: we can choose the direction of one of the axes. The simplest is to choose one of the axes parallel to the given line l. Considering the effect of the reflections in m and n leads directly to the proof of the theorem.

From this theorem we will now deduce a number of equalities for the angles that appear when two parallel lines intersect a third line. Let us first give a few definitions. Let m and n be two lines, and l a third line intersecting both m and n; see Fig. 2.19, on the left. We call the angles P_i and Q_i, $i = 1, \ldots, 4$,

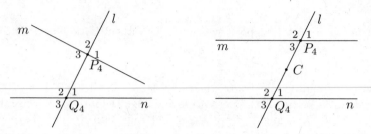

Fig. 2.19. Two lines meeting a third

corresponding angles. We call P_3 and Q_1 *alternate interior angles*; the pair P_4 and Q_2 are also alternate interior angles. We call P_1 and Q_3 *alternate exterior angles*; the pair P_2 and Q_4 are also alternate exterior angles. Since vertical angles are equal (Theorem 2.17), the following theorem also holds if we replace the alternate interior angles by alternate exterior angles.

Theorem 2.19. *Let m and n be two lines intersected by a third line l. If $m /\!/ n$, the corresponding angles and the alternate interior angles are equal. Conversely, if two corresponding angles or two alternate interior angles are equal, $m /\!/ n$.*

Proof. For the proof of the first statement we assume that $m /\!/ n$; see Fig. 2.19, on the right. We use the notation introduced above. Let C be the midpoint of the line segment $[PQ]$. Consider the effect of the reflection $\dot{S}_C$ in the point C: the point P is mapped to Q and the line l is mapped onto itself. The image of the line m is parallel to m (Theorem 2.18) and therefore coincides with n (Basic Assumption 1.20). Consequently, angle P_3 is mapped onto angle Q_1. Since congruent angles have the same measure, this prove the equality of any pair of alternate interior angles. Since $\angle P_3 = \angle P_1$ (Theorem 2.17), we also have $\angle P_1 = \angle Q_1$. This shows the equality of any pair of corresponding angles.

Let us now prove the converse. Let s be a line through P that is parallel to n. Compare the angles formed by l and s to the angles at Q: by the first part of the theorem, the corresponding angles and the alternate interior angles are equal. Compare the angles formed by l and m to the angles at Q: we have assumed that there are two equal corresponding angles or alternate interior angles. By Basic Assumption 2.12, property 4, m and s must coincide. Hence $m /\!/ n$.

This theorem has an important consequence for the angles of a triangle. When speaking of a triangle ABC, angle A stands for angle CAB, angle B

for angle ABC, and angle C for angle BCA. These are called the *interior angles* of the triangle. The intersection of the interior angles is the *interior* of the triangle. We can also describe the interior of $\triangle ABC$ in another way. Together, the lines AB, BC, and AC determine six open half-planes. Each of the points A, B, and C lies in exactly one open half-plane; the intersection of these three open half-planes is the interior of $\triangle ABC$.

The angles adjacent to the interior angles if we extend the sides of triangle ABC are called *exterior angles* of the triangle ABC. Each interior angle has two adjacent exterior angles. In Fig. 2.20, on the left, 2 indicates an exterior angle of angle C. The interior angles A and B are called *remote interior*

Fig. 2.20. $\angle C_2 = \angle A + \angle B$

angles of exterior angle C_2. Let l be a line through C that is parallel to AB; see Fig. 2.20, on the right. The line l divides angle C_2 into two angles, one equal to angle A (alternate interior angles), and the other one equal to angle B (corresponding angles). By Basic Assumption 2.12, it follows from this that $\angle C_2$ is equal to the sum of $\angle A$ and $\angle B$. The following theorem states this result.

Theorem 2.20. *In a triangle every exterior angle is equal to the sum of its two remote interior angles.*

The exterior angle C_2 and the interior angle C_1 add up to a straight angle. This leads to the following theorem:

Theorem 2.21. *The sum of the interior angles of a triangle is a straight angle.*

In other words, in $\triangle ABC$ we have $\angle A + \angle B + \angle C = \pi$ rad. This result is one of the most important theorems of plane geometry. In particular, it serves to differentiate the geometry of the plane from that of other surfaces [69]. Theorem 2.21 can also be useful for simpler things, for example for listing the congruence criteria. In addition to the two mentioned before, **SSS** (three equal sides) and **SAS** (two equal sides with equal included angle), we also have **SSA**, two equal sides and one equal angle opposite one of those sides plus a condition on the other angles, and **ASA**, two equal angles with equal included side. Let us first consider the criterion **ASA**. By Theorem 2.21, if two angles are equal, then so is the third. Because of this, criterion **AAS**, two equal angles and an equal side that does not lie between those angles, is equivalent to criterion **ASA**.

Congruence Criterion 4 (ASA). *Two triangles ABC and $A_1B_1C_1$ are congruent if $c = c_1$, $\angle A = \angle A_1$, and $\angle B = \angle B_1$.*

Proof. Since $c = c_1$, there is an isometry $\mathcal{F}$ with $\mathcal{F}(A) = A_1$ and $\mathcal{F}(B) = B_1$ (Sect. 2.1, Exercise 2.7). As shown in Fig. 2.21, $\mathcal{F}(C)$ and C_1 may lie in different open half-planes of the line A_1B_1. In that case we apply a reflection in the

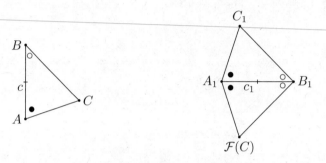

Fig. 2.21. The congruence criterion ASA

line A_1B_1 after $\mathcal{F}$; we denote the resulting map by $\mathcal{G}$. We have $\mathcal{G}(A) = A_1$ and $\mathcal{G}(B) = B_1$, and $\mathcal{G}(C)$ and C_1 lie in the same half-plane of the line A_1B_1. Since $\angle A = \angle A_1$, the line $\mathcal{G}(A)\mathcal{G}(C)$ must coincide with the line A_1C_1; otherwise there would be a conflict with Basic Assumption 2.12, property 4. Likewise, the line $\mathcal{G}(B)\mathcal{G}(C)$ must coincide with B_1C_1, and therefore $\mathcal{G}(C) = C_1$.

In order to state the last congruence criterion we first need two definitions. We call an angle A *acute* if $0 < \angle A < \pi/2$; we call it *obtuse* if $\pi/2 < \angle A < \pi$. The last congruence criterion concerns two equal sides and an equal angle opposite one of those sides. An additional condition is necessary to determine congruence. Because of this, this criterion is sometimes called the *ambiguous case* or **SSA** *condition* instead of *criterion*.

Congruence Criterion 5 (SSA, conditional). *Two triangles ABC and $A_1B_1C_1$ are congruent if $a = a_1$, $b = b_1$, and $\angle B = \angle B_1$, on the condition that the angles A and A_1 are either both acute, both right, or both obtuse.*

Proof. Consider Fig. 2.22. Under the assumptions of the theorem there are two complementary possibilities: either $c = c_1$ or $c \neq c_1$. In the first case, the triangles ABC and $A_1B_1C_1$ are congruent by **SSS**. In the second case, we may assume without loss of generality that $c < c_1$. In that case there is a point E on the line segment A_1B_1 such that $d(B_1, E) = d(B, A)$. Obviously this point differs from A_1. The triangles ABC and EB_1C_1 are congruent by **SAS**. Hence $\angle BAC = \angle B_1EC_1$. Moreover, $d(A, C) = d(E, C_1)$, so that

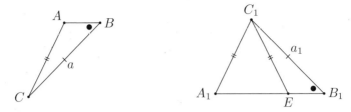

Fig. 2.22. The conditional congruence criterion SSA

$\triangle A_1EC_1$ is isosceles. It follows that the angles A_1 and A are supplementary; if one is acute, the other one is obtuse, and vice versa. This contradicts the extra condition on the angles A and A_1; the case $c \neq c_1$ therefore does not occur. This completes the proof.

The theorems treated in this section have many applications in geometry. Let us give an example; more can be found in the exercises.

Example 2.22. Let APB be an angle; see Fig. 2.23, on the left. By Basic Assumption 2.12, property 3, there is a point Q in the interior of APB such

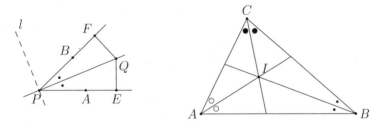

Fig. 2.23. The three angle bisectors of a triangle meet at one point

that $\angle APQ = \angle QPB = \angle APB/2$. The line PQ bisects angle APB and is called the *angle bisector* of APB. We sometimes also call this line an *interior angle bisector*. The bisector of the exterior angles of angle APB is called the *exterior angle bisector* of APB. In the figure, the line l is the exterior angle bisector of APB. Let E be the foot of the perpendicular through Q on PA, and F the foot of the perpendicular through Q on PB. The triangles PQE and PQF have the line segment $[PQ]$ in common; the triangles are congruent by **SAA**. Therefore $d(Q,E) = d(Q,F)$: the point Q is equidistant from the legs of the angle. This holds for every point of the angle bisector. In fact, this characterizes the angle bisector. If Q is equidistant from the legs of the angle, that is, $d(Q,E) = d(Q,F)$, then the triangles PQE and PQF are congruent by **SSA**; angle P is acute in both triangles, since the angles at E and F are right angles. Consequently, PQ is the angle bisector of angle APB. Conclusion: *a point Q in the interior of angle P lies on the interior angle* bisector of angle P if and only if Q is equidistant from the legs of angle P. Using this characterization, we immediately see that the three interior angle bisectors of $\triangle ABC$, Fig. 2.23, on the right, concur at the point I.

Exercises

2.16. Use **ASA** to prove that if in $\triangle ABC$ we have $\angle A = \angle B$, then $BC = AC$.
Conclusion: *a triangle is isosceles if and only if the base angles are equal.*

2.17. In any $\triangle ABC$, when comparing two angles and the sides opposite them, the angle lying opposite the longer side is the greater of the two. For example, if $a < b$, then $\angle A < \angle B$.
Hint: Choose a point D on CA (that is, side b) such that $d(C, D) = d(C, B)$, and compare $\angle BDC$ to the angles A and B.

2.18. In $\triangle ABC$, the side lying opposite the greater of two angles is the longer of the two. For example, if $\angle A < \angle B$, then $a < b$.
Hint: by contradiction.

2.19. A quadrilateral $ABCD$ is called a *trapezoid* if $AB \!\!\mathbin{/\mkern-5mu/} DC$ and $d(A, B) \neq d(D, C)$, where B and D lie on different sides of AC. The trapezoid is called *isosceles* if $d(A, D) = d(B, C)$. The angles DAB and CBA are called the *base angles*. A trapezoid is *isosceles* if and only if the base angles are equal.

2.20. Consider a triangle ABC. The interior angle bisector of A and the exterior angle bisectors of angles B and C are concurrent.

2.21. Let $ABCD$ be a quadrilateral with $d(A, B) = d(C, D)$ and $AB \!\!\mathbin{/\mkern-5mu/} DC$, where B and D lie on different sides of AC. Show that $ABCD$ is a parallelogram.

2.22. Let $ABCD$ be a quadrilateral with $d(A, B) = d(C, D)$ and $d(B, C) = d(A, D)$, where B and D lie on different sides of AC. Show that $ABCD$ is a parallelogram.

2.23. If the diagonals of a quadrilateral bisect each other, the quadrilateral is a parallelogram.

2.24. If the diagonals of a parallelogram are equal, the parallelogram is a rectangle.

2.25. Theorem *Let ACB be a triangle and D a point in its interior. The line AD meets the line segment $[BC]$.*

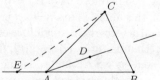

For the proof of the theorem, we choose a point E on AB such that $E \notin [AB]$ but $A \in [EB]$. Let us choose a coordinate system. As usual, we denote the vectors associated to points by the corresponding bold letters. Let $f(\mathbf{x}) = 0$ be the equation of the line **ac**; we may assume that $f(\mathbf{b}) > 0$.

(a) Show that $f(\mathbf{e}) < 0$ and that $f(\mathbf{y}) < 0$ for all Y in $[EC]$ other than C.
(b) Show that $f(\mathbf{z}) > 0$ for all Z other than A on the line AD that satisfy $Z \in [AD]$ or $D \in [AZ]$.
(c) Conclude from (a) and (b) that the line AD does not meet the line segment $[EC]$.
(d) The theorem now follows from Pasch's axiom applied to triangle EBC.

2.26. In triangle ABC, let D be the intersection point of the angle bisector of angle C and side AB. Then D divides side AB into line segments that are proportional to the adjacent sides.
Hint: Compute the ratio of the areas of the triangles ADC and BDC in two different ways.

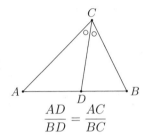

$$\frac{AD}{BD} = \frac{AC}{BC}$$

2.4 Translation and Rotation Groups

In the previous sections we stated various properties of isometries. In this section we will examine the set of all isometries of the plane. This study fits in with the discussion of Sect. 1.3. Let us recall the following: a map $\mathcal{F}$ from V to V is an isometry if it is surjective and distance-preserving. It is then also bijective.

We will use a number of results from algebra, in particular from group theory.

Definition 2.23 (Group). *A* group *is a nonempty set G with a group operation $*$ that to each pair of elements a, b of G associates an element $a * b$ of G. This association has the following properties:*

1. *Associativity: $a * (b * c) = (a * b) * c$ for all a, b, c in G.*
2. *There is an* identity element *e such that $e * a = a * e = a$ for all a in G.*
3. *For every a in G, there is an* inverse *a^{-1} with $a^{-1} * a = a * a^{-1} = e$.*

The standard example of a group is $\mathbb{R}$ under the addition operation $+$. The unit element is 0 and the inverse of a number a is $-a$. This terminology can be confusing: when speaking of addition, we often call the unit element the *zero element* and the inverse of an element its *negative*. However, we prefer to use the terms defined above. The group $\mathbb{R}$ is *abelian*, that is, for every a and b we have $a + b = b + a$.

Another example of a group is $\mathbb{R}^+$, the set of strictly positive real numbers, under the multiplication operation $\times$. In this example the unit element is 1. The inverse of a is $1/a$; note that $\mathbb{R}^+$ consists of strictly positive numbers, and that a can therefore not be zero.

These two examples are closely related. The exponential map $\exp \colon \mathbb{R} \to \mathbb{R}^+$ preserves the algebraic relations: $\exp(a+b) = (\exp a) \times (\exp b)$. In words: the image $\exp(a+b)$ of the sum $a+b$ under the map $\exp$ is equal to the product of the images $\exp a$ and $\exp b$. The logarithmic map $\log \colon \mathbb{R}^+ \to \mathbb{R}$ also preserves the algebraic relations: $\log(a \times b) = \log a + \log b$. In words: the image $\log(a \times b)$ of the product $a \times b$ is equal to the sum of the images $\log a$ and $\log b$. The maps log and exp are examples of group isomorphisms, that is, bijections that preserve the algebraic relations. More precisely: a *(group) isomorphism* from a group G_1 with operation $*_1$ to a group G_2 with operation $*_2$ is a bijection φ from G_1 to G_2 such that $\varphi(a *_1 b) = \varphi(a) *_2 \varphi(b)$ for all a and b in G_1. We call groups G_1 and G_2 *isomorphic* if there is a group isomorphism from G_1 to G_2.

The coordinate plane $\mathbb{R}^2$ introduced in Sect. 1.5 is a group; the group operation is the addition $+$ of vectors; the sum is taken componentwise:

$$(a_1, a_2) + (b_1, b_2) = (a_1 + b_1, a_2 + b_2)$$

for all (a_1, a_2) and (b_1, b_2) in $\mathbb{R}_2$. The unit element is **o** and the inverse of $\mathbf{a} = (a_1, a_2)$ is $(-a_1, -a_2)$. We call this group the *additive group of* $\mathbb{R}^2$. Scalar multiplication is an additional algebraic operation that makes $\mathbb{R}^2$ into a *vector space*.

Property 2 of the definition of a group says that there is a unit element. In the group $\mathbb{R}$ under the addition this is 0; in the group $\mathbb{R}^+$ under the multiplication this is 1. *In any group there is only one unit element.* We can show this as follows. Let e_1 and e_2 be two unit elements of G; in other words, both e_1 and e_2 satisfy the condition stated in property 2. By applying property 2 twice, once to e_1 and once to e_2, we find that

$$e_1 = e_1 * e_2 = e_2.$$

The unit element is therefore unique and is characterized by property 2. *An isomorphism maps the unit element of the first group onto the unit element of the second group.* This can be shown as follows. Let $\varphi \colon G_1 \to G_2$ be an isomorphism. Let e_i, $i = 1, 2$, be the unit element of G_i. Let d be an arbitrary element of G_2. Since the isomorphism φ is a bijection from G_1 to G_2, there is a b in G_1 such that $\varphi(b) = d$. Then

$$\varphi(e_1) *_2 d = \varphi(e_1) *_2 \varphi(b) = \varphi(e_1 *_1 b) = \varphi(b) = d\,.$$

This holds for every d in G_2. Likewise, we can show that $d *_2 \varphi(e_1) = d$ for every d in G_2. Consequently, $\varphi(e_1)$ is the unit element of G_2, and therefore $\varphi(e_1) = e_2$.

The third property of a group concerns the inverse. We can show that every element of the group has exactly one inverse: *the inverse is unique.* Let b and c be inverses of a in the group G with group operation $*$ and unit element e. We then have $b*a = a*b = e$ and $c*a = a*c = e$. It follows that

$$b = b*e = b*(a*c) = (b*a)*c = e*c = c,$$

whence $b = c$. Using the notation of previous paragraphs, we obtain, for every a in G_1,

$$\varphi(e_1) = \varphi\left(a^{-1} *_1 a\right) = \varphi\left(a^{-1}\right) *_2 \varphi(a) = e_2.$$

We see that the image $\varphi\left(a^{-1}\right)$ of the inverse of a under φ is the inverse of the image $\varphi(a)$.

Example 2.24 (Cyclic group). The groups C_n will play an important role in our story: C_n is the *cyclic group of n elements*. The elements of C_n are the integers $0, 1, \ldots, n-1$ and the group operation is addition modulo n: instead of taking the usual sum, we take its remainder after dividing by n. For example, C_3 consists of the elements 0, 1, 2, where $2+2 = 1$ because $2+2 = 4$ and 4 is has remainder 1 after dividing by 3. The unit element of C_n is 0 and, for example, the inverse of 1 in C_3 is 2, because $1+2 = 3$ and the remainder after dividing 3 by 3 is 0. In Fig. 2.24 (a), we see the addition table of C_4.

+	0	1	2	3
0	0	1	2	3
1	1	2	3	0
2	2	3	0	1
3	3	0	1	2

$*$	e	a	b	c
e	e	a	b	c
a	a	e	c	b
b	b	c	e	a
c	c	b	a	e

Fig. 2.24. (a) Addition table of C_4; (b) multiplication table of V_4

Another group that we will come across is the *Klein four-group* V_4. This group has four elements e, a, b, and c. Figure 2.24 (b) gives its multiplication table. The unit element is e. One of the particularities of this group is that every element x of V_4 satisfies $x*x = e$. The group C_4 does not have this property; the groups V_4 and C_4 are not isomorphic.

We will mostly consider *transformation groups*, groups whose elements are the isometries of a metric space; the group operation is the composition $\circ$, applying transformations one after the other. We stress that in general, transformation groups are *not* abelian; see Sect. 1.3, Exercises 1.14 and 1.15.

Theorem 2.25. *The set $\mathcal{I}(M)$ of all isometries of a metric space M is a group under the composition operation $\circ$.*

Proof. We mentioned the associativity of the composition in Remark 1.10. Let $\mathcal{F}$, $\mathcal{G}$, $\mathcal{H}$ be elements of $\mathcal{I}(M)$. For every X in M, we have

$$\mathcal{H} \circ (\mathcal{G} \circ \mathcal{F})(X) = \mathcal{H}((\mathcal{G} \circ \mathcal{F})(X))$$
$$= \mathcal{H}(\mathcal{G}(\mathcal{F}(X))) = (\mathcal{H} \circ \mathcal{G})(\mathcal{F}(X))$$
$$= (\mathcal{H} \circ \mathcal{G}) \circ \mathcal{F}(X) .$$

Consequently, $\mathcal{H} \circ (\mathcal{G} \circ \mathcal{F}) = (\mathcal{H} \circ \mathcal{G}) \circ \mathcal{F}$; the composition $\circ$ is therefore associative. The unit element of $\mathcal{I}(M)$ is the identity map id_M, which maps every element of M to itself. Every isometry is bijective (Theorem 1.6). The *inverse map* $\mathcal{F}^{-1}$ of an isometry $\mathcal{F}$ is defined by $\mathcal{F}^{-1}(Y) = X$ if and only if $\mathcal{F}(X) = Y$. Note that $\mathcal{F}^{-1}$ is well defined, because by the surjectivity of $\mathcal{F}$, for every Y there is an X with $\mathcal{F}(X) = Y$; moreover, by the injectivity of $\mathcal{F}$ there is only one such X. It is clear that

$$\mathcal{F}^{-1} \circ \mathcal{F} = \mathrm{id}_M = \mathcal{F} \circ \mathcal{F}^{-1} .$$

Furthermore, $\mathcal{F}^{-1}$ is an isometry. Indeed, since $\mathcal{F}$ is an isometry, all P and Q in M satisfy

$$d\left(\mathcal{F}^{-1}(P), \mathcal{F}^{-1}(Q)\right) = d\left(\mathcal{F}\left(\mathcal{F}^{-1}(P)\right), \mathcal{F}\left(\mathcal{F}^{-1}(Q)\right)\right) = d(P, Q) .$$

The map $\mathcal{F}^{-1}$ is therefore an isometry; it is the inverse of $\mathcal{F}$ in the group $\mathcal{I}(M)$.

Because of this result, we often write the *composition* of maps as a *product*. Let us now concentrate on the group $\mathcal{I}(V)$ of all isometries of the plane. We have already seen two properties of fundamental importance; they are stated in Theorem 2.4. Firstly, an isometry is completely determined by its action on three non-collinear points. Secondly, every isometry is the product of at most three reflections. The last property implies that every element of $\mathcal{I}(V)$ is the product of reflections; we say that the group $\mathcal{I}(V)$ is *generated* by the reflections. All information about the group $\mathcal{I}(V)$ is included in the properties of the set of reflections. This makes it possible to build up plane geometry from the notion of reflection [1].

Translation

We begin by studying the isometries that can be written as a product of two reflections. These preserve orientation; they are the translations and the rotations. We start with the translations.

Definition 2.26. *An isometry $\mathcal{F}$ of the plane is called a* translation *if for all points X and Y such that X, Y, and $\mathcal{F}(X)$ are noncollinear, the quadrilateral $XY\mathcal{F}(Y)\mathcal{F}(X)$ is a parallelogram.*

Based on this definition, id_V is also a translation, because we cannot find any X and Y such that X, Y, and $id_V(X) = X$ are noncollinear; the condition of the definition is therefore automatically satisfied. The following theorem gives a useful characterization of the translations among the isometries.

Theorem 2.27. *Let $\mathcal{F}$ be an isometry. If $\mathcal{F}$ is a translation, then in an arbitrary coordinate system, we have*

$$\mathcal{F}(\mathbf{x}) = \mathbf{x} + \mathcal{F}(\mathbf{o}) \tag{2.1}$$

for all $\mathbf{x}$ in the coordinate plane. The vector $\mathcal{F}(\mathbf{o})$ is called the translation vector *of $\mathcal{F}$ with respect to the coordinate system. If, conversely, there is a coordinate system such that (2.1) holds for every $\mathbf{x}$, then $\mathcal{F}$ is a translation.*

Proof. It is clear that (2.1) holds for the identity map. If $\mathcal{F}$ is a translation and $\mathcal{F} \neq id_V$, we choose a coordinate system such that $\mathbf{o} \neq \mathcal{F}(\mathbf{o})$. First choose a

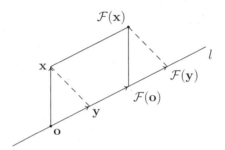

Fig. 2.25. A translation $\mathcal{F}$ satisfies $\mathcal{F}(\mathbf{x}) = \mathbf{x} + \mathcal{F}(\mathbf{o})$

point $\mathbf{x}$ outside of the line $\mathbf{o}\mathcal{F}(\mathbf{o})$; $\mathbf{x}$, $\mathbf{o}$, and $\mathcal{F}(\mathbf{o})$ are noncollinear. Since $\mathcal{F}$ is a translation, $\mathbf{o}\mathbf{x}\mathcal{F}(\mathbf{x})\mathcal{F}(\mathbf{o})$ is a parallelogram. Therefore $\mathcal{F}(\mathbf{x}) = \mathbf{x} + \mathcal{F}(\mathbf{o})$. Next consider a point $\mathbf{y}$ on the line $\mathbf{o}\mathcal{F}(\mathbf{o})$. Let $\mathbf{x}$ be as above. Then $\mathbf{x}$, $\mathcal{F}(\mathbf{x})$, and $\mathbf{y}$ are noncollinear, and therefore $\mathbf{x}\mathbf{y}\mathcal{F}(\mathbf{y})\mathcal{F}(\mathbf{x})$ is a parallelogram. Consequently, $\mathbf{y} + \mathcal{F}(\mathbf{x}) = \mathbf{x} + \mathcal{F}(\mathbf{y})$ (Sect. 1.5, Exercise 1.24). It immediately follows that

$$\mathcal{F}(\mathbf{y}) = \mathbf{y} + (\mathcal{F}(\mathbf{x}) - \mathbf{x}) = \mathbf{y} + \mathcal{F}(\mathbf{o}) ,$$

which we wanted to prove. We see that there is no fixed point. We can therefore choose the coordinate system arbitrarily.

For the proof of the second part of the theorem, we assume a coordinate system in which all $\mathbf{x}$ satisfy (2.1). For all $\mathbf{x}$ and $\mathbf{y}$ we then have $\mathcal{F}(\mathbf{x}) = \mathbf{x} + \mathcal{F}(\mathbf{o})$ and $\mathcal{F}(\mathbf{y}) = \mathbf{y} + \mathcal{F}(\mathbf{o})$. It follows that $\mathbf{y} + \mathcal{F}(\mathbf{x}) = \mathbf{x} + \mathcal{F}(\mathbf{y})$. Consequently, both $\mathbf{y}\mathbf{x}\mathcal{F}(\mathbf{x})\mathcal{F}(\mathbf{y})$ and $\mathbf{x}\mathbf{y}\mathcal{F}(\mathbf{y})\mathcal{F}(\mathbf{x})$ are parallelograms whenever $\mathbf{y}$, $\mathbf{x}$, and $\mathcal{F}(\mathbf{x})$ are noncollinear. We conclude that $\mathcal{F}$ is a translation.

Using this theorem we can describe the algebraic structure of the set of all translations. Let us discuss subgroups. Let G be a group with operation $*$ and unit element e. We call a subset H of G a *subgroup* if H under the operation $*$ is again a group, with unit element e. For example, $\mathbb{Z}$, the set of all integers, is a subgroup of $\mathbb{R}$ under the addition operation $+$. On the other hand, $\mathbb{N}\setminus\{\,0\,\}$ is not a subgroup of $\mathbb{R}^+$ under the multiplication operation $\times$ because $1/n$, the inverse of n in $\mathbb{R}^+$, is not an element of $\mathbb{N}$ for $n \geq 2$. We can use the following criterion to determine whether a subset H of a group G is a subgroup.

Lemma 2.28. *Let G be a group with group operation $*$. A subset H of G is a subgroup if and only if H is nonempty and all a, b in H satisfy $a * b^{-1} \in H$.*

Proof. If H is a subgroup of G, then H satisfies the conditions stated in the lemma. Indeed, H is nonempty because $e \in H$. Let a and b be elements of H. Then b^{-1} is also an element of H. Indeed, G contains a unique element, namely b^{-1}, that gives e when multiplied by b. Since H is a subgroup, it must contain an inverse of b, which must therefore be b^{-1}. But then $a * b^{-1}$ must also belong to H, because H is a group under the operation $*$. This proves the *only if* direction: a subset H of G is a subgroup only if H is nonempty and all a, b in H satisfy $a * b^{-1} \in H$. Let us now consider the *if* in the statement. First of all, we know that H is nonempty. Let $a \in H$. Then $e = a * a^{-1} \in H$; e will also be the unit element of H. All b in H satisfy $e * b^{-1} = b^{-1} \in H$, and b^{-1} will be the inverse of b in H. If a and b are elements of H, then so is $a * b = a * (b^{-1})^{-1}$; H is therefore closed under the operation $*$. Finally, H inherits the associativity from G. It follows that H is a group.

Theorem 2.29. *The translations form a subgroup of $\mathcal{I}(V)$. This subgroup is isomorphic to the additive group of $\mathbb{R}^2$.*

Proof. We use Theorem 2.27. Choose a coordinate system. If $\mathcal{T}_1$ and $\mathcal{T}_2$ are translations, the theorem tells us that

$$\mathcal{T}_2 \circ \mathcal{T}_1(\mathbf{x}) = \mathcal{T}_2(\mathbf{x} + \mathcal{T}_1(\mathbf{o})) = \mathbf{x} + \mathcal{T}_1(\mathbf{o}) + \mathcal{T}_2(\mathbf{o}) \qquad (2.2)$$

for all $\mathbf{x}$ in $\mathbb{R}^2$. If $\mathcal{T}$ is a translation, we define $\mathcal{T}'$ by $\mathcal{T}'(\mathbf{x}) = \mathbf{x} - \mathcal{T}(\mathbf{o})$. Then $\mathcal{T}'(\mathbf{o}) = -\mathcal{T}(\mathbf{o})$ and therefore $\mathcal{T}'(\mathbf{x}) = \mathbf{x} + \mathcal{T}'(\mathbf{o})$. The theorem implies that $\mathcal{T}'$ is a translation, and (2.2) shows that $\mathcal{T}'$ is the inverse of $\mathcal{T}$; therefore the inverse of $\mathcal{T}$ is a translation. For translations $\mathcal{T}_1$ and $\mathcal{T}_2$ we have

$$\mathcal{T}_1 \circ \mathcal{T}_2^{-1}(\mathbf{x}) = \mathbf{x} + \mathcal{T}_1(\mathbf{o}) - \mathcal{T}_2(\mathbf{o}) = \mathbf{x} + \mathcal{T}_1 \circ \mathcal{T}_2^{-1}(\mathbf{o}) \ .$$

The theorem implies that $\mathcal{T}_1 \circ \mathcal{T}_2^{-1}$ is a translation, and therefore that the translations form a subgroup. We define a map Φ from this subgroup to the coordinate plane by $\Phi(\mathcal{T}) = \mathcal{T}(\mathbf{o})$. For translations $\mathcal{T}_1$ and $\mathcal{T}_2$ we obtain

$$\Phi(\mathcal{T}_2 \circ \mathcal{T}_2) = \mathcal{T}_2 \circ \mathcal{T}_1(\mathbf{o}) = \mathcal{T}_1(\mathbf{o}) + \mathcal{T}_2(\mathbf{o}) = \Phi(\mathcal{T}_1) + \Phi(\mathcal{T}_2) \ .$$

Hence Φ is an isomorphism.

2.4 Translation and Rotation Groups 67

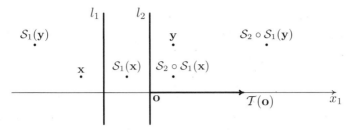

Fig. 2.26. Translation as product of two reflections

In particular, this theorem implies that the subgroup of the translations is abelian. Let us now show that every translation is the product of two reflections with parallel reflection axes. We first study the product of two reflections S_1 and S_2 with parallel axes l_1 respectively l_2. We choose the coordinate system such that the reflection axes are perpendicular to the x_1-axis; see Fig. 2.26. The second coordinate is then invariant under the reflections; we compute the effect on the first coordinate exactly as in Sect. 1.3. This shows that $S_2 \circ S_1$ is indeed a translation. Conversely, let $\mathcal{T}$ be a translation. We may assume without loss of generality that $\mathcal{T} \neq \mathrm{id}$. We choose the coordinate system as follows. Choose the origin at an arbitrary point O; next choose the axes such that the x_1-axis is the span of the vector $\mathcal{T}(\mathbf{o})$. Choose two reflection axes l_1 and l_2 perpendicular to the x_1-axis and at a distance $(1/2)\|\mathcal{T}(\mathbf{o})\|$ from each other. A well-chosen product of the corresponding reflections is equal to $\mathcal{T}$. Don't forget that the order of the reflections influences the result; see Sect. 1.3, Exercise 1.14. This concludes the proof of the following theorem. In the process we have just described, the order of the reflections, the direction of the reflection axes, and their mutual distance are determined by the translation that we want to write as a product; we do, however, still have one degree of freedom in the position of the axes. Let us summarize our results.

Theorem 2.30. *The product of two reflections with parallel axes is a translation. Every translation can be written as product of two reflections with parallel axes; the distance between the reflection axes is half the length of the translation vector. Every translation preserves orientation.*

Rotation

Having concluded our study of the structure of translations, let us now consider rotations. We will see that the product of two reflections with intersecting axes is a *rotation* about the intersection point. Reversing the order of the reflections changes the rotation direction. We must therefore use oriented angles when describing rotations. We tacitly assume that a coordinate system has been chosen with respect to which the orientation can be described. Here is the definition of a rotation.

Definition 2.31. *Let φ be a real number with $-\pi < \varphi \leq \pi$. Let C be a point. The* rotation $\mathcal{R}$ with rotation angle φ about the rotation center C *is the isometry given by*

1. *for $\varphi = 0$:* $\mathcal{R} = \mathrm{id}_V$;
2. *for $\varphi = \pi$:* $\mathcal{R} = \dot{\mathcal{S}}_C$;
3. *for φ with $0 < |\varphi| < \pi$: $\mathcal{R}(C) = C$ and for every X other than C, $\mathcal{R}(X)$ is the unique point such that $d(C, X) = d(C, \mathcal{R}(X))$, $\angle XC\mathcal{R}(X) = \varphi$, and angle $XC\mathcal{R}(X)$ has positive, respectively negative, orientation if φ is positive, respectively negative.*

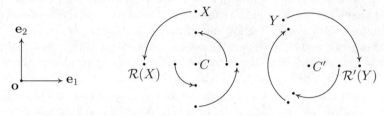

Fig. 2.27. (a) Angle $\mathbf{e}_1\mathbf{oe}_2$; (b) rotation over $\frac{1}{2}\pi$; (c) rotation over $-\frac{2}{3}\pi$

Let $\mathcal{R}$ be a rotation over an angle φ. To find the image $\mathcal{R}(X)$ of a point $X \neq C$, we first draw an angle of size φ with vertex C and one leg along CX, and then choose $\mathcal{R}(X)$ on the other leg of the angle such that $d(\mathcal{R}(X), C) = d(X, C)$. Figure 2.27 shows the geometric meaning of the sign of the rotation angle: a counterclockwise arrow indicates a positive orientation, a clockwise arrow a negative orientation. The sign of the rotation angle depends on the choice of coordinate system. By definition, the angle $\mathbf{e}_1\mathbf{oe}_2$ has a positive orientation; since we go counterclockwise to go from $\mathbf{e}_1$ to $\mathbf{e}_2$ along angle $\mathbf{e}_1\mathbf{oe}_2$, a counterclockwise arrow indicates a positive orientation. In the definition of a rotation, we have included the identity and the reflection in the point C as limiting cases; they can indeed be interpreted as rotations over respectively 0 and π rad.

In the last section we showed that the reflection in a point is the product of two reflections with perpendicular axes and that conversely, the product of two such reflections is a reflection in a point. We see that in this limiting case the rotation angle is twice the angle in which the rotation axes intersect. We will see that this holds more generally. Let l and m be two reflection axes that meet at the point C. Let ψ be one of the angles $\leq \pi/2$ formed by the intersection of l and m. We choose a point A on l and a point B on m such that $\psi = ACB$. Figure 2.28 shows that the effect of the map $\mathcal{S}_m \circ \mathcal{S}_l$, first $\mathcal{S}_l$ and then $\mathcal{S}_m$, is the same as that of a rotation over an angle φ, where $|\varphi| = 2\psi$. The sign of φ is determined by the orientation of angle ACB. For every point X other than C, the angles $XC\mathcal{R}(X)$ and ACB have the same orientation; this is because A lies on l, B lies on m, and $\mathcal{S}_l$ is carried out before $\mathcal{S}_m$. For

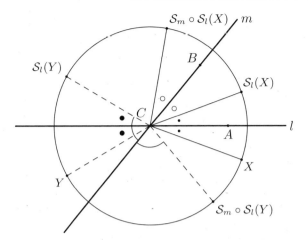

Fig. 2.28. A rotation is the product of two reflections

the map $\mathcal{S}_l \circ \mathcal{S}_m$, first $\mathcal{S}_m$ and then $\mathcal{S}_l$, it is the other way around; the sign of the rotation angle is determined by the orientation of angle BCA, which is the opposite of that of angle ACB. If $\psi = \pi/2$, we saw in the last section that the map $\mathcal{S}_m \circ \mathcal{S}_l$ is a reflection in the point C. If $l = m$, the map $\mathcal{S}_m \circ \mathcal{S}_l$ is the identity map. We have now shown that the product of two reflections with intersecting axes is a rotation. Conversely, every rotation can be obtained as follows: for a given rotation over an angle φ, $0 < |\varphi| < \pi$, and rotation center C we choose two reflection axes through C that meet under an angle of $(1/2)\varphi$. A well-chosen product of the corresponding reflections equals the given rotation. As with translations, the order of the reflections is important. We have shown the following.

Theorem 2.32. *The product of two reflections with intersecting axes is a rotation. Every rotation other than the identity map can be written as a product of two reflections whose axes meet at the center of the rotation; the size of the smallest angle between the reflection axes equals half the absolute value of the rotation angle. Every rotation preserves orientation.*

The composition of two rotations with about the same center is again a rotation about that same center; in determining the angle of the resulting rotation we work modulo 2π, as usual. This means that the sum of the rotation angles φ_1 and φ_2 is the unique number φ in $(-\pi, \pi]$ such that

$$\varphi = \varphi_1 + \varphi_2 + 2k\pi$$

for some integer k. Instead of this formula, we also write

$$\varphi = \varphi_1 + \varphi_2 \pmod{2\pi}.$$

Thus the composition of two rotations over an angle $2\pi/3$ is a rotation over an angle $2 \times 2\pi/3 - 2\pi = -2\pi/3$. Since for all φ satisfying $0 < |\varphi| < \pi$ the rotations over the angles φ and $-\varphi$ are each other's inverses, it follows without much difficulty that the rotations about a given point form a group. We note that the half-open interval $(-\pi, \pi]$ is an abelian group under the operation of *addition modulo* 2π. For example, the inverse of any number $x \neq \pi$ in the interval is $-x$, while the inverse of π is equal to π. Beware, though: addition modulo 2π differs from the usual addition of real numbers. Consequently, we do not have a subgroup of $\mathbb{R}$! By associating the rotation angle to each rotation, we define an isomorphism between the groups we have mentioned. The following theorem summarizes these results.

Theorem 2.33. *The rotations about a fixed center form a subgroup of $\mathcal{I}(V)$. This subgroup is isomorphic to the group of real numbers in the interval $(-\pi, \pi]$ with operation addition modulo 2π.*

Let us now consider the product of rotations with distinct rotation centers. Such a product is either a rotation or a translation, as we see in the following theorem.

Theorem 2.34. *Let $\mathcal{R}_1$ and $\mathcal{R}_2$ be rotations over respective angles φ_1 and φ_2. If $\varphi_1 + \varphi_2 = 0 \pmod{2\pi}$, the product $\mathcal{R}_2 \circ \mathcal{R}_1$ is a translation. Otherwise, the product $\mathcal{R}_2 \circ \mathcal{R}_1$ is a rotation over φ_0, where*

$$\varphi_0 = \varphi_1 + \varphi_2 \pmod{2\pi}.$$

Proof. Consider Fig. 2.29. The rotation center of $\mathcal{R}_i$ is C_i and the angle ψ_i

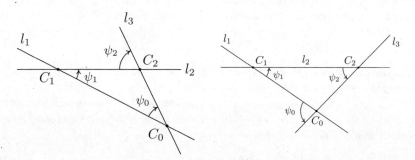

Fig. 2.29. The product of rotations

satisfies $\psi_i = (1/2)\varphi_i$, for $i = 1, 2$. The orientation of the angles is indicated by arrows: a counterclockwise arrow indicates a positive orientation, a clockwise arrow a negative orientation. The lines l_i are the reflection axes of reflections $\mathcal{S}_i$, $i = 1, 2, 3$. Let us describe the situation sketched in the figure on the left. The rotation $\mathcal{R}_1$ is the product of two reflections with axes l_1 and l_2. In this representation of $\mathcal{R}_1$ as a product of reflections, there is still a degree of freedom in the choice of the axes. We choose the line $C_1 C_2$ for l_2; this fixes l_1, because $\psi_1 = (1/2)\varphi_1$ and the angles are oriented. We have $\mathcal{R}_1 = \mathcal{S}_2 \circ \mathcal{S}_1$,

2.4 Translation and Rotation Groups

the reflection $\mathcal{S}_1$ followed by $\mathcal{S}_2$. Analogously, we write $\mathcal{R}_2$ as a product of reflections with axes l_2 and l_3: $\mathcal{R}_2 = \mathcal{S}_3 \circ \mathcal{S}_2$. We now have

$$\mathcal{R}_2 \circ \mathcal{R}_1 = (\mathcal{S}_3 \circ \mathcal{S}_2) \circ (\mathcal{S}_2 \circ \mathcal{S}_1) = \mathcal{S}_3 \circ (\mathcal{S}_2 \circ \mathcal{S}_2) \circ \mathcal{S}_1 = \mathcal{S}_3 \circ \mathcal{S}_1 \ .$$

This shows that the product $\mathcal{R}_2 \circ \mathcal{R}_1$ is a rotation about C_0, the intersection point of the reflection axes l_1 and l_3. We call the rotation angle ψ_0. By Theorem 2.20 we have $|\psi_2| = |\psi_1| + |\psi_0|$. If we also consider the orientations of the rotation angles, and therefore their signs, we obtain

$$\psi_0 = -|\psi_0| = |\psi_1| - |\psi_2| = \psi_1 + \psi_2 \ .$$

This reasoning doesn't quite work if $\psi_1 = -\psi_2$. In that case, $l_1 \mathbin{/\mkern-5mu/} l_3$ and $\mathcal{R}_2 \circ \mathcal{R}_1$ is a translation. The situation in the figure on the right is simpler, because all relevant angles are positive. It is useful to study as well the other cases that can occur.

We can now give an overview of the types of isometries that occur in $\mathcal{I}(V)$. Table 2.1 summarizes a number of things. The column on the right shows the least number of reflections that are necessary to represent an isometry of the type in question. We must of course work with minimal numbers; indeed, the identity can be written as a product of an arbitrarily large even number of reflections, since $\mathcal{S} \circ \mathcal{S} = \mathrm{id}_V$ for every reflection $\mathcal{S}$. The information in the

Table 2.1. The different types of isometries

isometry	number of fixed points	orientation- preserving	min. number of reflections
identity	∞	yes	0
reflection	∞	no	1
rotation*	1	yes	2
translation*	0	yes	2
glide reflection	0	no	3

* other than the identity

first four rows of the table follows from earlier results; we will come to the fifth row shortly. Note that the identity must be excluded from the rotations and translations, because the identity has infinitely many fixed points. The table is very useful for determining the type of an isometry (Exercises 2.27 and 2.28).

It is now time to study the isometries that can be written as a product of three reflections; this gives rise to a new type of isometry.

72   2 TRANSFORMATIONS

Definition 2.35. *The product of three reflections for which two axes are parallel but distinct and the third is perpendicular to the other two is called a glide reflection.*

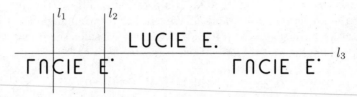

Fig. 2.30. The effect of a glide reflection (LUCIE E. is the first name and middle initial of the author's wife)

Let S_1 and S_2 be reflections whose respective axes l_1 and l_2 are parallel and distinct; see Fig. 2.30. If S_3 is a reflection whose axis l_3 is perpendicular to l_1 and l_2, then

$$S_3 \circ S_2 \circ S_1 = S_2 \circ S_3 \circ S_1 = S_2 \circ S_1 \circ S_3 \,,$$

because with reflections whose axes are perpendicular to each other, the order of the reflections may be inverted (Sect. 2.1, Exercise 2.4). This gives us a good idea of what a glide reflection is: *it is a translation $S_2 \circ S_1$ other than the identity followed by a reflection whose axis is parallel to the direction of the translation or, which is equivalent, a reflection followed by a translation.* The glide reflection has no fixed points; there is, however, a line, the *reflection axis*, that is mapped onto itself; in the notation above this is l_3. Theorem 2.4 and the following theorem imply that the inventory of the isometries we have given in Table 2.1 is complete.

Theorem 2.36. *Let $\mathcal{F}$ be an isometry that is the product of three distinct reflections. Then $\mathcal{F}$ is a reflection if and only if the three reflection axes are either concurrent or all parallel to one another. Moreover, if $\mathcal{F}$ is not a reflection, then $\mathcal{F}$ is a glide reflection.*

Proof. The first statement of the theorem is contained in Exercises 2.29 and 2.30. Let $\mathcal{F} = S_{l_3} \circ S_{l_2} \circ S_{l_1}$, and let us assume that $\mathcal{F}$ is not a reflection. Consider Fig. 2.31, on the left. Since $\mathcal{F}$ is not a reflection, it follows from the first statement of the theorem that the l_i are not all parallel. Assume that l_1 and l_2 intersect; the other cases are analogous. We can find l'_1 and l'_2 such that $S_{l'_2} \circ S_{l'_1} = S_{l_2} \circ S_{l_1}$ and l'_2 is perpendicular to l_3; see Fig. 2.31, in the middle. The intersection point of l_3 and l'_2 differs from the intersection point of l'_1 and l'_2 because the lines l_1, l_2, l_3 do not concur; otherwise, $\mathcal{F}$ would be a reflection. Next we choose perpendicular lines l''_2 and l'_3 such that $l''_2 \perp l'_1$ and $S_{l'_3} \circ S_{l''_2} = S_{l_3} \circ S_{l'_2}$; see Fig. 2.31, on the right. Then

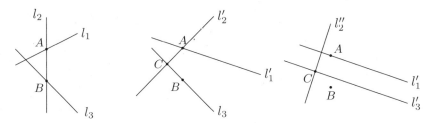

Fig. 2.31. The product of three reflections

$$\mathcal{F} = \mathcal{S}_{l_3} \circ \mathcal{S}_{l_2} \circ \mathcal{S}_{l_1} = \mathcal{S}_{l_3} \circ \mathcal{S}_{l'_2} \circ \mathcal{S}_{l'_1} = \mathcal{S}_{l'_3} \circ \mathcal{S}_{l''_2} \circ \mathcal{S}_{l'_1},$$

which implies that $\mathcal{F}$ is a glide reflection.

Digression 5 (Another Look at Orientation).
 In Sect. 2.2 we introduced the orientation of an angle. This depended on the choice of a coordinate system. We cannot explain what a positive orientation is without referring to the chosen coordinate system, or, more generally, without referring to something outside of mathematics, such as the rotational direction of the hands of a clock. But this direction depends on that of a sundial in the Northern Hemisphere; if the watch face had been invented in Australia, the rotational direction of the hands of the clock would probably be the other way around. We can also note that it is not truly important whether an angle has positive or negative orientation. What interests us is whether two angles have the same orientation, and also that a reflection inverses the orientation of an angle, while a rotation does not. Now that we know more about the transformations of the plane, we could treat orientation in a completely different manner. Let us present this briefly.

First we define when an isometry is direct, that is, preserves orientation, or is indirect or opposite, that is, reverses orientation. Only then do we define when two angles have the same orientation. We do not use coordinates in this definition. It is better, didactically, to use "distinct" and "opposite" rather than "preserves orientation" or "reverses orientation" until the meaning of the equality of two orientations is fixed. We already know that every isometry is the product of reflections.

Definition 2.37. *We call an isometry $\mathcal{F}$ direct if $\mathcal{F}$ is the product of an even number of reflections. An isometry $\mathcal{F}$ is indirect or opposite if $\mathcal{F}$ is the product of an odd number of reflections.*

We must of course show that these terms are well defined. For this we need to show that $\mathcal{F}$ cannot be both the product of an even number of reflections and the product of an odd number of reflections. It suffices to show that the identity is not equal to a product of an odd number of reflections. This is what the following lemma says.

Lemma 2.38. *If the identity is the product of reflections* $S_1, S_2, \ldots, S_k$, *then k is even.*

Proof. We have $\mathrm{id} = S_1 \circ S_2 \circ \cdots \circ S_k$ and must show that k is even. Let us first assume that $k \geq 4$. We will show that the identity can also be written as the product of $k - 2$ reflections. If this is true, we can repeatedly reduce the number of reflections by 2 in order to write the identity as product of m reflections, where $m < 4$ and k and m have the same parity. Theorem 2.36 then tells us that m must be even, which completes the proof.

Let us consider the product of the first three terms: $S_1 \circ S_2 \circ S_3$. By Theorem 2.36 this product is either a reflection or a glide reflection. In the first case, the number of reflections is reduced by 2. In the second case, by the definition of a glide reflection there exist reflections S_1', S_2', and S_3' such that

- $S_1 \circ S_2 \circ S_3 = S_1' \circ S_2' \circ S_3'$;
- the reflection axes l_2' of S_2' and l_3' of S_3' are parallel and distinct;
- the reflection axis l_1' of S_1' is perpendicular to l_2'.

See Fig. 2.32. If the reflection axis l_4 of S_4 is parallel to l_3', the product

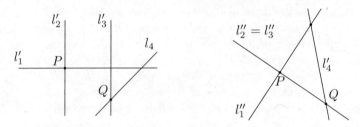

Fig. 2.32. Reduction of the number of reflections by 2

$S_2' \circ S_3' \circ S_4$ is a reflection; the number of reflections is then reduced by 2. In the other case, l_3' and l_4 have an intersection point, say Q. Let P be the intersection point of l_1' and l_2'. Then $P \neq Q$, because otherwise, l_2' and l_3' would coincide, which we have ruled out. Now let S_1'', S_2'', S_3'', and S_4'' be reflections satisfying

1. $S_1' \circ S_2' = S_1'' \circ S_2''$;
2. $S_3' \circ S_4 = S_3'' \circ S_4''$;
3. the reflection axes of S_2'' and S_3'' coincide with PQ.

Then $S_2'' \circ S_3'' = \mathrm{id}$ and the number of reflections is reduced by 2. This concludes the proof.

The proof of the following theorem follows directly from the definitions.

Theorem 2.39. *The product of two direct isometries is also a direct isometry. The product of two indirect isometries is a direct isometry. The product of a direct and an indirect isometry is an indirect isometry.*

With these results we define when two angles APB and CQD have the same orientation as follows. We may assume that $d(P, A) = 1$ and $d(P, B) = 1$, and also $d(Q, C) = 1$ and $d(Q, D) = 1$. There exists a direct isometry $\mathcal{F}$, the

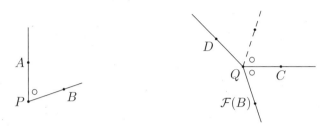

Fig. 2.33. Comparison of the orientation

product of two reflections, such that $\mathcal{F}(P) = Q$ and $\mathcal{F}(A) = C$. If the points D and $\mathcal{F}(B)$ lie in different half-planes of the line QC, as shown in Fig. 2.33, we say that the angles APB and CQD have *different orientations*; in the other case we speak of *equal orientations*.

Exercises

2.27. An isometry with exactly one fixed point is a rotation.

2.28. An indirect isometry without fixed point is a glide reflection.

2.29. Let $\mathcal{F}$ be an isometry that is the product of three reflections.
(a) If the three reflection axes concur, $\mathcal{F}$ is a reflection.
(b) If the three reflection axes are parallel, $\mathcal{F}$ is a reflection.

2.30. Let $\mathcal{S}$ be a reflection that is equal to a product $\mathcal{S}_3 \circ \mathcal{S}_2 \circ \mathcal{S}_1$ of three reflections; let l_i be the reflection axis of $\mathcal{S}_i$, for $i = 1, 2, 3$. The three axes l_i either concur or are all parallel.
Hint: $\mathcal{S}_3 \circ \mathcal{S}_2 = \mathcal{S} \circ \mathcal{S}_1$.

2.31. For $i = 1, 2$, let $\mathcal{S}_i$ be the reflection in the axis l_i. Assume that $l_1 \neq l_2$. Then $\mathcal{S}_1 \circ \mathcal{S}_2 = \mathcal{S}_2 \circ \mathcal{S}_1$ if and only if $l_1 \perp l_2$.
Hint: Sect. 2.1, Exercise 2.5.

2.32. In the coordinate plane we have two triangles, with respective vertices $(1, 1)$, $(2, 1)$, $(1, 3)$ and $(0, 0)$, $(2, 0)$, $(2, -1)$. Show that we need exactly three reflections to transform one of these triangles into the other.

2.33. The product of a translation and a rotation is a rotation over the same angle, but usually about another center.

2.34. Place two mirrors vertically so that they make a right angle with each other (think of a vanity table with a twofold mirror). How often do you see yourself in the mirrors? Are all images the same?

2.35. Place an asymmetric object (for example a mug) near the hinge of the mirror of a vanity table. What do you see as you open and close the mirror?

2.36. Let $\mathcal{R}_1$ and $\mathcal{R}_2$ be rotations. In general, $\mathcal{R}_1 \circ \mathcal{R}_2 \neq \mathcal{R}_2 \circ \mathcal{R}_1$. Give an example where inequality holds.
Hint: Take a reflection in a point for $\mathcal{R}_1$.

2.37. The product of two reflections in a point is a translation.

2.5 Dilation and Similarity

Theorem 2.18 implies that in a reflection in a point, every line is parallel to its image. Let us look more closely at this property.

Definition 2.40. *A bijection $\mathcal{F}$ from the plane to itself is called a* dilation *if for all X and Y with $X \neq Y$ we have $XY \mathbin{/\mkern-5mu/} \mathcal{F}(X)\mathcal{F}(Y)$.*

It is immediately clear that every translation and every reflection in a point is a dilation. A reflection (in a line) is not a dilation, and in general, neither is a rotation. As we will see, a dilation is not necessarily an isometry. One of the reasons dilations are important is the following theorem.

Theorem 2.41. *The image of a straight line under a dilation is again a straight line.*

Proof. Let $\mathcal{F}$ be a dilation. By definition, $\mathcal{F}$ is a bijection, and therefore has an inverse, which we will denote by $\mathcal{F}^{-1}$. (The existence of $\mathcal{F}^{-1}$ can be shown as in the proof of Theorem 2.25: $\mathcal{F}$ is surjective, so for every Y there is an X with $\mathcal{F}(X) = Y$; this X is unique because $\mathcal{F}$ is injective.) We can easily check that $\mathcal{F}^{-1}$ is also a dilation. If the distinct points X, Y, and Z lie on the line l, then the lines $\mathcal{F}(X)\mathcal{F}(Y)$ and $\mathcal{F}(X)\mathcal{F}(Z)$ are both parallel to l and therefore coincide. It follows that $\mathcal{F}(X)$, $\mathcal{F}(Y)$, and $\mathcal{F}(Z)$ are collinear. Consequently, the image of the line XY under the dilation $\mathcal{F}$ is contained in the line $\mathcal{F}(X)\mathcal{F}(Y)$. Since $\mathcal{F}^{-1}$ is also a dilation, the image of the line $\mathcal{F}(X)\mathcal{F}(Y)$ under the map $\mathcal{F}^{-1}$ must be contained in the line XY. It follows that the image of the line XY under $\mathcal{F}$ is the line $\mathcal{F}(X)\mathcal{F}(Y)$.

In the above proof we noted in passing that the inverse of a dilation is again a dilation. We can easily see that the composition of two dilations is also a dilation; the set of dilations is therefore closed under composition. Since the composition of maps is associative (Remark 1.10), we have proved the following theorem.

2.5 Dilation and Similarity

Theorem 2.42. *The set of all dilations is a group under the composition operation* ∘.

Let us now give an important example of a dilation, which in general is not an isometry.

Definition 2.43. *Let α be a real number other than 0. Let C be a point. A central dilation $\mathcal{V}$ with center C and scale factor α is given by $\mathcal{V}(C) = C$ and for any point X other than C, $\mathcal{V}(X)$ is the unique point with the properties (see Fig. 2.34)*

1. X, C, $\mathcal{V}(X)$ *are collinear;*
2. $d(C, \mathcal{V}(X))/d(C, X) = |\alpha|$;
3. X *and* $\mathcal{V}(X)$ *lie on the same side of C if $\alpha > 0$, while X and $\mathcal{V}(X)$ lie on different sides of C if $\alpha < 0$.*

We can easily check that if $\mathcal{V}$ is a central dilation with center C and scale factor α, then the central dilation with the same center but scale factor $1/\alpha$ is its inverse. A central dilation is therefore a bijective map. It immediately

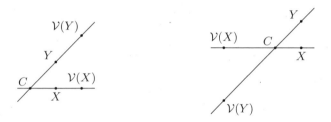

Fig. 2.34. Central dilations: (a) $\alpha > 0$; (b) $\alpha < 0$

follows from the definitions that the reflection in a point P is the same as the central dilation with center P and scale factor -1. The first statement of the following theorem implies that every central dilation is a dilation.

Theorem 2.44. *Let $\mathcal{V}$ be a central dilation with center C and scale factor α. For all X and Y with $X \neq Y$ we have*

1. $\mathcal{V}(X)\mathcal{V}(Y) // XY$;
2. $d(\mathcal{V}(X), \mathcal{V}(Y)) = |\alpha| d(X, Y)$;
3. $\angle XCY = \angle \mathcal{V}(X)C\mathcal{V}(Y)$ *if X, C, and Y are noncollinear.*

Proof. Consider Fig. 2.34. The statements of the theorem are fairly evident if X or Y is equal to C. Let us assume that this is not the case. We choose a coordinate system at the point C and choose the x_1-axis along the line XC. In this coordinate system C is the origin **o**, X is the vector $\mathbf{x} = (x_1, 0)$, and Y is the vector $\mathbf{y} = (y_1, y_2)$. The vector $(\alpha x_1, 0)$ corresponds to $\mathcal{V}(X)$, and the vector $(\alpha y_1, \alpha y_2)$ corresponds to $\mathcal{V}(Y)$. We find the following vector representations for the lines XY and $\mathcal{V}(X)\mathcal{V}(Y)$.

$$(z_1, z_2) = (1-\lambda)\mathbf{x} + \lambda \mathbf{y} = \mathbf{x} + \lambda(\mathbf{y} - \mathbf{x})$$
$$= (x_1, 0) + \lambda(y_1 - x_1, y_2),$$

respectively

$$(z_1, z_2) = (\alpha x_1, 0) + \lambda\alpha(y_1 - x_1, y_2).$$

These lines are therefore parallel (Sect. 1.5, Exercise 1.23); this proves the first statement. We check the correctness of the second statement with a simple computation:

$$d(X, Y) = \sqrt{(x_1 - y_1)^2 + y_2^2},$$
$$d(\mathcal{V}(X), \mathcal{V}(Y)) = \sqrt{(\alpha x_1 - \alpha y_1)^2 + (\alpha y_2)^2} = |\alpha|\sqrt{(x_1 - y_1)^2 + y_2^2}.$$

Now the third statement. This is evident if $\alpha > 0$. Let us therefore assume that $\alpha < 0$. The angles XCY and $\mathcal{V}(X)C\mathcal{V}(Y)$ are vertical angles, and therefore of equal size. The orientation of angle XCY depends on the sign of the determinant $\det((x_1, 0)(y_1, y_2))$; this determinant is equal to $x_1 y_2$. The orientation of angle $\mathcal{V}(X)C\mathcal{V}(Y)$ depends on the sign of $\det((\alpha x_1, 0)(\alpha y_1, \alpha y_2))$, which is equal to $\alpha^2 x_1 y_2$. These determinants have the same sign, since α^2 is always positive. Consequently, the angles have the same orientation.

The following theorem says that the only dilations are the central dilations and the translations.

Theorem 2.45. *A bijective map from the plane to itself is a dilation if and only if it is a central dilation or a translation.*

Proof. We have already shown that central dilations and translations are dilations. Let $\mathcal{F}$ be a dilation; we will show that $\mathcal{F}$ is either a central dilation or a translation. Since the identity map is a translation, we will assume that $\mathcal{F}$ is not the identity map. Choose a point P such that $\mathcal{F}(P) \neq P$. We are now in one of the following cases:

1. Every point Q satisfies $Q\mathcal{F}(Q) \,/\!/\, P\mathcal{F}(P)$.
2. There is a point Q such that the lines $Q\mathcal{F}(Q)$ and $P\mathcal{F}(P)$ intersect; see Fig. 2.35.

Let us first consider case 1. Take an arbitrary point Q outside the line $P\mathcal{F}(P)$. Since $\mathcal{F}$ is a dilation, we have $PQ \,/\!/\, \mathcal{F}(P)\mathcal{F}(Q)$. It follows that $PQ\mathcal{F}(Q)\mathcal{F}(P)$ is a parallelogram. Hence $\mathcal{F}$ is a translation.

Now case 2. Take a point Q such that the lines $l = P\mathcal{F}(P)$ and $m = Q\mathcal{F}(Q)$ intersect. Let C be the intersection point. The points P and Q cannot both coincide with C; otherwise, the lines l and m would coincide. Assume that $P \neq C$. Since $\mathcal{F}$ is a dilation, we have $PC \,/\!/\, \mathcal{F}(P)\mathcal{F}(C)$. Then $\mathcal{F}(P)\mathcal{F}(C) \,/\!/\, l$

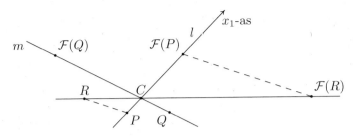

Fig. 2.35. Construction of a central dilation

and therefore $\mathcal{F}(C) \in l$. If we had $Q = C$, then we would have $\mathcal{F}(Q) \in l$, so l and m would coincide, which is not the case. Consequently, $Q \neq C$, and in the same way as above, we can show that $\mathcal{F}(C) \in m$. Therefore $\mathcal{F}(C) = C$: the dilation $\mathcal{F}$ has a fixed point C. Next let R be an arbitrary point, distinct from C. Since $\mathcal{F}$ is a dilation and $C = \mathcal{F}(C)$, we have $CR \mathbin{/\mkern-5mu/} C\mathcal{F}(R)$, so R, C, and $\mathcal{F}(R)$ are collinear. First assume that R does not lie on l. Since $\mathcal{F}$ is a dilation, we have $PR \mathbin{/\mkern-5mu/} \mathcal{F}(P)\mathcal{F}(R)$. Let us choose a coordinate system with origin at C and x_1-axis along l. The point P has coordinates $(p_1, 0)$. The point $\mathcal{F}(P)$ has coordinates $(\alpha p_1, 0)$ for some well-chosen α; $\alpha > 0$ if P and $\mathcal{F}(P)$ lie on the same side of C, while $\alpha < 0$ if this is not the case. Let (r_1, r_2) be the coordinates of R. The point $\mathcal{F}(R)$ is the intersection point of the line CR and the line through $\mathcal{F}(P)$ that is parallel to RP. The equations of these lines are respectively

$$(z_1, z_2) = \mu(r_1, r_2)$$

and

$$(z_1, z_2) = (\alpha p_1, 0) + \lambda(r_1 - p_1, r_2) .$$

Simple computations show that the intersection point $\mathcal{F}(R)$ has coordinates $(\alpha r_1, \alpha r_2)$. What we have just shown holds in particular for $R = Q$. If the point R lies on l, we can do the same deduction, with Q replaced by P. Result: for the point R with coordinates (r_1, r_2), $\mathcal{F}(R)$ has coordinates $(\alpha r_1, \alpha r_2)$. The map $\mathcal{F}$ is therefore the central dilation with center C and scale factor α.

A direct consequence of this theorem and of Theorem 2.42 is that the composition of two central dilations, with possibly different centers, is either a central dilation or a translation. We easily see that if the composition of two central dilations is again a central dilation, the scale factor of the composition is the product of the scale factors of the two original central dilations. Let us illustrate the theory with an example.

Example 2.46 (Nine-point circle). Consider Fig. 2.36. In triangle ABC, the points D, E, F are the respective midpoints of sides a, b, c. The feet of the altitudes from A, B, C are respectively P, Q, R. The orthocenter of the triangle is H. The midpoints of the line segments $[AH]$, $[BH]$, $[CH]$ are respectively U, V, W. Euler (1707–1783) showed that the six points D, E,

F, P, Q, R lie on one circle. In 1820, Brianchon (1783–1864) and Poncelet (1778–1867) showed that this circle also contains the points U, V, and W [6]. They also showed that the nine-point circle is tangent to the incircle and to the three excircles of ABC. This same result was proved independently by Feuerbach (1800–1834) in 1822 and is now known as Feuerbach's theorem. The name *nine-point circle* was introduced by Terquem in 1842 [70].

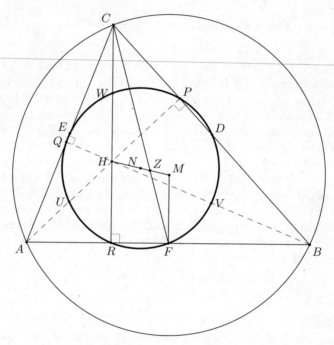

Fig. 2.36. The nine-point circle

Before proving that all nine points line on one circle, let us first make some remarks. In Sect. 1.4, Exercise 1.20, we showed that the perpendicular bisectors of $\triangle ABC$ are concurrent; in the figure their intersection point is M. The point M is the unique point that is equidistant from the three vertices of ABC. There is therefore exactly one circle through A, B, and C; this circle is called the circumcircle of the triangle; we will study it in Sect. 4.1. In Example 1.28 we showed that the medians of triangle ABC concur at a point Z and that the lengths of the line segments $[CZ]$ and $[ZF]$ are in the ratio $2 : 1$. Let $\mathcal{F}$ be the central dilation $\mathcal{F}$ with center Z and scale factor -2; then $\mathcal{F}(F) = C$. For the moment, let us write X for $\mathcal{F}(M)$. Since $\mathcal{F}$ is a dilation, MF and XC are parallel. Hence X lies on the altitude CR from C. Likewise, we can show that X lies on the other altitudes, so $X = H$, the orthocenter. Thus we have given a new proof of the concurrence of the three altitudes. We also see that the *orthocenter H, the centroid Z, and the center*

2.5 Dilation and Similarity

of the circumcircle M lie on one line, which is called the *Euler line*. Moreover, we have $d(Z, H) = 2d(Z, M)$. Let us now consider the central dilation $\mathcal{G}$ with center H and scale factor $1/2$. We have $U = \mathcal{G}(A)$, $V = \mathcal{G}(B)$, and $W = \mathcal{G}(C)$. Let $N = \mathcal{G}(M)$. The point N lies on the Euler line and is the midpoint of the line segment $[HM]$. The circumcircle is mapped by $\mathcal{G}$ onto the circle Γ with center N passing through the points U, V, and W.

Let us show that Γ is the nine-point circle. For this, we first consider the dilation $\mathcal{H} = \mathcal{F}^{-1} \circ \mathcal{G}^{-1}$; $\mathcal{G}^{-1}$ is the central dilation with center H and scale factor 2, and $\mathcal{F}^{-1}$ is the central dilation with center Z and scale factor $-1/2$. Consequently, $\mathcal{H}$ is a central dilation with scale factor -1. Note that $\mathcal{G}^{-1}(N) = M$ and $\mathcal{H}(N) = \mathcal{F}^{-1}(M) = N$; the center of $\mathcal{H}$ is therefore N. It follows that $\mathcal{H}(\Gamma) = \Gamma$; $\mathcal{H}$ maps every point of Γ onto the point diametrically opposite it on Γ. Moreover, we easily verify that $\mathcal{H}(U) = D$, $\mathcal{H}(V) = E$, and $\mathcal{H}(W) = F$. Therefore D, E, and F also lie on Γ. We must still prove that P, Q, and R lie

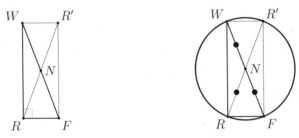

Fig. 2.37. R also lies on the nine-point circle

on Γ. We will prove this for R; the proofs for the other points are analogous. From the above we know that the triangle WRF (see also Fig. 2.37 on the left) has a right angle at R, and that N is the midpoint of the line segment WF. Choose the point R' such that $WR' \parallel RF$ and $FR' \parallel RW$. Then $WRFR'$ is a rectangle, so the diagonals RR' and WF have the same length (Sect. 1.6, Exercise 1.32) and bisect each other (Sect. 1.5, Exercise 1.25). It follows that $d(R, N) = d(F, N)$; see the figure on the right. Therefore R lies on Γ.

Remark 2.47. If H is the orthocenter of triangle ABC, we also know that A is the orthocenter of triangle BCH, B is the orthocenter of triangle $\triangle CAH$, and C is the orthocenter of triangle ABH. Because of this property we call $\{A, B, C, H\}$ an *orthocentric system*: each of the four points is the orthocenter of the triangle formed by the other three. The four triangles have the same nine-point circle; see Fig. 2.38. It is interesting to study the roles played by the nine points in the four triangles.

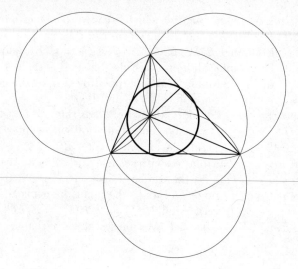

Fig. 2.38. An orthocentric system has one nine-point circle

Similarity

We saw earlier that congruence is an important notion; the notion of similarity is nearly as important.

Definition 2.48. *We call two figures F_1 and F_2 in the plane similar, denoted by $F_1 \sim F_2$, if there exist an isometry $\mathcal{H}$ of the plane and a central dilation $\mathcal{V}$ such that $\mathcal{V} \circ \mathcal{H}(F_1) = F_2$. The composition $\mathcal{V} \circ \mathcal{H}$ is called a similarity.*

By definition, a similarity is a composition $\mathcal{V} \circ \mathcal{H}$, where the central dilation $\mathcal{V}$ follows the isometry $\mathcal{H}$. The composition $\mathcal{H} \circ \mathcal{V}$, where the isometry $\mathcal{H}$ follows the central dilation $\mathcal{V}$, is also a similarity; see Exercise 2.42. If $\mathcal{G}$ is a similarity and $\mathcal{G} = \mathcal{V} \circ \mathcal{H}$ as in the definition, then

$$d(\mathcal{G}(X), \mathcal{G}(Y)) = |\alpha| d(X, Y) \text{ for all } X, Y,$$

where α is the scale factor of the central dilation $\mathcal{V}$. Because of this we call $|\alpha|$ the *scale factor* of $\mathcal{G}$. There is a clear analogy between the theory of similarity and that of congruence. We will list a number of properties of similarity. For some of the proofs we will refer to the exercises. It is immediately clear that a similarity maps straight lines onto straight lines. The size of angles is also invariant under similarities; see Exercise 2.39.

Theorem 2.49. *Let $\mathcal{G}_1$ and $\mathcal{G}_2$ be similarities. If there exist three noncollinear points A, B, and C such that $\mathcal{G}_1(A) = \mathcal{G}_2(A)$, $\mathcal{G}_1(B) = \mathcal{G}_2(B)$, and $\mathcal{G}_1(C) = \mathcal{G}_2(C)$, then $\mathcal{G}_1 = \mathcal{G}_2$.*

Exercise 2.38 gives hints for the proof of this theorem. More succinctly, the theorem states that a similarity is determined by its action on three noncollinear points. The following theorem gives a very useful method for determining whether figures are similar.

Theorem 2.50 (Similarity criterion AAA). *Two triangles ABC and $A_1B_1C_1$ are similar if $\angle A = \angle A_1$, $\angle B = \angle B_1$, and $\angle C = \angle C_1$.*

Proof. We are going to construct a similarity $\mathcal{H}$ satisfying $\mathcal{H}(A) = A_1$, $\mathcal{H}(B) = B_1$, and $\mathcal{H}(C) = C_1$. Since $\angle A = \angle A_1$, there is an isometry $\mathcal{F}_1$ that maps angle A to angle A_1; see Fig. 2.39. If $\mathcal{F}_1(B)$ and B_1 lie on different

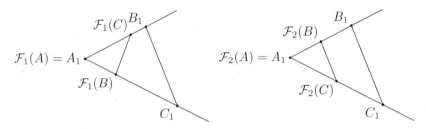

Fig. 2.39. On the similarity criterion **AAA**

legs of angle A_1, as sketched in the figure on the left, we compose $\mathcal{F}_1$ with the reflection in the bisector of angle A_1. We call the composition $\mathcal{F}_1$ followed by the reflection $\mathcal{F}_2$; see the figure on the right. Note that by the equality of the corresponding angles and Theorem 2.19 we know that the line $\mathcal{F}_2(B)\mathcal{F}_2(C)$ is parallel to B_1C_1. In the obvious way we construct a central dilation $\mathcal{V}$ with center A_1 that maps $\mathcal{F}_2(B)$ onto B_1. This $\mathcal{V}$ maps the point $\mathcal{F}_2(C)$ onto C_1, because $\mathcal{V} \circ \mathcal{F}_2(C)$ must lie both on the line A_1C_1 and on the line through B_1 that is parallel to $\mathcal{F}_2(B)\mathcal{F}_2(C)$. The map $\mathcal{V} \circ \mathcal{F}_2$ is the similarity $\mathcal{H}$ we were looking for.

It is not immediately clear from the definition of similarity that the similarity relation is symmetric, that is, that if $F_1 \sim F_2$, then $F_2 \sim F_1$. We will now prove that the relation is indeed symmetric.

Theorem 2.51. *If $F_1 \sim F_2$, then $F_2 \sim F_1$.*

Proof. Since $F_1 \sim F_2$, there is a similarity $\mathcal{G}_1$ such that $\mathcal{G}_1(F_1) = F_2$. Choose three noncollinear points A_1, B_1, C_1. Let $\mathcal{G}_1(A_1) = A_2$, $\mathcal{G}_1(B_1) = B_2$, $\mathcal{G}_1(C_1) = C_2$; see Fig. 2.40. By Exercise 2.39, we have $\angle A_1 = \angle A_2$, $\angle B_1 = \angle B_2$, $\angle C_1 = \angle C_2$. Using Theorem 2.50 we can find a similarity $\mathcal{G}_2$ such that $\mathcal{G}_2(A_2) = A_1$, $\mathcal{G}_2(B_2) = B_1$, and $\mathcal{G}_2(C_2) = C_1$. Consider $\mathcal{H} = \mathcal{G}_2 \circ \mathcal{G}_1$. We have $\mathcal{H}(A_1) = A_1$, $\mathcal{H}(B_1) = B_1$, and $\mathcal{H}(C_1) = C_1$. It follows that the product of the factors of $\mathcal{G}_1$ and $\mathcal{G}_2$ is equal to 1, so that $\mathcal{H}$ is an isometry. Theorem 2.3 now tells us that $\mathcal{H}$ is the identity. Consequently, $\mathcal{G}_2(F_2) = F_1$.

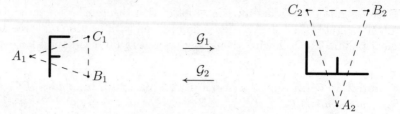

Fig. 2.40. The similarity relation is symmetric

The following is another criterion for the similarity of triangles. The proof is left to the reader.

Theorem 2.52. *Two triangles ABC and $A_1B_1C_1$ are similar if $\angle A = \angle A_1$ and $d(A_1, B_1) : d(A, B) = d(A_1, C_1) : d(A, C)$.*

Exercises

2.38. Let $\mathcal{G}_i = \mathcal{V}_i \circ \mathcal{H}_i$ for $i = 1, 2$, where $\mathcal{H}_i$ is an isometry and $\mathcal{V}_i$ is a central dilation. For the three noncollinear points A, B, and C we have $\mathcal{G}_1(A) = \mathcal{G}_2(A)$, $\mathcal{G}_1(B) = \mathcal{G}_2(B)$, and $\mathcal{G}_1(C) = \mathcal{G}_2(C)$. Then

(a) If α_i is the scale factor of $\mathcal{V}_i$ for $i = 1, 2$, then $|\alpha_1| = |\alpha_2|$.
(b) $\mathcal{V}_2^{-1} \circ \mathcal{V}_1$ is an isometry.
(c) $\mathcal{H}_2 = \mathcal{V}_2^{-1} \circ \mathcal{V}_1 \circ \mathcal{H}_1$.
(d) $\mathcal{G}_1 = \mathcal{G}_2$.

2.39. If $\mathcal{G}$ is a similarity and A, B, and C are three noncollinear points, then $\mathcal{G}(A)$, $\mathcal{G}(B)$, $\mathcal{G}(C)$ are noncollinear and $\angle A = \angle \mathcal{G}(A)$, $\angle B = \angle \mathcal{G}(B)$, $\angle C = \angle \mathcal{G}(C)$.
Hint: First prove this for a dilation.

2.40. Prove Theorem 2.52. You might start as in the proof of Theorem 2.50.

2.41. In triangle ABC, let D, E, F be the feet of the altitudes from respectively A, B, and C. Triangle DEF is called the *orthic triangle* of ABC.

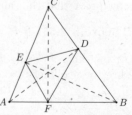

(a) Show that $\triangle BDA \sim \triangle BFC$.
(b) We have $d(B, D) : d(B, F) = d(A, B) : d(C, B)$.
    Show this and deduce from it that
    $d(B, D) : d(A, B) = d(B, F) : d(C, B)$.
(c) $\triangle BDF \sim \triangle BAC$, and in particular $\angle BDF = \angle BAC$ and $\angle BFD = \angle BCA$. Because of the equality of the angles we call the lines AC and DF *antiparallel* with respect to angle B.

(d) The altitudes of triangle ABC coincide with the angle bisectors of the orthic triangle. Using Example 2.22, we can now prove that the three altitudes concur.

2.42. Let $\mathcal{H}$ be an isometry and let $\mathcal{V}$ be a central dilation.

(a) $\mathcal{H} \circ \mathcal{V}$ is a similarity.
Hint: We have $(\mathcal{H} \circ \mathcal{V})^{-1} = \mathcal{V}^{-1} \circ \mathcal{H}^{-1}$; this map is a similarity. See also the proof of Theorem 2.51.
(b) In general, $\mathcal{H} \circ \mathcal{V}$ and $\mathcal{V} \circ \mathcal{H}$ are different transformations.

2.43. The similarity relation is transitive: if $F_1 \sim F_2$ and $F_2 \sim F_3$, then $F_1 \sim F_3$.
Hint: First assume that F_1 is a triangle and use Theorem 2.50. Another method is to use Exercise 2.42.

2.44. In Example 2.46, we showed that in a right triangle WRF with right angle at R the length of the median from R onto the hypotenuse is equal to half the length of the hypotenuse WF.

(a) Deduce from this that the sum of the angles of $\triangle WRF$ is equal to π.
(b) Use this result to show that in every triangle the sum of the angles is equal to π.
(c) If, moreover, $d(R, F) = d(W, F)/2$, then $\angle WFR = \pi/3$.

2.45. Here is another similarity criterion. Two triangles ABC and $A_1B_1C_1$ are similar if

$$d(A_1, B_1) : d(A, B) = d(A_1, C_1) : d(A, C) = d(B_1, C_1) : d(B, C) .$$

Hint: Apply a suitable central dilation $\mathcal{V}$ such that $\triangle \mathcal{V}(A)\mathcal{V}(B)\mathcal{V}(C)$ is congruent to $\triangle A_1B_1C_1$.

2.46. Let n_1, n_2, n_3 be three distinct lines. Let l be a line that intersects these three in respectively A_1, A_2, A_3. Let m be a line that intersects the three in respectively B_1, B_2, B_3.

(a) If $n_1 // n_2 // n_3$, then $d(A_1, A_2) : d(A_2, A_3) = d(B_1, B_2) : d(B_2, B_3)$.
Hint: Draw the line l' through B_1 and parallel to l.
(b) If $d(A_1, A_2) : d(A_2, A_3) = d(B_1, B_2) : d(B_2, B_3)$ and $n_1 // n_2$, then the lines n_1, n_2, n_3 are parallel to each other.

2.47. Let ABC be a triangle. We want to draw a square whose four vertices lie on the three sides of the triangle. Is this possible?
Hint: First try to draw three vertices on two sides.

2.48. The golden section. Let $[AB]$ be a line segment. Let C be a point cutting the line segment $[AB]$ according to the *golden section*, that is,

$$d(A,B) : d(A,C) = d(A,C) : d(C,B) .$$

We call the ratio $d(A,B) : d(A,C)$ the *golden ratio*; it is seen as the ideal ratio for lengths in many objects, for example paintings.

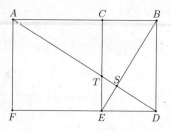

(a) We set $d(A,C) = 1$ and $d(C,B) = \tau$. The golden ratio is then $1 + \tau$. We have $1 + \tau = 1/\tau$ and $\tau^2 + \tau - 1 = 0$. Hence $\tau = (-1 + \sqrt{5})/2$.
Let us now consider the *golden* rectangle $AFDB$ with $d(A,F) = d(A,C)$ and $CE \parallel AF$. Let S be the intersection point of AD and EB, and T that of AD and EC.
(b) We have $\triangle ABD \sim \triangle BDE$.
(c) It follows that $\angle ASB = \pi/2$.
(d) We also have $d(S,B) : d(S,A) = \tau$.
(e) The rectangle $BDEC$ arises from $ABDF$ by a rotation over $-\pi/2$ about S and a central dilation with center S and scale factor τ. The figure shows the effect of iterations of this similarity. See also [33].

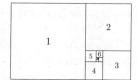

2.6 Fractals, Dimension, and Measure

Though examples of fractals had already been studied previously, in 1975 Mandelbrot (b. 1924) coined the term fractal and gave a mathematical definition of such a set. However, this definition is incomplete, since it excludes certain sets one would consider fractals. One of the most important properties of a fractal is its *self-similarity*. We will illustrate this with examples.

Example 2.53. The first example is the *Sierpiński triangle*. Sierpiński (1882–1969) described some of its properties in 1916 in connection with the topological study of curves. We start with a full triangle ABC, that is, a triangle ABC plus its interior; see Fig. 2.41 (1). First we divide the full triangle into four congruent pieces by adding the mid-parallels DE, EF, and FD, and remove the interior of the middle triangle EFD. What is left consists of three congruent triangles; see (2). Next we apply the same process to each of these triangles; the result is drawn in (3). A further iteration of the process gives the set sketched in (4). Figures (1) through (4) give an increasingly accurate approximation of the Sierpiński triangle we want to construct. By continuing the process of removing the middle triangles, we obtain the Sierpiński triangle as a limit. A closer look at the figures shows us that (2) is made up of

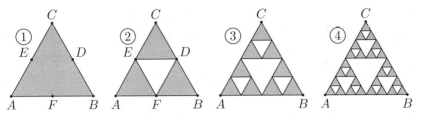

Fig. 2.41. The Sierpiński triangle

three copies of (1) after the application of a central dilation with scale factor $1/2$; (3) is made up of three copies of (2) after the application of a central dilation with scale factor $1/2$; and so on. This suggests that the Sierpiński triangle has the following property: the central dilations $\mathcal{V}_1$, $\mathcal{V}_2$, and $\mathcal{V}_3$, each with scale factor $1/2$, and with the vertices as centers (in (1) these are A, B, and C), map the Sierpiński triangle onto a subset of itself; the set is the union of the three images. We call this property the *self-similarity* of the Sierpiński triangle. In this context people sometimes speak of the *Droste effect*, a Dutch term for such recursive figures. On the boxes of Droste cocoa there is an image of a nurse carrying a tray with a wholesome cup of hot chocolate and a box of Droste cocoa; on that box we again see the whole image at a smaller scale. In our example this repetitive effect is repeatedly carried out in triplicate. The Sierpiński triangle was invented long before the term *fractal* came into use. The following example is even older; it was constructed as an example of a curve that is nowhere differentiable.

Example 2.54. This example is called the *Koch curve*, after its discoverer von Koch (1870–1924). We begin with a line segment, see Fig. 2.42 (1). To fix our thoughts, let us assume that this segment goes from $(0,0)$ to $(1,0)$. The

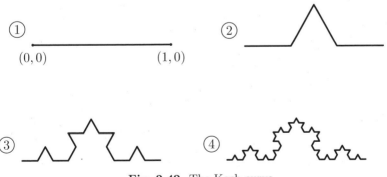

Fig. 2.42. The Koch curve

middle third of this line segment is replaced by the legs of an equilateral triangle with the middle third as basis; see (2). In the next step each of the four line segments of (2) is replaced by a copy of (2) scaled by a factor $1/3$, and so on. By repeating this process we obtain the Koch curve K as "limit."

Consider the following similarities:

$\mathcal{G}_1$ the central dilation with center $(0,0)$ and scale factor $1/3$,

$\mathcal{G}_2$ the map $\mathcal{G}_1$ followed first by a translation with vector $(1/3, 0)$ and then by a rotation about $(1/3, 0)$ with rotation angle $\pi/3$,

$\mathcal{G}_3$ the map $\mathcal{G}_1$ followed first by a translation with vector $(1/3, 0)$ and then by a rotation about $(2/3, 0)$ with rotation angle $-\pi/3$,

$\mathcal{G}_4$ the map $\mathcal{G}_1$ followed first by a translation with vector $(2/3, 0)$.

Together, the images $\mathcal{G}_1(K)$, $\mathcal{G}_2(K)$, $\mathcal{G}_3(K)$, and $\mathcal{G}_4(K)$ cover K, that is,

$$K = \mathcal{G}_1(K) \cup \mathcal{G}_2(K) \cup \mathcal{G}_3(K) \cup \mathcal{G}_4(K).$$

Iterated Function Systems, IFS

The constructions of Examples 2.53 and 2.54 display a common pattern. This leads to the definition of the notion *IFS*, *iterated function system*, introduced by Hutchinson in 1981 [34]. We will limit ourselves to sets in the plane; a thorough and more general description, with many applications, can be found in [2] or [54].

Definition 2.55. *An IFS is a set $\{\mathcal{G}_1, \ldots, \mathcal{G}_n\}$, $n > 1$, of n distinct similarities with respective scale factors $r_1, \ldots, r_n$, where $r_i < 1$ for all i.*

The following theorem is one of the most important statements concerning IFSs; for a proof, see [2]. In the statement of the proof, and therefore also in the definition of invariant set that it contains, we use certain topological notions that we will not explain further.

Theorem 2.56. *For every IFS $\{\mathcal{G}_1, \ldots, \mathcal{G}_n\}$ there is exactly one nonempty bounded closed set K such that*

$$K = \mathcal{G}_1(K) \cup \cdots \cup \mathcal{G}_n(K).$$

This set K is called the invariant set *of the IFS.*

In Example 2.53, the Sierpiński triangle is the invariant set of the IFS $\{\mathcal{V}_1, \mathcal{V}_2, \mathcal{V}_3\}$. In Example 2.54, the Koch curve is the invariant set of the IFS $\{\mathcal{G}_1, \mathcal{G}_2, \mathcal{G}_3, \mathcal{G}_4\}$. In general the invariant sets of IFSs are fractals. Figure 2.43 gives two more examples of fractals. The example on the left is the *Sierpiński carpet*. This can be described as the invariant set of an IFS consisting of eight similarities, each with scale factor $1/3$. The other example, on the right, is called the *Menger sponge* (1902–1985). In contrast to the other examples, this one is three-dimensional. This fractal was devised in the 1920s in connection with the topological study of curves. The Menger sponge is the invariant set of an IFS consisting of twenty similarities, each with scale factor $1/3$. Here is a

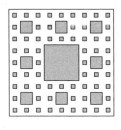
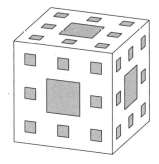

Fig. 2.43. (a) Sierpiński carpet; (b) Menger sponge

more graphic description. In your mind, divide a cube into twenty-seven cubes, by dividing each edge into three equal pieces. Remove the central column of cubes, from top to bottom, and the central rows, both from front to back and from left to right. In all, seven cubes are removed (the central cube is part of all columns and rows that are removed). Repeat this process with each of the remaining cubes, and so on.

How can we find the invariant set of a given IFS $\{\mathcal{G}_1, \ldots, \mathcal{G}_n\}$? We can present the answer to this question in the form of general theorems. However, we choose to limit ourselves to giving a description in a simple case. Let R_0 be a rectangle together with its interior. We then define $R_1, R_2, \ldots$ by

$$R_1 = \mathcal{G}_1(R_0) \cup \cdots \cup \mathcal{G}_n(R_0),$$
$$R_2 = \mathcal{G}_1(R_1) \cup \cdots \cup \mathcal{G}_n(R_1),$$

and in general,

$$R_{k+1} = \mathcal{G}_1(R_k) \cup \cdots \cup \mathcal{G}_n(R_k). \tag{2.3}$$

We can prove that regardless of the choice of R_0, the series of sets R_0, R_1, $R_2, \ldots$ has the invariant set as "limit." In this way, we obtain the invariant set as the limit of an *iteration*: we begin with a set R_0, the *initial value*. Instead of a rectangle we can take an arbitrary set. From this set the operation described in (2.3) forms the set R_1, then R_2, and so on. The resulting sets "converge" to the invariant set of the IFS. We illustrate this further with two examples.

Example 2.57. Consider the IFS consisting of two central dilations, each with scale factor $1/3$, one with center $(0,0)$, the other with center $(1,0)$. See Fig. 2.44. As initial value we take the rectangle R_0 with sides parallel to the axes and vertices $(0, 1/2)$ and $(1, 2/3)$. The figure shows R_1, R_2, R_3, and R_4. The invariant set of the IFS is the "limit" of the series $R_0, R_1, R_2, \ldots$. This is a subset of the line segment between $(0,0)$ and $(1,0)$. This set was first described by Cantor (1845–1918) in 1883, and is therefore called the *Cantor set*. It is a set that has "as many" points as the real line, but contains no interval. Hence there are many "holes" in the set. The set is sometimes also

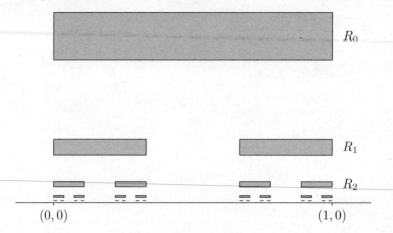

Fig. 2.44. The Cantor set

called the *Cantor discontinuum*. Its structure is completely different from that of the real line, which in Cantor's time was called the continuum. The name continuum refers to the connectedness of the real axis: it has no holes.

Example 2.58. Let us now describe a variant of the last example. Unlike the previous examples, the similarities of this IFS have different scale factors. The IFS consists of two central dilations. One has scale factor $1/3$ and center $(0,0)$, the other has scale factor $1/2$ and center $(1,0)$; see Fig. 2.45. The initial set

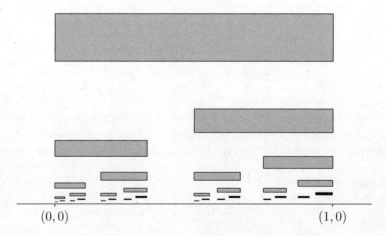

Fig. 2.45. An IFS with distinct scale factors

of the iteration is the same as in the last example. Again the invariant set of the IFS is a subset of the real line. It resembles the Cantor set very much.

The iteration provides a method for rapidly obtaining an idea of what the fractal looks like. The applications of IFSs are based on this property. In general, the fractals are generated using iterations. As soon as one knows which similarities are involved, one can make an image of the fractal. Whenever an image can be made up of fractals, it can be constructed in a simple manner. This is why fractals are used in image compression; see [2].

Dimension and Measure

The word fractal is derived from the Latin adjective *fractus*, which in turn is related to the verb *frangere*, which means to break. The name fractal refers to a characteristic property of fractals, the *broken dimension*. The dimension function in question was introduced by Hausdorff (1869–1942) in 1918 and is related to the manner in which the measure of a figure changes under central dilations. For most fractals, the value of this dimension function is not an integer. Let us first give a simple, but important, example.

Example 2.59. Consider Fig. 2.46. When we apply a central dilation with scale

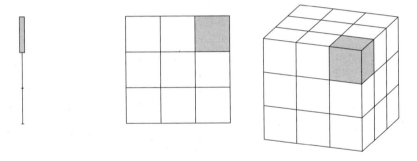

Fig. 2.46. Length, area, and volume

factor $1/k$ for an integer k greater than 0 to an interval, the *length* of the image is $1/k$ times the length of the original interval. Moreover, k copies of the image together cover the original interval. When we apply a central dilation with scale factor $1/k$ to a square, the *area* of the image is $1/k^2$ times the area of the original square, and k^2 copies of the image taken together cover the original square. In Sect. 1.1, when we discussed the proof of the Pythagorean theorem by Einstein, we stated this property more generally: if two figures are similar, the areas of the figures are in the same ratio as the squares of the lengths of corresponding line segments. When we apply a central dilation with scale factor $1/k$ to a cube, the *volume* of the image equals $1/k^3$ times the volume of the original cube, and k^3 copies of the image fill up the original cube. This tells us how the length, area, and volume behave under scaling (central dilation), something we possibly remember from elementary school: 1

meter is 10 decimeters, 1 square meter is 100 square decimeters, and 1 cubic meter is 1000 cubic decimeters.

Let us once more consider the Sierpiński triangle of Example 2.53. What is its area? If the area of triangle ABC in Fig. 2.41 is a, then the areas in (2), (3), and (4) are equal to respectively $(3/4)a$, $(3/4)^2 a = (9/16)a$, and $(3/4)^3 a = (27/64)a$. The series $a, (3/4)a, (3/4)^2 a, (3/4)^3 a, \ldots$ converges to 0. Consequently, the Sierpiński triangle has area 0. How long is the Sierpiński triangle? It is not clear how we should measure the length. But it is plausible that the length must be greater than the sum of the perimeters of the smaller triangles in any approximation; indeed, the sides of the triangles are part of the Sierpiński triangle. If the perimeter of the triangle ABC is equal to s, then the sums of the perimeters of the smaller triangles in subsequent approximations are respectively s, $(4/3)s$, $(4/3)^2 s$, $(4/3)^3 s$, and so on. We see that the sums of the perimeters grow without bound and that the length of the Sierpiński triangle is "infinite." It seems that on the one hand, the size of the Sierpiński triangle is too small for the area measure, while on the other hand, it is too large for the length measure.

Fractal Dimension

In an article in 1919, Hausdorff pointed out the connection between measure and dimension. We measure the size of a 1-dimensional figure such as the interval using the length; consequently, length is sometimes called a 1-dimensional measure. We measure the size of a 2-dimensional figure such as the square using the area; consequently, the area is sometimes called a 2-dimensional measure. We measure the size of a 3-dimensional figure such as the cube using the volume; consequently the volume is sometimes called a 3-dimensional measure. In Example 2.59, we illustrated the following fundamental property of the n-dimensional measure, for $n = 1$, 2, and 3: when we apply a central dilation with scale factor $1/k$ for k an integer ≥ 1 to an n-dimensional figure, the n-dimensional measure of the image is equal to $(1/k)^n$ times the n-dimensional measure of the original figure. This property is called the *scaling property* of the n-dimensional measure. We recognize the dimension n in the exponent of $(1/k)^n$. Hausdorff also gave a definition of D-dimensional Hausdorff measures, where D is a positive number that is not necessarily an integer. This measure has the scaling property mentioned above: when we apply a central dilation with scale factor $1/k$ for k an integer ≥ 1 to a figure, the D-dimensional Hausdorff measure of the image is equal to $(1/k)^D$ times the D-dimensional Hausdorff measure of the original figure.

Let us again consider the Sierpiński triangle S of Example 2.53. We have seen that the 2-dimensional measure of S equals 0, while the 1-dimensional measure is "infinite." Is there a number D such that the D-dimensional Hausdorff measure of S is greater than 0 but less than "infinite"? What conditions must such a D satisfy? Assume that the Sierpiński triangle has a positive D-dimensional Hausdorff measure m. We have seen that S is the union of three copies $\mathcal{V}_1(S)$, $\mathcal{V}_2(S)$, and $\mathcal{V}_3(S)$ of itself. Each copy arises from S through a central dilation with scale factor $1/2$. Hence $m(\mathcal{V}_i(S)) = (1/2)^D m(S)$ for $i = 1, 2, 3$ (scaling property). Since any two copies have only one common point, it is likely that we must have

$$m(S) = m(\mathcal{V}_1(S)) + m(\mathcal{V}_2(S)) + m(\mathcal{V}_3(S)) = 3\left(\tfrac{1}{2}\right)^D m(S).$$

But then $3 = 2^D$. There is only one number that satisfies this, namely $D = \log_2 3$, the number to which power you must take 2 to obtain 3. This number is called the *fractal* or *Hausdorff* dimension of S. The Hausdorff dimension is a notion that exists for more general settings than only fractals. In the case of fractals, simpler computation rules hold; we then speak of the *fractal dimension*. In general, we define the fractal dimension corresponding to an IFS as follows.

Definition 2.60. *The* fractal dimension *of the IFS $\{\mathcal{G}_1, \ldots, \mathcal{G}_n\}$ with respective scale factors $r_1, \ldots, r_n$, $0 < r_i < 1$ for all i, is the unique number D satisfying $\sum_{i=1}^{n} r_i^D = 1$.*

Table 2.2 gives the fractal dimensions of the IFSs that have been mentioned in this section.

Table 2.2. The fractal dimensions of a number of IFSs

IFS	fractal dimension
Sierpiński triangle	$\log_2 3 \approx 1.585$
Koch curve	$\log_3 4 \approx 1.262$
Sierpiński carpet	$\log_3 8 \approx 1.893$
Menger sponge	$\log_3 20 \approx 2.727$
Cantor set	$\log_3 2 \approx 0.631$
Example 2.58	≈ 0.788

Exercises

2.49. Consider the IFS of Example 2.57.

(a) Make a sketch of the effect of three iterations with initial value the line segment between $(0, 0)$ and $(1, 0)$.
(b) Show that the Cantor set does not contain any interval.

2.50. On the real line $\mathbb{R}$, we study the IFS consisting of the following maps: $f_1(x) = (1/3)x - 2/3$, $f_2(x) = (1/3)x$, and $f_3(x) = (1/3)x + 2/3$.

(a) The invariant set is the line segment between $(-1,0)$ and $(1,0)$.
    Hint: Theorem 2.56.
(b) What is the fractal dimension of this IFS?
    Consider the IFS consisting of the following maps: $g_1(x) = (2/3)x - 1/3$, $g_2(x) = (2/3)x$, and $g_3(x) = (2/3)x + 1/3$.
(c) What is the invariant set of the IFS $\{g_1, g_2, g_3\}$?
(d) What is the fractal dimension of the IFS $\{g_1, g_2, g_3\}$?

Remark: The answers to (a) and (c) are the same, but those to (b) and (d) are different. Apparently, the fractal dimension depends not only on the invariant set, but also on the maps that occur in the IFS.

2.51. Show that there is no IFS whose invariant set is a circle.

2.52. Figure 2.47 (a) is an approximation of the invariant set of an IFS with fractal dimension $\log_3 4$.

(a) Describe such an IFS.
(b) Show that the invariant set does not contain any line segment.

2.53. Figure 2.47 (b) is an approximation of the invariant set of an IFS with fractal dimension $\log_4 8 = 1.5$. Describe such an IFS.

Fig. 2.47. Figures for (a) Exercise 2.52; (b) Exercise 2.53

3
SYMMETRY

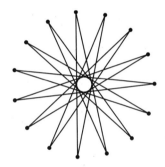

In this chapter we take a closer look at symmetry. Its importance in art is evident in the work of Escher. In the Alhambra, that monument of Moorish art that so inspired Escher [62], all decorations consist of symmetric tilings. The booklet [40] is filled with examples of regular structures in art.

Mathematically, we describe the symmetry properties of a figure using its *symmetry group*, the group of all isometries of the figure. A *symmetry property* is characterized by a, usually simple, subgroup of the group $\mathcal{I}(V)$ of all isometries of the plane.

In Sect. 3.1 we first discuss *rotational symmetry*. Then we look at the symmetry groups of a number of figures. Finally, we determine which groups can occur as finite subgroups of $\mathcal{I}(V)$. In the next section we analyze the structure of frieze patterns by studying their symmetry groups. We first define when two frieze patterns are the same and then give a classification of all frieze patterns. This is a good preparation for the study of periodic tilings. In Sect. 3.3 we introduce *Voronoi cells*, an important tool for the analysis of periodic tilings. Our first use of Voronoi diagrams is the definition of the conics as *front lines*. We then proceed, in the final three sections, to classify the periodic tilings. The form of the Voronoi cell associated to a periodic tiling gives a good indication of its symmetry. We will use this in our enumeration of the possible tilings. Here too the classification is based on symmetry properties.

3 SYMMETRY

3.1 What Is Symmetry?

In the previous chapter we discussed the notions of reflection symmetry and point symmetry. In this section we continue our study of symmetry. We will often use the results of Sect. 2.4, in particular Table 2.1.

To study the symmetry of a figure F, we consider all isometries that map F onto itself; the idea is that this set of isometries is a good reflection of the isometries of F.

Definition 3.1. *Let F be a subset of the plane. The* symmetry group *of F is the group $\mathcal{I}(F)$ of all isometries of the plane that map F onto itself.*

To justify this definition we need to check that $\mathcal{I}(F)$ is a subgroup of $\mathcal{I}(V)$. We studied this latter group in detail in Sect. 2.4. We know that if $\mathcal{H}$ is an isometry that maps F onto itself, $\mathcal{H}^{-1}$ also maps F onto itself. Moreover, the composition of two isometries that map F onto itself also maps F onto itself. It immediately follows that $\mathcal{I}(F)$ is a subgroup of $\mathcal{I}(V)$.

In addition to reflection symmetry and point symmetry, we also have *rotational symmetry*. A set F is *rotationally symmetric* and has an *n-fold rotational symmetry*, where n is a positive integer, if $\mathcal{I}(F)$ contains a rotation $\mathcal{R}$ over an angle $2\pi/n$. Figure 3.1 shows three figures with 12-fold rotational symmetry.

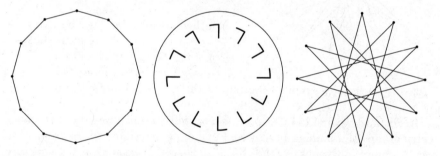

Fig. 3.1. 12-fold rotational symmetry

On the left we see the regular 12-gon. For $n \geq 3$, the regular n-gon is defined as follows. Consider a rotation $\mathcal{R}$ over an angle $2\pi/n$. Choose a point X other than the center of the rotation. Let L be the line segment $[X\mathcal{R}(X)]$. The figure

$$L \cup \mathcal{R}(L) \cup \mathcal{R}^2(L) \cup \cdots \cup \mathcal{R}^{n-1}(L)$$

is a *regular n-gon*. In the formula, $\mathcal{R}^k$ denotes the k-fold composition of $\mathcal{R}$:

$$\mathcal{R}^k = \underbrace{\mathcal{R} \circ \mathcal{R} \circ \cdots \circ \mathcal{R}}_{k \text{ times}}.$$

The regular 12-gon clearly has a 12-fold rotational symmetry. It also has reflection symmetry. There are 12 reflections that map the 12-gon onto itself; the reflection axes are the six joins of diametrically opposite vertices and the six common perpendicular bisectors of diametrically opposite sides. The middle picture in Figure 3.1 also has 12-fold rotational symmetry, but no reflection symmetry: the hooks in the figure, which have no symmetry, prevent its occurrence. The figure on the right is the $\{12/5\}$ star polygon. More generally, for any two integers n and d with $n-1 > d > 1$ and $\gcd(n,d) = 1$, the $\{n/d\}$ *star polygon* is defined as follows [10]: let $\mathcal{R}$ be a rotation over an angle $2\pi/(n/d)$. Let L be the line segment $[X\mathcal{R}(X)]$, where X is not the rotation center. Then $L \cup \cdots \cup \mathcal{R}^{n-1}(L)$ is the $\{n/d\}$ star polygon. The $\{12/5\}$ star polygon has "the same" reflection symmetry as the regular 12-gon.

Rotational symmetry is very common. The sixfold symmetry of the honeycomb is well known. Less known, but particularly nice, is the fivefold symmetry of an apple; you can see it by cutting an apple crosswise.

Definition 3.2. *We say that a rotation $\mathcal{R}$ has* order n *if n is the smallest strictly positive integer such that $\mathcal{R}^n = \mathrm{id}$.*

Thus rotations over angles $\pi/12$, $5\pi/12$, $7\pi/12$, and $11\pi/12$ all have order 12. Let $\mathcal{R}$ be a rotation of order n. We consider the set

$$H = \left\{ \mathrm{id}, \mathcal{R}, \mathcal{R}^2, \ldots, \mathcal{R}^{n-1} \right\}.$$

The elements of this set are all distinct; indeed, if $\mathcal{R}^k = \mathcal{R}^l$ for some $l \leq k$, then $\mathcal{R}^{k-l}$ is the identity, whence $k = l$, because n is the smallest positive integer for which $\mathcal{R}^n = \mathrm{id}$.

If k and l are positive integers with $k < n$ and $l < n$, the product $\mathcal{R}^k \circ \mathcal{R}^l$ is equal to $\mathcal{R}^s$ for $s = k+l \pmod{n}$; by this we mean that $s = k+l$ if $k+l < n$ and $s = k+l-n$ if $k+l \geq n$. By convention, $\mathcal{R}^0 = \mathrm{id}$. The inverse of $\mathcal{R}^k$ is equal to $\mathcal{R}^{n-k}$. It follows that H is a subgroup of $\mathcal{I}(V)$. This subgroup is isomorphic to C_n, the cyclic group with n elements. We obtain the isomorphism by mapping $\mathcal{R}^k$ to k, for $0 \leq k < n$. In the subgroup H, $\mathcal{R}^{n-1}$ is the inverse of $\mathcal{R}$; the rotation angle of $\mathcal{R}^{n-1}$ is therefore equal to 0 if $n = 1$, π if $n = 2$, and $-2\pi/n$ otherwise.

Symmetry groups play an important role in science. In chemistry, for example, symmetry groups of molecules give insight into the possible vibrations and rotations of molecules. This is used to make predictions concerning the color spectra of substances. In crystallography, symmetry groups are used to obtain an overview of all possible crystal forms. Since we can determine the symmetry groups of crystals using x-ray techniques, this is an important tool for identifying substances.

Let us first consider the symmetry groups of a number of simple figures.

Example 3.3 (A & Z). See Fig. 3.2. When treating symmetry groups, we will use symbols in our figures that are borrowed from crystallography. The letter A, on the left in the figure, has only reflection symmetry. The reflection axis is denoted by a line. The letter Z, on the right in the figure, has only

Fig. 3.2. The symmetry groups of the letters A & Z

point symmetry, or, which amounts to the same, twofold rotational symmetry. The rotation center is indicated in the figure by a lens (a regular 2-gon). The symmetry group of the letter Z is isomorphic to C_2, the cyclic group of two elements. The symmetry group of the letter A also consists of two elements, the identity and a reflection $\mathcal{S}$; we denote this group by D_1. The group D_1 is isomorphic to C_2. There is, however, a good reason to use different notations for the symmetry groups of A and Z: the symmetry group of Z consists of orientation-preserving transformations, while the symmetry group of A contains the orientation-reversing reflection $\mathcal{S}$.

Example 3.4 (Δ). Let us now consider the symmetry group of an equilateral triangle Δ; see Fig. 3.3. We immediately see that Δ has threefold rotational

Fig. 3.3. The symmetry group of an equilateral triangle

symmetry. Which isometries are elements of the symmetry group of Δ? First there are the rotations over $2\pi/3$ and $4\pi/3$; together with the identity these form the rotation subgroup, which is isomorphic to C_3. The small triangle in the figure indicates the common center of the rotations. Next there are three reflections, whose axes are indicated in the figure. Note that rotating over $\pi/3$ maps the reflection axes onto each other. Let us now show that these are all the elements of the symmetry group. Every isometric map that leaves Δ invariant induces a permutation of the vertices of Δ. We can check that every permutation of the vertices is induced by one of the maps that we have listed. Since an isometry is determined by its action on three noncollinear points (Theorem 2.4), we have found all isometries that map Δ onto itself. The symmetry group of Δ is denoted by D_3. More generally, the symmetry group of the regular n-gon is denoted by D_n. It contains n reflections, and n rotations that form a subgroup isomorphic to C_n.

Example 3.5 (H). Let us study the symmetry group of the letter H; see Fig. 3.4 (**a**). The letter H has twofold rotational symmetry. The rotations

3.1 What Is Symmetry?

Fig. 3.4. (a) The letter H; (b) a windmill

form a subgroup of the symmetry group that is isomorphic to C_2. There are two reflections, whose axes are perpendicular to each other. In the figure these are indicated by thin lines. The symmetry group of H consists of four transformations; we denote it by D_2. The windmill in (b) has a fourfold rotational symmetry. The small square indicates the rotation center. There is no reflection symmetry. The symmetry group of the windmill is isomorphic to C_4. We note that C_4 and D_2 both have four elements. Nevertheless, the groups are not isomorphic: in D_2 every element $\mathcal{F}$ satisfies $\mathcal{F}^2 = \text{id}$, which is not the case in C_4. The group D_2 is isomorphic to the Klein four-group V_4 from Example 2.24.

Many of the symmetry groups we have seen contain both orientation-preserving and orientation-reversing maps. The following theorem examines this phenomenon more closely.

Theorem 3.6. *Let G be a subgroup of $\mathcal{I}(V)$. Let G^+ be the set of direct isometries of G, and let G^- be the set of indirect isometries of G. Then G^+ is a subgroup of G. Moreover, if $G^- \neq \emptyset$, say $\mathcal{H} \in G^-$, then the map ι_r given by $\mathcal{F} \mapsto \mathcal{F} \circ \mathcal{H}$ is a bijection from G^+ to G^-. Likewise, the map ι_l given by $\mathcal{F} \mapsto \mathcal{H} \circ \mathcal{F}$ is a bijection from G^+ to G^-.*

Proof. We can easily check that G^+ is a subgroup of G. Let $\mathcal{H} \in G^-$. To prove the injectivity of ι_r, we must show that for two distinct elements $\mathcal{F}_1$ and $\mathcal{F}_2$ of G^+, we have $\mathcal{F}_1 \circ \mathcal{H} \neq \mathcal{F}_2 \circ \mathcal{H}$. We will do this by contradiction. If $\mathcal{F}_1 \circ \mathcal{H} = \mathcal{F}_2 \circ \mathcal{H}$, multiplying both members of the equality by $\mathcal{H}^{-1}$ on the right side gives

$$\mathcal{F}_1 \circ \mathcal{H} \circ \mathcal{H}^{-1} = \mathcal{F}_2 \circ \mathcal{H} \circ \mathcal{H}^{-1}, \text{ and therefore } \mathcal{F}_1 = \mathcal{F}_2,$$

a contradiction. Now the surjectivity of ι_r. Let $\mathcal{F} \in G^-$; then $\mathcal{F} \circ \mathcal{H}^{-1} \in G^+$ and we have $(\mathcal{F} \circ \mathcal{H}^{-1}) \circ \mathcal{H} = \mathcal{F}$. Hence ι_r is surjective. We can prove the statements concerning ι_l in the same manner.

If we apply the theorem to the group D_n, the symmetry group of the regular n-gon, G^+ is the rotation group of the regular n-gon, which is isomorphic to C_n. The set G^- consists of reflections. The reflection axes all pass through the common center of the rotations.

We can use these theorems to list all finite isometry groups, that is, all isometry groups with a finite number of elements. The result of the following theorem is said to have already been known to Leonardo da Vinci, who valued it highly in connection with architectural designs. The groups D_n play a dominant role in many constructions (Aachen Cathedral, Castel del Monte, Pentagon, pyramids).

Theorem 3.7. *If G is a finite subgroup of $\mathcal{I}(V)$, the elements of G have a common fixed point P. The group G is isomorphic to either C_n or D_n, for some positive integer n.*

Proof. Let us first consider the subgroup G^+ of G; see Theorem 3.6. If G^+ contains only the unit element, G^+ is isomorphic to C_1. Let us assume for the moment that G^+ has more than one element. We will prove that G^+ is isomorphic to the cyclic group C_n for some $n \geq 2$. Let us first note that there is no translation in G^+ other than the identity map. If there were, say $\mathcal{T} \in G^+$, we would also have $\mathcal{T}^n \in G^+$ for all integers n, which is impossible since G is finite and $\mathcal{T} \neq \text{id}$. Hence G^+ contains only rotations.

Moreover, all rotations in G^+ have the same center P. We will prove this by contradiction. Let $\mathcal{R}_1$, $\mathcal{R}_2$ be two rotations other than the identity, with respective centers C_1 and C_2. Assume that $C_1 \neq C_2$. The rotation angle of $\mathcal{R}_1$ is equal to $\pm 2\pi/n$ for some $n \geq 2$; see Fig. 3.5. Let $C_3 = \mathcal{R}_2(C_1)$. Since

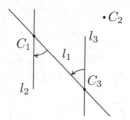

Fig. 3.5. More than one rotation center

$C_1 \neq C_2$ and $\mathcal{R}_2 \neq \text{id}$, we have $C_1 \neq C_3$. Let $\mathcal{H} = \mathcal{R}_2 \circ \mathcal{R}_1 \circ \mathcal{R}_2^{-1}$. Then $\mathcal{H}$ has a fixed point in C_3 and preserves orientation, since it is a product of direct isometries. It cannot equal the identity because $\mathcal{R}_1 \neq \text{id}$; $\mathcal{H}$ is therefore a rotation with center C_3. Theorem 2.34 implies that the rotation angles of $\mathcal{R}_1$ and $\mathcal{H}$ are equal. In Fig. 3.5, $\mathcal{R}_1$ is represented as the product of reflections in the axes l_1 and l_2, where l_1 is the line joining the points C_1 and C_3. We use the same notation as in the proof of Theorem 2.34: the clockwise arrow in the figure indicates that the rotation angle of $\mathcal{R}_1$ is negative. The transformation $\mathcal{H}^{-1}$ is then the product of the reflections in the axes l_3 and l_1, where $l_3 \mathbin{//} l_2$. The composition of $\mathcal{H}^{-1}$ and $\mathcal{R}_1$ is a translation other than the identity; this follows from Theorem 2.34. But we have already shown that G^+ contains only rotations, and no translations. This contradiction comes from the assumption that $C_1 \neq C_2$. Our conclusion is that all rotations in G^+ have the same center, which we call P.

Since G^+ is finite, we can find a rotation $\mathcal{R}$ in G^+ such that the absolute value of the angle of $\mathcal{R}$ is minimal. It follows from the finiteness of G^+ that there is an n such that $\mathcal{R}^n = \text{id}$. We take the smallest n with this property. The minimality of the rotation angle now implies that

$$G^+ = \{\text{id}, \mathcal{R}, \mathcal{R}^2, \ldots, \mathcal{R}^{n-1}\};$$

in other words, G^+ is isomorphic to C_n. If $G^- = \emptyset$, this completes the proof. Let us now assume that $\mathcal{F} \in G^-$. Then $\mathcal{F}$ is either a reflection or a glide reflection. The latter is excluded because the product of a glide reflection with itself is a translation, and there are no nontrivial translations in G. Moreover, the reflection axis of $\mathcal{F}$ must pass through P. Indeed, if this were not the case, the product of $\mathcal{F}$ and $\mathcal{R}$ would be a glide reflection, which is impossible. By Theorem 3.6 we find that G is isomorphic to D_n; see Exercise 3.27. If $G^+ = C_1$ and $G^- \neq \emptyset$, we can easily show that $G = D_1$.

Using Theorem 2.6, we can draw the following conclusion from the last theorem.

Corollary 3.8. *Let F be a subset of the plane containing three noncollinear points. If G is a finite subgroup of the symmetry group of F, G is isomorphic to either C_n or D_n for some n.*

Exercises

3.1. We place two mirrors vertically at an angle of $\pi/3$ radians, and we place an asymmetric object on the angle bisector. What do we see, a C_6 or a D_3?

3.2. Any two groups with two elements are isomorphic. The symmetry groups of A and Z are therefore isomorphic.

3.3. Let G be a group with three elements. Let a be an element of G other than the unit element e. Then $G = \{e, a, a^2\}$.
After this, prove that any two groups with three elements are isomorphic.

3.4. Prove that the groups D_2 and V_4 are isomorphic.

3.5. Let G be a subgroup of $\mathcal{I}(V)$ with $2n$ elements such that G^+ has n elements.
(a) If $\mathcal{S} \in G^-$ and $\mathcal{R} \in G^+$, then $\mathcal{S} \circ \mathcal{R} = \mathcal{R}^{-1} \circ \mathcal{S}$.
Hint: We know from the proof of Theorem 3.7 that $\mathcal{S}$ is a reflection and $\mathcal{R}$ is a rotation.
(b) Show that G is isomorphic to D_n.
Hint: It is not difficult to make an isomorphism from G^+ to C_n. Next take a reflection from G and map it to a reflection in D_n. Finally, use Theorem 3.6 to make an isomorphism from G to D_n.

3.2 There Are Seven Types of Frieze Patterns

Figure 3.6 gives an example of a frieze pattern. This figure has the characteristic property that it is not invariant under arbitrary translations but only under powers of a "base translation" $\mathcal{T}$. Let us rephrase this as a mathemat-

Fig. 3.6. A frieze pattern

ical definition. For a translation $\mathcal{T}$ and a positive integer k, we let $\mathcal{T}^k$ denote the k-fold composition of $\mathcal{T}$:

$$\mathcal{T}^k = \underbrace{\mathcal{T} \circ \mathcal{T} \circ \cdots \circ \mathcal{T}}_{k \text{ times}}.$$

Moreover, by convention: $\mathcal{T}^0 = \mathrm{id}$ and $\mathcal{T}^{-k} = \left(\mathcal{T}^{-1}\right)^k$ for $k > 0$.

Definition 3.9. *A frieze pattern D is a subset of the plane for which the set of translations in the symmetry group $\mathcal{I}(D)$ is equal to $\{\mathcal{T}^k : k \in \mathbb{Z}\}$ for some translation $\mathcal{T}$. We say that $\mathcal{T}$ generates the frieze pattern.*

If $\mathcal{T}$ generates the frieze pattern D, then so does $\mathcal{T}^{-1}$. A natural question is, how many different frieze patterns are there? From a practical point of view the answer is infinitely many. We can always slightly change the motif in a frieze pattern; Escher applied this principle often. If we want to answer the question from a mathematical point of view, we must first determine when two patterns are equivalent.

Definition 3.10. *We say that the frieze patterns D_1 and D_2 have the same type if there is an isomorphism φ from $\mathcal{I}(D_1)$ to $\mathcal{I}(D_2)$ that maps $\mathcal{I}(D_1)^+$ onto $\mathcal{I}(D_2)^+$. We will say that two frieze patterns are* essentially different *if they do not have the same type.*

For example, the frieze patterns in Fig. 3.7 have the same type. For every pattern D in the figure, there is a translation $\mathcal{T}$ such that every translation that maps D onto itself is equal to $\mathcal{T}^k$ for some $k \in \mathbb{Z}$, while every translation $\mathcal{T}^k$, $k \in \mathbb{Z}$, maps D onto itself. We have

$$\mathcal{I}(D) = \mathcal{I}(D)^+ = \{\mathcal{T}^k : k \in \mathbb{Z}\}.$$

This group is isomorphic to $\mathbb{Z}$; to obtain an isomorphism we map $\mathcal{T}^k$ to k, for $k \in \mathbb{Z}$. The mathematical question is now, *how many essentially different frieze patterns are there?* We will show that there are seven essentially different frieze patterns.

3.2 There Are Seven Types of Frieze Patterns

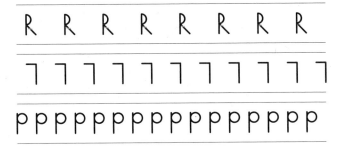

Fig. 3.7. Frieze patterns of the same type

We first list the different types of frieze patterns that have only direct isometries—either translations or rotations. The translations are all of the form $\mathcal{T}^k$, $k \in \mathbb{Z}$, where $\mathcal{T}$ is a translation that generates the frieze pattern. If a frieze pattern has rotational symmetry, this can only be twofold rotational symmetry.

Theorem 3.11. *Every rotation other than the identity in the symmetry group of a frieze pattern is of order* 2.

Proof. Let $\mathcal{T}$ be a translation that generates the frieze pattern. Let $\mathcal{R}$ be a rotation in the symmetry group, with center C. Consider the map $\mathcal{F} = \mathcal{R} \circ \mathcal{T} \circ \mathcal{R}^{-1}$. It follows from Sect. 2.4, Exercise 2.33 and Theorem 2.34 that $\mathcal{F}$ is a translation. We have

$$\mathcal{F}(C) = \mathcal{R} \circ \mathcal{T} \circ \mathcal{R}^{-1}(C) = \mathcal{R} \circ \mathcal{T}(C) \, .$$

Since $\mathcal{F} \in \mathcal{I}(D)$ and $\mathcal{F}$ is a translation, $\mathcal{F} = \mathcal{T}^k$ for some k, so

$$\mathcal{R} \circ \mathcal{T}(C) = \mathcal{F}(C) = \mathcal{T}^k(C) \, . \tag{3.1}$$

Since $\mathcal{R}$ is a rotation with center C, we obtain

$$d\left(C, \mathcal{T}^k(C)\right) = d\left(C, \mathcal{R} \circ \mathcal{T}(C)\right) = d\left(C, \mathcal{T}(C)\right) \, .$$

Consequently, $k = 1$ or $k = -1$. The first case, $k = 1$, does not occur. Indeed, (3.1) implies that in this case, $\mathcal{T}(C)$ is the rotation center of $\mathcal{R}$, whence $\mathcal{T}(C) = C$. This is impossible because a nontrivial translation has no fixed points. In the second case the rotation angle of $\mathcal{R}$ equals π, and $\mathcal{R}$ is of order 2.

We have found two types of frieze patterns, drawn on the left in Fig. 3.8. These can be seen as a succession of tiles. These tiles may be infinitely "high," but for practical reasons we will draw them with finite height. Together, the images of a single tile under the translations $\mathcal{T}^n$ cover the pattern, where the tiles have only boundary points in common. In the patterns on the left such a tile has been drawn. On the right the tile has been drawn again, now

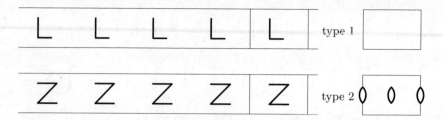

Fig. 3.8. Frieze patterns without indirect isometries

without motif. Instead, we have indicated the symmetry elements. The length of a tile, measured in the direction of the translation vector of the generating translation $\mathcal{T}$, is equal to the length of that vector.

A pattern of type 1 allows only translations. In the rest of this section we will denote the symmetry group of the frieze pattern of type 1 by H_1, and that of type 2 by H_2. The symmetry groups of frieze patterns are also called *frieze groups*. A pattern of type 2 allows both translations and rotations over π. The translations are of the form $\mathcal{T}^k$ with $k \in \mathbb{Z}$, where $\mathcal{T}$ is a generating translation. Note that $\mathcal{I}(D)$ being a group implies that as soon as there is one rotation $\mathcal{R}$ over an angle π that maps a frieze pattern D into itself, there are also other rotations with this property.

Let us examine this further. We find the rotations as follows; see Fig. 3.9. We write the rotation $\mathcal{R}$, with center C, as a product $\mathcal{S}_{l_2} \circ \mathcal{S}_{l_1}$ of two reflections

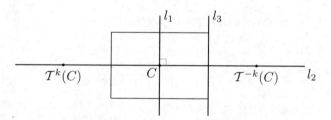

Fig. 3.9. $\mathcal{R} = \mathcal{S}_{l_2} \circ \mathcal{S}_{l_1}$, $\mathcal{T}^k = \mathcal{S}_{l_1} \circ \mathcal{S}_{l_3}$, $\mathcal{R} \circ \mathcal{T}^k = \mathcal{S}_{l_2} \circ \mathcal{S}_{l_3}$

with perpendicular axes l_1 and l_2, where l_2 is parallel to the span of $\mathcal{T}$, say $\mathcal{R} = \mathcal{S}_{l_2} \circ \mathcal{S}_{l_1}$. Next we write the translation $\mathcal{T}^k$, $k \neq 0$, as a product of two reflections with parallel axes l_1 and l_3 that are perpendicular to the span of $\mathcal{T}$, say $\mathcal{T}^k = \mathcal{S}_{l_1} \circ \mathcal{S}_{l_3}$; first $\mathcal{S}_{l_3}$, then $\mathcal{S}_{l_1}$. The product $\mathcal{R} \circ \mathcal{T}^k$ is the rotation $\mathcal{S}_{l_2} \circ \mathcal{S}_{l_3}$ with center the midpoint of the line segment $[C\mathcal{T}^{-k}(C)]$. From this we can easily deduce the rotation centers, indicated in Fig. 3.8 by small lenses. For later use we note that in the same way, using Exercise 2.31 it follows that

$$\mathcal{R} \circ \mathcal{T} = \mathcal{T}^{-1} \circ \mathcal{R} \quad \text{and} \quad \mathcal{R} \circ \mathcal{T}^{-1} = \mathcal{T} \circ \mathcal{R}. \tag{3.2}$$

Theorem 3.12. *Consider a frieze pattern D with generating translation $\mathcal{T}$ with translation vector $\mathbf{t}$. If $\mathcal{R}_1$ and $\mathcal{R}_2$ are rotations in $\mathcal{I}(D)$ with distinct centers C_1 and C_2, respectively, there is a natural number k such that $d(C_1, C_2) = (k/2)\|\mathbf{t}\|$.*

Proof. The map $\mathcal{R}_1 \circ \mathcal{R}_2$ is a translation; the length of the translation vector is equal to $2d(C_1, C_2)$. This latter must be an integral multiple of $\|\mathbf{t}\|$. Therefore there is a natural number k such that $2d(C_1, C_2) = k\|\mathbf{t}\|$.

We continue the discussion we started before the theorem. We were studying which rotations can occur in the symmetry group of a pattern of type 2. We indicated these schematically in Fig. 3.8, in the lower right corner. The theorem we just proved implies that we have already found all rotations that can occur. Thus we arrive at the following classification of frieze patterns with only direct isometries.

Theorem 3.13. *If the symmetry group of a frieze pattern contains only direct isometries, it has type 1 or 2.*

Proof. Let $\mathcal{T}_i$, for $i = 1, 2$, be a generating translation of the frieze pattern of type i. In addition to translations, the symmetry group H_2 of the frieze pattern of type 2 also contains a rotation, which we denote by $\mathcal{R}_2$.

Above we saw that we can write every rotation in H_2 as a product $\mathcal{R}_2 \circ \mathcal{T}_2^k$ with $k \in \mathbb{Z}$. Let D be a frieze pattern with generating translation $\mathcal{T}$. If $\mathcal{I}(D)$ does not contain any rotations, $\mathcal{I}(D) = \{\mathcal{T}^k : k \in \mathbb{Z}\}$. By associating $\mathcal{T}^k$ to $\mathcal{T}_1^k$, for $k \in \mathbb{Z}$, we find an isomorphism from $\mathcal{I}(D)$ to H_1. If $\mathcal{I}(D)$ does contain a rotation, say $\mathcal{R}$, then in the same manner as above we find that the elements of $\mathcal{I}(D)$ are of the form $\mathcal{T}^k$ and $\mathcal{R} \circ \mathcal{T}^k$, $k \in \mathbb{Z}$. We define an isomorphism from $\mathcal{I}(D)$ to H_2 as follows: for every $k \in \mathbb{Z}$, we associate $\mathcal{T}^k$ to $\mathcal{T}_2^k$ and $\mathcal{R} \circ \mathcal{T}^k$ to $\mathcal{R}_2 \circ \mathcal{T}_2^k$. This map is indeed an isomorphism; to prove this we can use the equations in (3.2), which hold both in H_2 and in $\mathcal{I}(D)$. For example, the image of $\mathcal{R}_2 \circ \mathcal{T}_2^{-8} = \mathcal{T}_2^5 \circ (\mathcal{R}_2 \circ \mathcal{T}_2^{-3})$ is $\mathcal{T}^5 \circ (\mathcal{R} \circ \mathcal{T}^{-3}) = \mathcal{R} \circ \mathcal{T}^{-8}$.

To find the other types of frieze patterns, we will add indirect isometries to the symmetry groups H_1 and H_2. This method is the converse of what is described in Theorem 3.6. In that theorem we started out with a group of transformations and studied the place of the direct isometries in the group. Now we have a group H of direct isometries, where $H = H_1$ or $H = H_2$, and we try to expand this group to a group G in such a way that H is exactly the subgroup G^+. For every element $\mathcal{F}$ in the group G we want to construct, we have $\mathcal{F} \circ \mathcal{F} \in H$. This is because $\mathcal{F} \circ \mathcal{F}$ is always a direct isometry, regardless of whether $\mathcal{F}$ is direct. Our strategy is to repeatedly look for an indirect isometry $\mathcal{F}$ such that $\mathcal{F}^2$ is an element of H, and add this to H.

Extensions of H_1. Let us start with the symmetry group H_1. We have $H_1 = \{\mathcal{T}^k : k \in \mathbb{Z}\}$, where $\mathcal{T}$ is the generating translation of the type-1

106   3 SYMMETRY

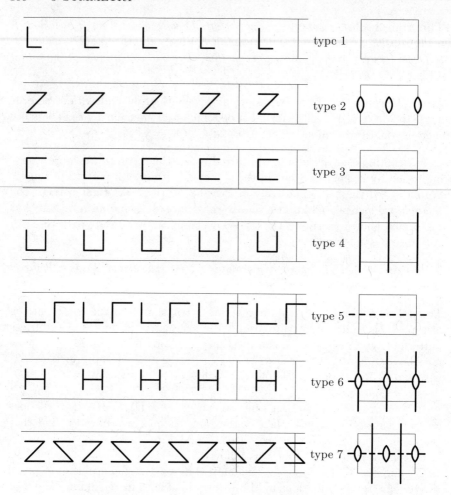

Fig. 3.10. The seven essentially different frieze patterns

frieze pattern. Which indirect isometries can we add to this? Since id $= \mathcal{S}^2$ for every reflection $\mathcal{S}$, we first study which reflections we can add to H_1. The only reflections that qualify are those whose reflection axis is either perpendicular or parallel to the translation vector of $\mathcal{T}$; see Exercise 3.9. There can be only one of the latter type in the symmetry group, because otherwise we could find a composition that is a translation in a direction perpendicular to that of $\mathcal{T}$, which is excluded by the definition of a frieze pattern.

Let us first consider the extension of H_1 by a reflection with axis parallel to the translation vector of $\mathcal{T}$. This leads to the symmetry group H_3 of a type-3 frieze pattern; see Fig. 3.10. This has the following elements:

1. Translations $\mathcal{T}^n$, $n \in \mathbb{Z}$ (the elements of H_1);
2. A reflection $\mathcal{S}$ whose axis is parallel to the translation vector of $\mathcal{T}$;

3.2 There Are Seven Types of Frieze Patterns

3. Glide reflections, namely products of S and translations from H_1.

In the figure representing the tile, on the right, the reflection S is indicated by a continuous straight line. The symmetry group H_3 of the type-3 frieze pattern is one of the extensions of the translation group we had in mind.

We obtain a second extension of H_1 by adding a reflection whose axis is perpendicular to the translation vector of T. The products of this reflection with the elements of H_1 are reflections. Figure 3.10 shows a type-4 frieze pattern. The symmetry group H_4 of this strip is exactly the extension of H_1 we just described. The representation of the tile, on the right, shows which reflections are in H_4.

There exists yet another extension of H_1. For this we write the translation T as the square of a glide reflection G. The symmetry group H_5 of the type 5 frieze pattern contains G, the translations, and other glide reflections. The square of such a glide reflection is a translation T^n with n odd, and every translation T^n with n odd can be obtained in this way. It follows that we have now found all extensions of H_1. The axis of a glide reflection is commonly indicated with a dashed line; see Fig. 3.10.

Extensions of H_2. We now turn our attention to extensions of the symmetry group H_2 of the type-2 frieze pattern; see Fig. 3.8. The symmetry group H_6 of the type-6 frieze pattern, Fig. 3.10, contains the elements of H_2 plus other reflections. Note that all these reflections can be obtained as product of one fixed reflection and the point reflections or translations from H_2.

The symmetry group H_7 of the type-7 frieze pattern contains both the elements of H_2 and a glide reflection G whose square is equal to T. Since we are dealing with a group, we also find reflections in the symmetry group. We can, for example, find a reflection as shown in Fig. 3.11. The result is indicated

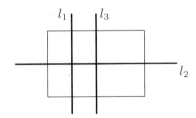

Fig. 3.11. $R = S_{l_3} \circ S_{l_2}$, $G = S_{l_1} \circ S_{l_2} \circ S_{l_3}$, $G \circ R = S_{l_1}$

schematically on the tile, on the right in Fig. 3.10; the dashed line indicates a glide reflection, the thick continuous lines indicate reflections. We have now found all possible extensions of H_2.

Theorem 3.14. *Every frieze pattern has one of the types 1 through 7.*

Proof. Let D be a frieze pattern. The group $\mathcal{I}(D)^+$ is isomorphic to either H_1 or H_2. If $\mathcal{I}(D) = \mathcal{I}(D)^+$, D must have type 1 or 2. If $\mathcal{I}(D) \neq \mathcal{I}(D)^+$, $\mathcal{I}(D)$ results from $\mathcal{I}(D)^+$ by adding a reflection or a glide reflection. The group $\mathcal{I}(D)$ therefore corresponds to one of the extensions of H_1 or H_2 described above. The proof of Theorem 3.13 shows how to make an isomorphism from $\mathcal{I}(D)$ to one of the symmetry groups H_3 through H_7.

In the following example we look back at Definition 3.10 of type. We will show why we need to bother with the direct isometries and may not simply say that D_1 and D_2 have the same type if there is an isomorphism from $\mathcal{I}(D_1)$ to $\mathcal{I}(D_2)$.

Example 3.15. We consider the symmetry group H_2 of the type-2 frieze pattern and the symmetry group H_4 of the type-4 frieze pattern. We will show that H_2 and H_4 are isomorphic. Nevertheless, types 2 and 4 are different, since $H_2^+ = H_2$ and $H_4^+ \neq H_4$.

We assume that the translations in both groups are powers of the same translation $\mathcal{T}$. We can choose coordinates in such a way that the centers of the rotations in H_2 are equal to $(n, 0)$, $n \in \mathbb{Z}$. We let $\mathcal{R}_n$ denote the rotation over π with center $(n, 0)$. Likewise, we can choose coordinates such that the axes of the reflections in H_4 have equations $x_1 = n$. We let $\mathcal{S}_n$ denote the reflection in the line $x_1 = n$. Note that in both cases, the "width" of the tile is equal to 2 and the norm of the translation vector is also equal to 2: $\|\mathcal{T}(\mathbf{o})\| = 2$. We can easily show that $\mathcal{R}_n = \mathcal{T}^n \circ \mathcal{R}_0$ and $\mathcal{S}_n = \mathcal{T}^n \circ \mathcal{S}_0$ for all integers n. To define the isomorphism from H_2 to H_4, we map $\mathcal{T}^n$ to $\mathcal{T}^n$ and $\mathcal{R}_n$ to $\mathcal{S}_n$, for all n; we can easily check that this defines an isomorphism.

Exercises

3.6. Show that the symmetry group of the type-1 and type-5 frieze patterns are isomorphic, whereas the types are different.

3.7. The symmetry groups of the type-2, -4, and -7 frieze patterns are isomorphic; nevertheless, the types are different.

3.8. Try to use the group properties to show that the symmetry groups of the type-3, -4, and -5 frieze patterns are not isomorphic.
Hint: The symmetry group of the type-5 frieze pattern contains no transformation whose square is the identity, other than the identity itself.

3.9. Let D be a frieze pattern with generating translation $\mathcal{T}$. If the reflection $\mathcal{S}$ is an element of the symmetry group $\mathcal{I}(D)$, the reflection axis of $\mathcal{S}$ is either parallel or perpendicular to the translation vector of $\mathcal{T}$.
Hint: Consider the map $\mathcal{S} \circ \mathcal{T} \circ \mathcal{S}$.

3.3 Voronoi Diagrams

Voronoi diagrams have many interesting applications in geometry. We will use them to classify the periodic tilings. We first introduce some notions from the theory of Voronoi diagrams.

Definition 3.16. *Consider a finite set E of points in the plane, containing at least two points. (A set is called finite if it has only a finite number of points.) Let Q be a point of E. The* Voronoi cell $V(Q)$ *of Q is the set of points in the plane that lie closer to Q than to any other point of E; see Fig. 3.12. This can be expresses as a formula:*

$$V(Q) = \{ X : d(X,Q) \leq d(X,R) \text{ for all } R \text{ in } E \text{ with } R \neq Q \}.$$

How do we find the Voronoi cell of Q? For every point $R \in E$ with $R \neq Q$

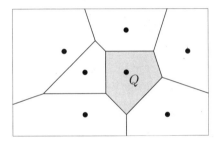

Fig. 3.12. The Voronoi cell of Q

we draw the perpendicular bisector of $[RQ]$ and determine the closed half-plane H_R containing Q; the intersection of all the resulting half-planes is the Voronoi cell of Q. Thus a Voronoi cell is a convex set; that is, the line segment joining any two points of the cell lies entirely inside the cell. More generally, we call a subset D of the plane *convex* if for any two points A and B of D, the line segment $[AB]$ lies entirely inside D; see also Exercise 3.11. We say that two different Voronoi cells *are adjacent to each other* or *are neighbors* if their intersection contains more than one point; in that case the intersection is called an *edge* of the cells. The edge of the Voronoi cells of P and Q is a subset of the perpendicular bisector of $[PQ]$. Since both cells are convex, the edge also is. Because of this there exist points A and B such that the edge is equal to one of the following sets:

1. The line segment $[AB]$;
2. The *half-line* $\{ X : X \in [AB] \text{ or } B \in [AX] \}$;
3. The line AB.

Together, the edges of a Voronoi cell form a broken line; we call this broken line the *boundary* of the cell. We call a point that belongs to different boundaries a *vertex*. Together, the boundaries of all cells form the *Voronoi diagram* of the set E. The points of E are called the *sites* of the Voronoi diagram.

Example 3.17. Our first example of a Voronoi diagram concerns rainfall measurement. In a given region we measure the amount of rainfall in rain gauges set at a number of fixed points. The Voronoi diagram with these places as sites gives a partition of the region. We assume that in each cell the amount of rainfall is approximately that measured at the corresponding site. This allows us to give a global estimate of the amount of rainfall for the whole region. In meteorology a Voronoi diagram is sometimes called a *Thiessen polygon*.

The following theorem can be proved from the definitions without much difficulty.

Theorem 3.18. *If the line l is the extension of the edge of cells with sites P and Q, the reflection in the line l maps P and Q onto each other.*

The deduction of the following theorem is more subtle.

Theorem 3.19. *Every vertex H of a Voronoi diagram belongs to at least three different cells; the sites of these cells lie on a circle with center H. There are no sites inside this circle.*

Proof. Figure 3.13 shows part of the Voronoi diagram. By definition, the vertex H belongs to at least two boundaries. Let H be a vertex that is part of the edge of the cells with sites P and Q. This edge is a subset of the perpendicular

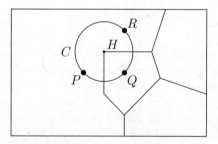

Fig. 3.13. Edges meeting at a vertex

bisector of the line segment $[PQ]$. The vertex H also belongs to a second edge. Without loss of generality we may, and do, assume that the angle between these two edges contains no other edge with vertex H. The points P and Q lie on a circle C with center H. No other site lies inside the circle, since the position of H would otherwise be incorrect. It follows from the definition of cells that reflecting Q in the second edge gives another site R. This site also lies on C. Since there are no sites inside the circle, there must be at least one edge separating R from P. This edge also has vertex H. Consequently, at least three edges meet at H.

When studying the Voronoi diagram of a set E, it is sometimes useful to determine the convex envelope of E. Consider Fig. 3.14. We join any two

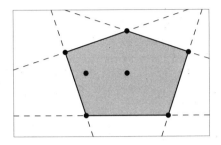

Fig. 3.14. The convex hull and convex envelope of E

points of E by a straight line. Each of these lines determines two closed half-planes. For some of these lines all points of E lie in the same closed half-plane; these lines are dashed in the figure. For each of the dashed lines we take the closed half-plane that contains E. The intersection of these half-planes is the *convex hull* of E (shaded in the figure), and the boundary of the convex hull of E is called the *convex envelope* of E. The convex hull of E is convex; see Exercise 3.11.

Up to now, we have restricted the figures we were considering in this section to a frame. What happens to the figures outside the frame? The following theorem tells us more about this.

Theorem 3.20. *Let E be a finite set. The Voronoi cell of any point on the convex envelope of E is unbounded.*

Proof. If the points of the set E are collinear, the theorem follows directly from the figure. Let us assume that the points of E are noncollinear. Let P be a point on the convex envelope of E, and let A and B be the "neighbors" of P on this convex envelope; see Fig. 3.15. We first assume that the points A, P, and B are noncollinear. The points of E lie in the shaded angle at P. The Voronoi

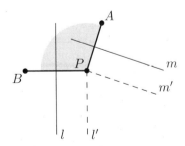

Fig. 3.15. The cell of a point on the convex envelope

cell of P lies in the same closed half-plane of the perpendicular bisector l of $[BP]$ as P. It also lies in the same closed half-plane of the perpendicular bisector m of $[AP]$ as P. Let l' and m' be the lines obtained by translating l and m, respectively, such that both pass through P. These lines determine four angles at P. One of those lies inside the same half-planes of l and m as P. We can easily check that this angle lies in the interior of the Voronoi cell of P. This implies that the cell of P is unbounded.

If the points A, P, and B are collinear, we prove in an analogous manner that the cell of P contains a half-line, and is therefore unbounded.

Let us return for the moment to Example 3.17. It concerned Voronoi diagrams used in the study of rainfall in a given region. Such a diagram can be incomplete, for example because the positions of a number of rain gauges are lost. These incomplete diagrams must be completed to Voronoi diagrams; we call this the *reconstruction problem*. Let us first give an example.

Example 3.21. Figure 3.16 (a) contains a figure that is part of a Voronoi diagram. Can you find the Voronoi diagram? Figure 3.16 (b) shows a solution of

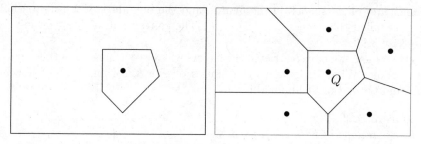

Fig. 3.16. (a) A partial Voronoi diagram; (b) its reconstruction

the reconstruction problem. Based on the properties of Voronoi diagrams we see that at least five sites and five edges must be added. Note that Fig. 3.12 also gives a solution of the reconstruction problem; such a solution is not unique.

Example 3.22. A more general reconstruction problem concerns finding the sites of a given Voronoi diagram. We discuss the simplest case. Figure 3.17, on the left, shows part of a Voronoi diagram, namely the neighborhood of a vertex. We call the edges that meet at the vertex l_1, l_2, and l_3. The question is whether we can find the position of the corresponding sites. The figure on the right contains an analysis of the problem. Suppose that we knew the location of a site; call this site P. Let $\mathcal{S}_i$ be the reflection in the line l_i. We reflect the point P in the line l_1; since P is a site, its image $\mathcal{S}_1(P)$ is also one. Hence $\mathcal{S}_2 \circ \mathcal{S}_1(P)$ is also a site. Consequently, $\mathcal{S}_3 \circ \mathcal{S}_2 \circ \mathcal{S}_1(P) = P$. The effect of the

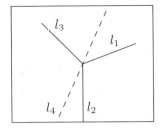 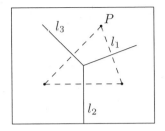

Fig. 3.17. Where are the sites?

reflection S_1 followed by S_2 is equal to the clockwise rotation over twice the angle between l_1 and l_2.

Next draw the line l_4 (dashed in the figure on the left) such that the reflection S_4 in l_4 satisfies $S_2 \circ S_1 = S_3 \circ S_4$, and therefore

$$S_3 \circ S_2 \circ S_1 = S_3 \circ S_3 \circ S_4 = S_4.$$

Then $S_4(P) = P$, whence P must lie on l_4. To further determine the position of P we need more information on the Voronoi diagram.

Generalized Voronoi Diagrams

By replacing the finite set of points E by a finite set of sets, we obtain an extension of the theory of Voronoi diagrams; see for example Fig. 3.18, where the sets replacing E are indicated with thick line segments. The Voronoi diagram

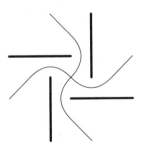

Fig. 3.18. A Voronoi diagram (drawn with thin lines)

now consists of four gracefully curved lines.

Definition 3.23. *Let U and W be sets in the plane. The (generalized) Voronoi diagram of U and W is the set of all points X of the plane that are equidistant from U and W.*

3 SYMMETRY

We must of course still define the distance from the point X to the set U. If U is a straight line, this is the distance from X to the foot of the perpendicular from X to the line; this distance is less than the distance from X to any other point on the line. More generally, we would like to define the distance from X to the set U as the minimal value of the distances $d(X, Y)$ for $Y \in U$. This definition cannot always be used, because there need not be any Y in U with minimal $d(X, Y)$. In the situations we consider, though, it works well.

If U and W both consist of exactly one point, say $U = \{A\}$ and $W = \{B\}$, the Voronoi diagram of U and W is the perpendicular bisector of $[AB]$. If U and W are the sides AB and AC of a triangle, the Voronoi diagram of U and W is the bisector of angle CAB.

In Sect. 4.4 we will discuss conics in details. At this point we will just show that the *conics* can be introduced in an elegant manner as Voronoi diagrams. Let us start with the parabola. We begin with a line l, the *directrix*, and a point F, the *focus*. We now make the Voronoi diagram of the directrix and the focus. This diagram consists of exactly one boundary, the *parabola*; see Fig. 3.19. That is, the parabola is the Voronoi diagram of the directrix l and

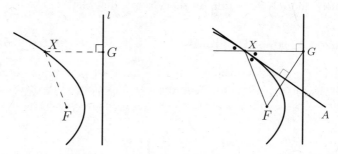

Fig. 3.19. (a) The parabola as a Voronoi diagram; (b) the tangent construction

the the focus F. Every point X of the parabola satisfies $d(X, F) = d(X, G)$, where G is the foot of the perpendicular from X on the directrix l.

It is immediately clear from this definition of the parabola how to draw the *tangents* to the parabola. The figure on the right shows the tangent to the parabola at X. Let G be the foot of the perpendicular from X to the directrix. Then $d(X, F) = d(X, G)$; hence $\triangle GXF$ is isosceles. Consequently, the line XA, the bisector of angle FXG, is also the perpendicular bisector of $[FG]$. It follows without much trouble that every point Y of the line XA other than X lies outside the parabola. Namely, we have

$$d(Y, F) = d(Y, G) > d(Y, l) .$$

The line XA therefore has exactly one point in common with the parabola. It follows that XA is the *tangent* to the parabola at X. In the figure we see, moreover, the well-known reflection property of the parabola. The straight line through F and perpendicular to l is called the *axis* of the parabola. The reflection property says that light falling parallel to the axis onto a parabolic mirror is reflected through the focus. This property explains the name *focus*. Conversely, a light source placed at the focus gives a bundle of reflected light parallel to the axis.

Let us now define the ellipse; see Fig. 3.20 (a). We begin with a circle C,

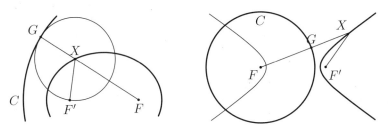

Fig. 3.20. (a) The ellipse as a Voronoi diagram; (b) the hyperbola as a Voronoi diagram

the *directrix circle*, with center F and radius r, and a point F' other than F, the *focus*, which lies inside C. The Voronoi diagram of the directrix circle and the focus consists of a single boundary, the *ellipse*. That is, the ellipse is the Voronoi diagram of the directrix circle C and the focus F'. Every point X of the ellipse satisfies $d(X, F') = d(X, G)$, where G is the intersection point of FX and the circle C. Indeed, every point Y of the circle other than G satisfies $d(F, Y) = d(F, G)$; it easily follows that $d(X, Y) > d(X, G)$, and therefore that the distance from X to the circle C is equal to $d(X, G)$. We immediately deduce from the figure that every point X of the ellipse satisfies

$$d(X, F') + d(X, F) = r\,,$$

and conversely, every point with this property lies on the ellipse. The ellipse is often defined using this property. By reflecting the figure in the perpendicular bisector of $[FF']$, we see that in the construction of the ellipse we can interchange the roles of F and F' (for a different directrix circle, namely the circle with center F' and radius r); F is consequently also called a focus of the ellipse.

Let us now define the hyperbola; see Fig. 3.20 (b). We begin with a circle C, the *directrix circle*, with center F and radius r, and a point F', the *focus*, which now lies outside the circle C. The *hyperbola* is the Voronoi diagram of the focus F' and the directrix circle C. Every point X of the hyperbola satisfies $d(X, F') = d(X, G)$, where G is the intersection point of FX and the circle C. Indeed, every point Y of the circle C other than G satisfies $d(F, Y) = d(F, G)$; it easily follows that $d(X, Y) > d(X, G)$. Hence the distance from X to the circle C is equal to $d(X, G)$. In the figure we see that every point X of the hyperbola satisfies

$$d(X,F) - d(X,F') = r\,,$$

and, conversely, every point with this property lies on the hyperbola. In fact, we obtain only one branch of the hyperbola this way. By interchanging the roles of F and F' we obtain the other branch (with another directrix circle, with center F' and radius r).

Exercises

3.10. Let us discuss the Voronoi diagram of a finite set E. A vertex of a Voronoi cell is defined to be a point that belongs to at least three cells. Prove that a vertex of a Voronoi cell is equidistant from the sites of all cells the vertex belongs to. We call this distance r. There are no sites inside the circle with center at the vertex and radius r.

3.11. We prove, among other things, that the convex hull of a finite set E is indeed convex.

(a) A closed half-plane H associated to a line l is convex.
 Hint: Choose a coordinate system, with one axis along l.
(b) The intersection of any number of convex sets is convex.
(c) A Voronoi cell is a convex set.
(d) The convex hull of a finite set E is convex.

3.12. Let A and B be points of a finite set E that are neighbors on the convex envelope of E. The common edge of the Voronoi cells of A and B is unbounded. In particular, it is not a line segment but either a half-line or a straight line.

3.13. Let K, L, M be three sets in the plane. If the Voronoi diagram of K and L and that of L and M have a point in common, that point also lies on the Voronoi diagram of K and M.

3.14. The definitions of the ellipse and hyperbola give a method for drawing these figures point by point if the directrix circle $C = \odot(F,r)$ and the focus F' are given.

(a) First assume that F' lies inside the directrix circle and that $F \neq F'$; see Fig. 3.21 **(a)**. For every point G of the directrix circle, the line segment $[FG]$ contains exactly one point X of the ellipse with directrix circle $\odot(F,r)$ and focus F'. Show that X is the intersection point of the line GF and the perpendicular bisector of the line segment $[GF']$.
(b) Which Voronoi diagram do we obtain if $F' = F$?
(c) Now assume that F' lies outside the directrix circle; see Fig. 3.21 **(b)**. For every point G of the directrix circle satisfying $\angle FGF' > \pi/2$ the line FG contains exactly one point X of the hyperbola with directrix circle $\odot(F,r)$ and focus F'. Prove this and show how we can find the point X.

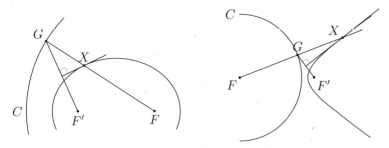

Fig. 3.21. Drawing point by point: (a) the ellipse; (b) the hyperbola

(d) If F' lies on the circle $\odot(F, r)$, what is the set of points equidistant from F' and the circle?

3.15. Let K be a parabola with focus F and directrix l. Let S be the foot of the perpendicular on l through F. The midpoint T of FS is called the *vertex* of K.

(a) Show that the perpendicular bisector m of $[FS]$ has exactly one point in common with K, namely T; m is the *tangent* to the parabola at its vertex.
(b) Let P be an arbitrary point of K. Let Q be the foot of the perpendicular from P onto m. Let R be the intersection point of m and the tangent to K at P. Show that $d(T, R) = d(R, Q)$.

3.4 Voronoi Cells in Lattices

In the last section we defined Voronoi diagrams of finite sets. In this section we define Voronoi diagrams of lattices, and use these to study periodic tilings. Let us first define a lattice.

Definition 3.24. *Let P be a point in the plane. We take two translations T_1 and T_2 whose translation vectors have different spans. The set*

$$\{\, T_1^n \circ T_2^m(P) : n \in \mathbb{Z}, m \in \mathbb{Z} \,\}$$

is called the lattice of P. *We call T_1 and T_2 the* generating translations *of the lattice.*

Let us describe the lattice of P in more detail. For this we first introduce a coordinate system at P. Using Theorem 2.27 and the proof of Theorem 2.29, we see that

$$T_1^n \circ T_2^m(P) = T_1^n \circ T_2^m(\mathbf{o}) = T_1^n\left(m T_2(\mathbf{o})\right) = n T_1(\mathbf{o}) + m T_2(\mathbf{o}) \,.$$

In the local coordinate system, the lattice of P is the set

118 3 SYMMETRY

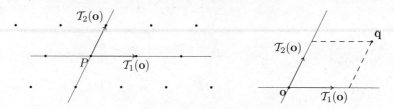

Fig. 3.22. The lattice of P

$$\{ nT_1(\mathbf{o}) + mT_2(\mathbf{o}) : n \in \mathbb{Z},\ m \in \mathbb{Z} \} ;$$

see Fig. 3.22, on the left. Since the origin $\mathbf{o}$ is chosen at a point of the lattice, namely P, we see that the lattice is a subgroup of the additive group of $\mathbb{R}^2$. We check this using Lemma 2.28. It is clear that the lattice is nonempty. If n, m, n', and m' are in $\mathbb{Z}$, then

$$(nT_1(\mathbf{o}) + mT_2(\mathbf{o})) - (n'T_1(\mathbf{o}) + m'T_2(\mathbf{o})) = (n - n')T_1(\mathbf{o}) + (m - m')T_2(\mathbf{o}) .$$

The right-hand side clearly represents a lattice point. This shows that the lattice is a subgroup of the additive group of $\mathbb{R}^2$. Let us pause to deduce a number of properties of this lattice. These properties are not as simple as they may seem at first. It may help to consider the results in the explicit case of the translations mentioned in Exercise 3.16. We first note that every vector $\mathbf{q}$ of the plane can be written in a unique way as *linear combination* of $T_1(\mathbf{o})$ and $T_2(\mathbf{o})$; that is, $\mathbf{q} = \lambda T_1(\mathbf{o}) + \mu T_2(\mathbf{o})$. Indeed, we see in Fig. 3.22, on the right, that we can write $\mathbf{q}$ as the sum of two vectors, one of which has the same span as $T_2(\mathbf{o})$; by forming the parallelogram, $\mathbf{q}$ is decomposed along the vectors $T_1(\mathbf{o})$ and $T_2(\mathbf{o})$. Therefore there are λ and μ such that $\mathbf{q} = \lambda T_1(\mathbf{o}) + \mu T_2(\mathbf{o})$. There is a unique way to do this. Namely, if we also have $\mathbf{q} = \rho T_1(\mathbf{o}) + \sigma T_2(\mathbf{o})$, then

$$(\lambda - \rho)T_1(\mathbf{o}) = (\sigma - \mu)T_2(\mathbf{o}) . \tag{3.3}$$

Since $T_1(\mathbf{o})$ and $T_2(\mathbf{o})$ have different spans, both sides of (3.3) must equal the zero vector; hence $\lambda = \rho$ and $\mu = \sigma$. It follows that there is only one way to write $\mathbf{q}$ as linear combination of $T_1(\mathbf{o})$ and $T_2(\mathbf{o})$. We note that $P = \mathbf{o}$ is the only lattice point in the interior of the parallelogram with successive vertices $T_1(\mathbf{o}) + T_2(\mathbf{o})$, $-T_1(\mathbf{o}) + T_2(\mathbf{o})$, $-T_1(\mathbf{o}) - T_2(\mathbf{o})$, $T_1(\mathbf{o}) - T_2(\mathbf{o})$; see Fig. 3.23. The unique decomposition of vectors in the plane along the spans of $T_1(\mathbf{o})$ and $T_2(\mathbf{o})$ implies that a point $\mathbf{q}$ inside the parallelogram in question satisfies $\mathbf{q} = \lambda T_1(\mathbf{o}) + \mu T_2(\mathbf{o})$ with $-1 < \lambda < 1$ and $-1 < \mu < 1$, and that this decomposition is unique. If $\mathbf{q}$ is a lattice point, λ and μ must be integers. Hence $\mathbf{q} = \mathbf{o}$ and P is the only lattice point inside the parallelogram. Likewise, for every point of the lattice we can find a parallelogram congruent to the one in Fig. 3.23 such that the given point is the only lattice point in the interior. We can state this property as follows: *the lattice of P is discrete*. This property

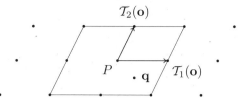

Fig. 3.23. The lattice is discrete

implies that there is a number $d > 0$ such that the distance between any two distinct lattice points is greater than or equal to d. For d we can take, for example, the radius of a circle inscribed in the parallelogram. In that case the disks with radius $d/3$ around the lattice points are pairwise disjoint; see Fig. 3.24. A computation shows that this implies that inside any circle, of any size, there are only finitely many lattice points.

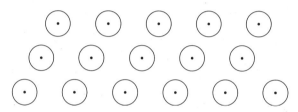

Fig. 3.24. Disjoint disks

Voronoi Cells

The following definition is almost literally Definition 3.16. Let Q be a point of the lattice of P. The *Voronoi cell* $V(Q)$ of Q is the set of all points of the plane that lie closer to Q than to any other point of the lattice. Expressed as a formula,

$$V(Q) = \{\, X \in V : d(X, Q) \leq d(X, R) \text{ for all lattice points } R \neq Q \,\}.$$

As we have just seen, the lattice is discrete. Because of this a Voronoi cell is not flat (1-dimensional), but contains a disk; see Fig. 3.25. For the construction of a Voronoi cell we need to consider only a finite number of lattice points. We will come back to this later. Let us first prove that the Voronoi cells of the points of a lattice are all congruent.

Theorem 3.25. *Consider the lattice L of a point P. The Voronoi cells of any two lattice points are congruent.*

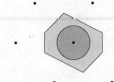

Fig. 3.25. A lattice with Voronoi cell

Proof. To begin, we choose a coordinate system at P, that is, $P = \mathbf{o}$. Let Q be a point of L. As usual, we denote the vectors associated with the points by the corresponding bold lowercase letters. The Voronoi cell of $Q = \mathbf{q}$ is

$$V(\mathbf{q}) = \{\, \mathbf{x} : \|\mathbf{x} - \mathbf{q}\| \leq \|\mathbf{x} - \mathbf{r}\| \text{ for all } \mathbf{r} \in L \text{ with } \mathbf{r} \neq \mathbf{q} \,\}.$$

Let $\mathcal{T}$ be the translation with $\mathcal{T}(\mathbf{q}) = \mathbf{o}$. Since $\mathcal{T}$ is an isometry, we can rewrite the above as follows:

$$\begin{aligned} V(\mathbf{q}) &= \{\, \mathbf{x} : \|\mathcal{T}(\mathbf{x}) - \mathcal{T}(\mathbf{q})\| \leq \|\mathcal{T}(\mathbf{x}) - \mathcal{T}(\mathbf{r})\| \text{ for all } \mathbf{r} \in L \text{ with } \mathbf{r} \neq \mathbf{q} \,\} \\ &= \{\, \mathbf{x} : \|\mathcal{T}(\mathbf{x}) - \mathbf{o}\| \leq \|\mathcal{T}(\mathbf{x}) - \mathbf{s}\| \text{ for all } \mathbf{s} \in L \text{ with } \mathbf{s} \neq \mathbf{o} \,\} \\ &= \{\, \mathbf{x} : \mathcal{T}(\mathbf{x}) \in V(\mathbf{o}) \,\} = \mathcal{T}^{-1}\left(V(\mathbf{o})\right). \end{aligned}$$

We see that the Voronoi cell of an arbitrary lattice point $\mathbf{q}$ is the image of the Voronoi cell $V(\mathbf{o})$ under the translation $\mathcal{T}^{-1}$, and is therefore congruent to $V(\mathbf{o})$.

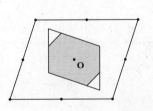

Fig. 3.26. A Voronoi cell

Let us now consider the form of the Voronoi cell of a lattice; see Fig. 3.26. By the previous theorem we only need to study the form of the cell of the origin $\mathbf{o}$ of the coordinate plane. It will turn out that in general, the Voronoi cell is a point-symmetric hexagon; in the figure the cell of $\mathbf{o}$ is shaded. We haven't reached that point yet, though. We will first show that the Voronoi

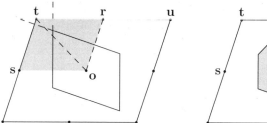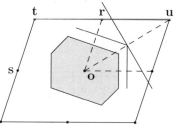

Fig. 3.27. Details for the study of the Voronoi cell

cell of **o** is contained in the inner parallelogram of Fig. 3.26. Below we describe how to find this parallelogram. Consider Fig. 3.27, on the left. Let $\mathbf{r} \neq \mathbf{o}$ be a lattice point with minimal norm $\|\mathbf{r}\|$; that is, $\mathbf{r}$ is chosen as close as possible to **o**. Let **s** be a lattice point outside the span of **r** with minimal norm among such points. That is, if we disregard the points of the span of **r**, the point **s** is the closest possible to **o**. Let $\mathbf{t} = \mathbf{r} + \mathbf{s}$ and $\mathbf{u} = \mathbf{r} - \mathbf{s}$. After, if necessary, replacing **s** by $-\mathbf{s}$, we may, and do, assume that

$$\|\mathbf{r}\| \leq \|\mathbf{s}\| \leq \|\mathbf{t}\| \leq \|\mathbf{u}\| \ .$$

We call the parallelogram **orts** (*shaded* in Fig. 3.27 on the left) a *fundamental parallelogram* of the lattice. We can check that there are no lattice points inside the fundamental parallelogram; this is due to the choice of **r** and **s**. The easiest way to see this is by dividing the fundamental parallelogram into four using the diagonals; the distance from a point inside the fundamental parallelogram to a suitable vertex is *less* than the maximum of $\|\mathbf{r}\|$ and $\|\mathbf{s}\|$. The triangle **ort** satisfies $\angle \mathbf{t} \leq \angle \mathbf{o} \leq \angle \mathbf{r}$; see Sect. 2.3, Exercise 2.17. The perpendicular bisector of [**or**] will therefore meet the line segment **ot**. This perpendicular bisector determines two closed half-planes. The point **o** and the Voronoi cell of **o** lie in one of these half-planes; the line segment **tr** lies in the other. In a similar manner we see that the point **o** lies in one of the closed half-planes of the perpendicular bisector of [**os**], while the line segment **st** lies in the other. This reasoning finally leads us to the realization that the Voronoi cell lies inside the inner parallelogram. From this we can now deduce that we need to consider only finitely many lattice points to construct the Voronoi cell of Q; see Exercise 3.17. We can now prove the following theorem.

Theorem 3.26. *Let Q be a point of the lattice of P. The Voronoi cell of Q is bounded by either a hexagon or a rectangle that is point-symmetric with respect to Q.*

Proof. The lattice is point-symmetric; every lattice point can be used as a reflection point. We see this by setting a coordinate system at the point in question. It immediately follows that the Voronoi cell of Q is point-symmetric with respect to Q. Because of the previous theorem we only need to prove this theorem for the lattice point P. We fix a coordinate system at P, the point **o** in Fig. 3.27, on the right. We choose the points **r, s, t,** and **u** as before the theorem. Let us first assume that the angle **ros** is obtuse. After some puzzling we see that the Voronoi cell of **o** (*shaded*) is bounded by a broken line formed by line segments that lie, successively, on the perpendicular bisectors of [**ro**], [(−**s**)**o**], [(−**t**)**o**], [(−**r**)**o**], [**so**], [**to**]. The points inside this broken line lie closer to **o** than to **r,** −**s,** −**t,** −**r, s,** or **t.** Let us prove that this broken line is indeed the boundary of the Voronoi cell.

We need to show that the other lattice points have no influence on the form of the Voronoi cell. As an example we take the lattice point **u.** Figure 3.27 shows the perpendicular bisector of the line segment [**ou**], and the perpendiculars from the midpoint of [**ou**] onto the lines **o**(−**s**) and **or.** We can deduce from the positions of the lines that **u** has no influence on the form of the cell. See also Exercise 3.17. If **ros** is a right angle, the Voronoi cell is a rectangle; see for example Fig. 3.28, on the left.

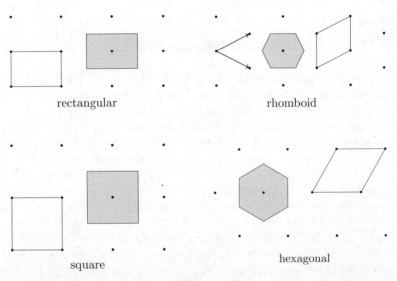

Fig. 3.28. Lattices with more regularity: the Voronoi cells are shaded, the fundamental parallelograms are open

Figure 3.28 shows lattices with some regularity. Let us explain the adjectives placed under the lattices. In the *rectangular* and *square* lattices, the fundamental parallelograms are a rectangle and a square, respectively. In the *rhomboid* lattice, the fundamental parallelogram is not necessarily a rhombus, but it can be generated by two translations with translation vectors of equal length, which therefore span a rhombus. The Voronoi cell is an irregular hexagon. In the *hexagonal* lattice, the fundamental parallelogram is a particular rhombus, the union of two equilateral triangles. The word *hexagonal* refers to the form of the Voronoi cell, a regular hexagon. This list of possible lattices will be useful in the construction of the different periodic tilings in the next section.

Quadratic Forms

In number theory, lattices are studied in connection with quadratic forms with integer coefficients. The original work of Voronoi (1868–1908) concerns the theory of n-dimensional lattices. Since Dirichlet (1805–1859) also studied them, Voronoi cells are sometimes called *Dirichlet regions*. We will illustrate the connection between Voronoi diagrams and quadratic forms with a simple example.

We are looking for *integers* x and y that satisfy the equation

$$5x^2 + 6xy + 2y^2 = 1 \, . \tag{3.4}$$

We might be able to find the solutions quickly by educated guessing. Let us show how the solutions can be found systematically. What is the attraction of equations of this type? The school of Pythagoras showed much interest in integer solutions of the equation $x^2 + y^2 = z^2$. The triples (x, y, z) of integers satisfying this equation can be described using simple formulas. They are called Pythagorean triples; see Sect. 1.1. They are closely related to the Pythagorean theorem. Later, Diophantus (ca. 246–ca. 330) studied equations of the type $x^2 + y^2 = x^3$. As was usual at that time, he looked for rational solutions. In time, the term *Diophantine equation* has come to mean an equation where one looks for integer solutions. Fermat (1601–1665) studied Diophantine equations of the type $x^2 + y^2 = p$, where p is a prime number. Later, Euler and Lagrange (1736–1813) would extend this study to Diophantine equations of the form $Ax^2 + 2Bxy + Cy^2 = n$, where A, B, C are integers, n is a positive integer, and the form $Ax^2 + 2Bxy + Cy^2$ is positive definite, that is, for all nonzero x and y, the expression is positive.

If the form $Ax^2 + 2Bxy + Cy^2$ is positive definite, then necessarily $A > 0$ and $C > 0$; if, for example, $A = 0$, the form would be 0 for $(x, y) = (1, 0)$, which is excluded. Since

$$Ax^2 + 2Bxy + Cy^2 = A\left(x + \tfrac{B}{A} y\right)^2 + \tfrac{1}{A}\left(AC - B^2\right) y^2 \, ,$$

we see that we must also have $AC - B^2 > 0$. Conversely, if $A > 0$ and $AC - B^2 > 0$, the form $Ax^2 + 2Bxy + Cy^2$ is positive definite. Gauss found a link between positive definite forms and inner products. To explain this link we return to (3.4). Choose vectors **p** and **q** in the plane such that $\|\mathbf{p}\| = \sqrt{5}$, $\|\mathbf{q}\| = \sqrt{2}$, and the positive angle ϑ between the vectors satisfies $\cos \vartheta = 3/\sqrt{10}$; see Fig. 3.29.

If, more generally, we consider the positive definite form $Ax^2 + 2Bxy + Cy^2$, we choose vectors with lengths $\sqrt{A}$ and $\sqrt{C}$ whose enclosed angle satisfies $\cos \vartheta = B/\sqrt{AC}$; this angle is well defined because the form is positive definite,

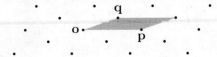

Fig. 3.29. The lattice for the equation $5x^2 + 6xy + 2y^2 = 1$

and therefore $AC > B^2$. See also Sect. 4.2, Exercise 4.14 (b). If the x and y in our example are integers, we obtain

$$\|x\mathbf{p} + y\mathbf{q}\|^2 = \langle x\mathbf{p} + y\mathbf{q}, x\mathbf{p} + y\mathbf{q}\rangle = 5x^2 + 6xy + 2y^2 \ .$$

The vector $x\mathbf{p} + y\mathbf{q}$ is the position vector of a point in the lattice

$$E = \{\, n\mathbf{p} + m\mathbf{q} : n \in \mathbb{Z}, m \in \mathbb{Z}\,\}\ ,$$

and the formula we just found tells us that the square of the norm of $x\mathbf{p} + y\mathbf{q}$ is an integer. We can now state the original question as follows: are there lattice points in E with norm 1? Since the square of the norm of a point in E is always an integer, it suffices to show there are lattice points with norm less than $\sqrt{2}$.

The fundamental parallelogram spanned by $\mathbf{p}$ and $\mathbf{q}$ has area 1. If, more generally, we consider the positive definite form $Ax^2 + 2Bxy + Cy^2$, the area of the fundamental parallelogram is $\sqrt{AC - B^2}$. In this section we have studied the Voronoi diagram of such a lattice. The Voronoi cells are hexagons; the area of such a Voronoi cell equals the area of the fundamental parallelogram, namely 1. Opposite sides of these hexagons are parallel and the distance between such sides is equal to the distance between two lattice points. How great is this distance? The minimal distance is maximal if the hexagon is regular. If a is the distance between opposite sides of a regular hexagon, the area of the hexagon is $a^2\sqrt{3}/2$. By setting this equal to 1, we find that the maximal minimal distance is $\sqrt{2}/\sqrt[4]{3}$, which is less than $\sqrt{2}$. It follows that (3.4) has an integer solution. If we consider the lattice again, we can find it easily.

The search for solutions of Diophantine equations using lattices and Voronoi cells is part of the *geometry of numbers*. See, for example, [8], [67].

Exercises

3.16. Let T_1 and T_2 be translations in the coordinate plane with translation vectors $T_1(\mathbf{o}) = (2,3)$ and $T_2(\mathbf{o}) = (3,5)$. Show that the lattice of $\mathbf{o}$ equals $\{\,(n,m) : n \in \mathbb{Z}, m \in \mathbb{Z}\,\}$.
Hint: $(1,0) = 5T_1(\mathbf{o}) - 3T_2(\mathbf{o})$.

3.17. Let Q be a lattice point. We know that the Voronoi cell $V(Q)$ of Q lies inside the circle with radius r and center Q. Consequently, for every lattice point R with $d(R, Q) \geq 2r$, the perpendicular bisector of $[RQ]$ is disjoint

from $V(Q)$.
Conclusion: a lattice point R with $d(R,Q) \geq 2r$ has no influence on the form of the Voronoi cell of Q.

3.18. Consider Fig. 3.29. The parallelogram spanned by **p** and $2\mathbf{p}+\mathbf{q}$ has area 1. The same holds for the parallelogram spanned by $\mathbf{p}+\mathbf{q}$ and **q**.

3.19. What are the integer solutions of the equation $5x^2 + 26xy + 34y^2 = 1$?

3.5 Periodic Tilings

A frieze pattern is generated by a single translation. We will now consider translations in different directions. We call a subset D of the plane a *periodic tiling* if there exist linearly independent translations $\mathcal{T}_1$ and $\mathcal{T}_2$, that is, translations whose vectors have different spans, such that the set of all translations in the symmetry group $\mathcal{I}(D)$ is equal to $\{\mathcal{T}_1^n \circ \mathcal{T}_2^m : n \in \mathbb{Z}, m \in \mathbb{Z}\}$. We say that $\mathcal{T}_1$ and $\mathcal{T}_2$ are *generating* translations of D. Periodic tilings are common; think of wall tilings or wallpaper. As we did for frieze patterns, we can ask how many different periodic tilings there are. Let us first define when two tilings are equal.

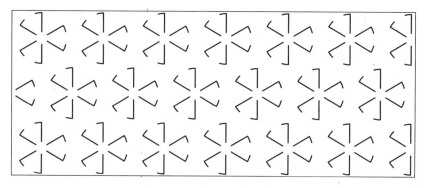

Fig. 3.30. A periodic tiling

Definition 3.27. *We say that two periodic tilings D_1 and D_2 have the same type if there exists an isomorphism φ from $\mathcal{I}(D_1)$ to $\mathcal{I}(D_2)$ that maps $\mathcal{I}(D_1)^+$ onto $\mathcal{I}(D_2)^+$. We say that two periodic tilings are* different *if they do not have the same type.*

The two tilings in Fig. 3.31 have the same type. If we consider either of these tilings, we see that its symmetry group $\mathcal{I}$ contains only translations. In particular, all isometries that map the tiling onto itself are direct; consequently, $\mathcal{I} = \mathcal{I}^+$. There are two generating translations, which we call $\mathcal{T}_1$

and $\mathcal{T}_2$. We see that $\mathcal{I} = \{\, \mathcal{T}_1^n \circ \mathcal{T}_2^m : n \in \mathbb{Z}, m \in \mathbb{Z} \,\}$. This is isomorphic to $\mathbb{Z}^2$, the subgroup of $\mathbb{R}^2$ consisting of all vectors (n, m) with n and m in $\mathbb{Z}$. The isomorphism is defined by mapping $\mathcal{T}_1^n \circ \mathcal{T}_2^m$ to (n, m). The two tilings have the same symmetry group.

Again we ask *how many different periodic tilings* there are. In the next section we will deduce that there are exactly 17. Broadly speaking, the deduction is the same as that of the existence of 7 frieze patterns.

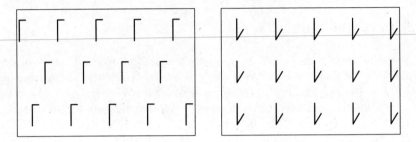

Fig. 3.31. Tilings of the same type

Let D be a periodic tiling with generating translations $\mathcal{T}_1$ and $\mathcal{T}_2$. We associate a lattice to D: for any point P in the plane, the set $\{\, \mathcal{T}_1^n \circ \mathcal{T}_2^m (P) : n \in \mathbb{Z}, m \in \mathbb{Z} \,\}$ is the *lattice* of P. Of course, the lattice of P depends on the choice of the point P. Two lattices associated to D can always be transformed into each other by a translation, which, of course, in general does not belong to the symmetry group of D. We consider the Voronoi cells of the points of the lattice of P. Since the Voronoi cells cover the plane, as in Fig. 3.32, we can use these cells for the construction of periodic tilings; to design a periodic tiling it suffices to design the motif for a single Voronoi cell.

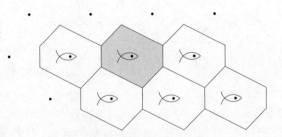

Fig. 3.32. Tiling with Voronoi cells

The symmetry of the Voronoi cells reflects that of the lattice. Figure 3.44 shows 17 Voronoi cells that form the tiles of the same number of periodic tilings. We now come to an important question: *which rotational symmetry can occur in periodic tilings?* The following theorem gives an answer. To understand the theorem we first need to understand what the order of a rotation is. By Definition 3.2, the order of a rotation is n if n is the smallest strictly positive integer such that $\mathcal{R}^n = \text{id}$.

Theorem 3.28. *Let D be a periodic tiling. Let $\mathcal{R}$ be a rotation other than the identity in the symmetry group $\mathcal{I}(D)$ of D. The order of $\mathcal{R}$ is 2, 3, 4, or 6. Moreover, $\mathcal{R}$ maps the lattice of its rotation center onto itself.*

Proof. We will deduce this from Theorem 3.26. We begin with the second result. Choose a coordinate system with origin at the rotation center C of $\mathcal{R}$. We first prove that the rotation $\mathcal{R}$ maps the lattice of $C = \mathbf{o}$ onto itself. Let $\mathbf{q}$ be an arbitrary lattice point. The translation $\mathcal{T}$ determined by $\mathcal{T}(\mathbf{o}) = \mathbf{q}$ is an element of $\mathcal{I}(D)$. The map $\mathcal{R} \circ \mathcal{T} \circ \mathcal{R}^{-1}$ is also an element of $\mathcal{I}(D)$ and is a translation; this last statement follows from Sect. 2.4, Exercise 2.33, and Theorem 2.34. Hence $\mathcal{R} \circ \mathcal{T} \circ \mathcal{R}^{-1}(\mathbf{o})$ is an element of the lattice. We have

$$\mathcal{R} \circ \mathcal{T} \circ \mathcal{R}^{-1}(\mathbf{o}) = \mathcal{R} \circ \mathcal{T}(\mathbf{o}) = \mathcal{R}(\mathbf{q}) \ .$$

Consequently, $\mathcal{R}(\mathbf{q})$ is an element of the lattice; therefore $\mathcal{R}$ maps the lattice onto itself. This proves the second result of the theorem. Since $\mathcal{R}^{-1}$ is also a rotation other than the identity in $\mathcal{I}(D)$, it likewise maps the lattice onto itself, whence it follows that $\mathcal{R}$ is bijective. The order of $\mathcal{R}$ is then a divisor of the number of sides of the Voronoi cell. In the last section we saw that the Voronoi cell is either a hexagon or a rectangle. This implies the first result of the theorem.

Digression 6. *For a long time, this last theorem was seen as the most important result on periodic tilings. It forms an essential element of the theory of the three-dimensional lattices on which crystallography is based. The symmetry group of a three-dimensional lattice also cannot contain a rotation of order 5; see Theorem 5.66. In the 1980s, the discovery of semiregular spatial tilings with fivefold symmetry in so-called* quasicrystals *revolutionized crystallography. Of course, there is no lattice associated to a semiregular spatial tiling; this follows from the previous theorem. For more on this subject see [65].*

Let us now begin the classification of the periodic tilings.

We first list the types of periodic tilings whose symmetry group contains only direct isometries, that is, translations or rotations.

By Theorem 3.28 we need to look only for the presence of rotations of orders 2, 3, 4, and 6. We will number the types we find this way 1 through 5 and will denote the corresponding symmetry groups by H_1 through H_5.

Type 1. To begin, there is the periodic tiling whose symmetry group contains only translations; it has type 1. Each row of Fig. 3.33 shows, on the left, a Voronoi cell with a motif, and on the right the same cell, this time with its symmetry elements. In this case there are no symmetry elements; the point-symmetry of the lattice is destroyed by the asymmetry of the motif.

128   3 SYMMETRY

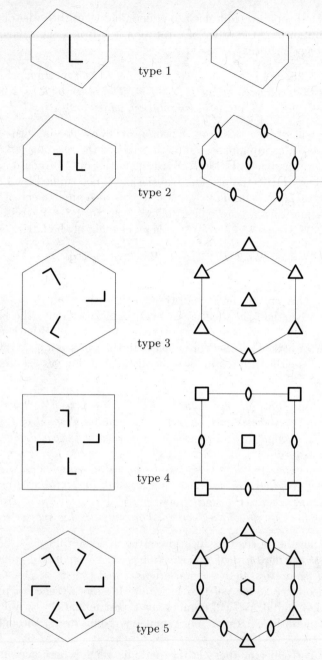

Fig. 3.33. Periodic tilings of types 1 through 5

3.5 Periodic Tilings

Note that lattices can differ geometrically even though their translation groups are equal algebraically; see Fig. 3.31. All periodic tilings whose symmetry group contains only translations have the same type, namely type 1. The symmetry group H_1 of such a tiling is isomorphic to $\mathbb{Z}^2$. The symmetry groups of periodic tilings are also called *wallpaper groups*.

Type 2. Next there is the periodic tiling whose symmetry group contains translations and a rotation of order 2, and no rotations of higher order; this has type 2; see the second row of Fig. 3.33.

Let us study the lattice of a rotation center. The top figure shows a Voronoi cell associated to this lattice. All lattice points are centers of rotations of order 2, as are many other points; see Exercise 3.20. The rotation centers are shown in the figure on the right. The rotations arise by composing rotations with centers at the lattice points and translations that map the lattice onto itself.

Fig. 3.34. Creation of a new rotation

Figure 3.34 shows a rotation with a lattice point as center as the product of two reflections with perpendicular axes. One of the translations in the symmetry group is represented as the product of two reflections with parallel axes. Together, the rotation and the translation give a rotation of order 2 with center on the boundary of the Voronoi cell. This method gives six rotation centers on the boundary of the Voronoi cell. If the Voronoi cell is a rectangle, this same method leads to new rotation centers. In particular, we obtain all rotation centers this way. We can prove this as follows. Let C be the center of a rotation of order 2. Let Q be the center of a Voronoi cell containing C. Composing the two rotations with centers C and Q gives a translation; this must be one of the translations of the lattice. It is not difficult to deduce from this that C is one of the rotation centers we just found.

We can make a periodic tiling of type 2 by filling the plane with copies of this Voronoi cell, as shown in Fig. 3.35. All periodic tilings whose symmetry group contains only translations and rotations of order 2 have the same type. We can see this as follows. Given two such periodic tilings, we begin by making an isomorphism between the translation subgroups of the symmetry groups. Then we choose a rotation of order 2 in each of the symmetry groups; these are mapped to each other. Any other element of one of the symmetry groups is the composition of the rotation and a suitable translation. As in the proof of Theorem 3.13, we can use this to extend the isomorphism between the translation subgroups to an isomorphism between the symmetry groups. We denote the corresponding symmetry group by H_2.

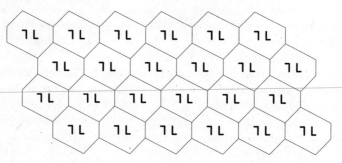

Fig. 3.35. How a tile gives rise to a tiling

Type 3. Let us now consider the case that in addition to the translations, the symmetry group of a periodic tiling also contains a rotation of order 3, and no rotations of higher order. This is type 3; see the third row of Fig. 3.33. We consider the lattice of a rotation center: all lattice points are centers of rotations of order 3; see Exercise 3.20. Theorem 3.28 implies that the lattice, and therefore also the Voronoi cell, has threefold rotational symmetry. The lattice is therefore hexagonal: a rhomboid lattice has threefold rotational symmetry only if it is hexagonal. We add a motif with threefold rotational symmetry to the Voronoi cell.

It turns out that the vertices of the Voronoi cell are also centers of rotations of order 3. We can easily prove this by representing such a rotation with a lattice point as center as the product of two reflections whose axes meet at a lattice point under an angle of $\pi/3$, and then composing this rotation with one of the translations, written as the product of two reflections with parallel axes. This proof is very similar to the proof we gave for type 2.

There are no other rotations of order 3. We prove this as follows. Let C be the center of a rotation of order 3, and let Q be the center of a Voronoi cell containing C. The composition of the rotation over an angle $2\pi/3$ about C and the rotation over an angle $-2\pi/3$ about Q is a translation; the length of the translation vector is $\sqrt{3}d(Q,C)$. This translation must be one of the translations of the lattice. It follows without much trouble that C must be one of the centers we found before. We furthermore note that there are no rotations of order 2. Indeed, a composition of a rotation over an angle π with a rotation over an angle $-2\pi/3$ is a rotation of order 6, which does not occur in the case that we are studying.

As in the type-2 case, these considerations lead to the following conclusion: all periodic tilings whose symmetry group contains translations, a rotation of order 3, and no rotations of higher order have the same type, namely type 3. We denote the symmetry group of this type by H_3.

Type 4. Let us now consider a periodic tiling whose symmetry group contains a rotation of order 4, but no rotation of higher order; see the fourth row of Fig. 3.33.

Again we consider the lattice of a rotation center and notice that all lattice points are centers of rotations of order 4. Since the Voronoi cells must also have fourfold rotational symmetry, we have a square lattice. We add a motif with fourfold rotational symmetry to the Voronoi cell.

It turns out that the vertices of the Voronoi cell are also centers of rotations of order 4 and that the midpoints of the edges are centers of rotations of order 2. Indeed, every lattice point is also the center of a rotation of order 2. By composing this rotation with a translation of the lattice we again obtain a rotation of order 2.

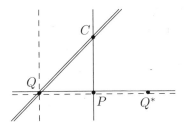

Fig. 3.36. There are no other rotations of order 4

There are no other rotations of order 4. We prove this as follows; see Fig. 3.36. Let C be the center of a rotation of order 4, and let Q be the center of a Voronoi cell containing C. The composition of the rotation over an angle $\pi/2$ about Q and the rotation over an angle $\pi/2$ about C is a rotation of order 2 about P. The composition of the point reflections $\dot{S}_Q$ followed by $\dot{S}_P$ (the rotation of order 2 we just found) is a translation. We find that the image Q^* of Q under $\dot{S}_P$ must be a lattice point. But then P lies in the middle of the edge of the cell of Q, and C lies at a vertex of that cell. We see that we have already counted both P and C in our inventory of the rotation centers. Moreover, there can be no rotations of order 3, as we will see in Exercise 3.21. Likewise, there are no rotations of order 2 other than those we already have.

We come to the following conclusion: all periodic tilings whose symmetry group contains translations, a rotation of order 4, and no rotations of higher order have the same type, namely type 4. We denote the corresponding symmetry group by H_4.

Type 5. Let us now consider a tiling whose symmetry group contains a rotation of order 6; see the bottom row of Fig. 3.33. The lattice and the Voronoi cell now have sixfold rotational symmetry. The vertices of the Voronoi cell are centers of rotations of order 3 and the midpoints of the edges are centers of rotations of order 2. As above we can prove that there are no other rotations of order 6, 3, or 2, and also no rotations of order 4.

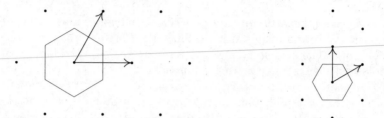

Fig. 3.37. How to construct an isometry for a tiling of type 5

Given two periodic tilings whose symmetry groups contain a rotation of order 6, we can construct an isomorphism between the symmetry groups as follows; see Fig. 3.37. We choose a rotation of order 6 in each of the symmetry groups, and fix a coordinate system at each of the rotation centers. The lattices of the centers are both hexagonal. We can see the lattice points as the vectors of the translations in the symmetry groups. We can find a similarity that maps one lattice onto the other, and, moreover, maps the origins of the coordinate systems onto each other. Furthermore, we can choose generating translations in the symmetry groups that correspond to each other under the similarity; these translation vectors are drawn in the figure.

Since the other elements in the symmetry group arise by multiplying the chosen rotation and powers of the generating translations, this fixes the isomorphism between the symmetry groups. We can almost read the isomorphism in the similarity between the lattices; the transformation maps rotation centers onto rotation centers. We conclude that all periodic tilings whose symmetry group contains translations and a rotation of order 6 have the same type, namely type 5. We denote the symmetry group by H_5.

This completes the classification of the periodic tilings with only direct isometries in the symmetry group.

Theorem 3.29. *If the symmetry group $\mathcal{I}(D)$ of a periodic tiling D contains only direct isometries, that is, if $\mathcal{I}(D) = \mathcal{I}(D)^+$, then D has type 1, 2, 3, 4, or 5.*

Exercises

3.20. Let $\mathcal{T}$ be a translation. If $\mathcal{R}$ is a rotation with center P, then $\mathcal{T}(P)$ is the center of the rotation $\mathcal{T} \circ \mathcal{R} \circ (\mathcal{T})^{-1}$. These rotations have the same order.

3.21. The symmetry group of a periodic tiling D can never contain both a rotation of order 3 and a rotation of order 4.
Hint: How many sides does the Voronoi cell of the lattice of D have?

3.22. Let $\mathcal{R}$ be a rotation of order 6 in the symmetry group of a periodic tiling of type 5. Let $\mathcal{T}$ be a generating translation in the same symmetry group. Determine the rotation centers and the orders of the compositions of $\mathcal{T}$ and $\mathcal{R}^k$ for $k = 1, \ldots, 5$.

3.6 All Types of Periodic Tilings

In this section we turn our attention to periodic tilings with both direct and indirect isometries. This will complete our classification of the periodic tilings. Our method is very similar to that of the second part of Sect. 3.2. The idea is to add indirect isometries to the groups H_1 through H_5. As we observed before, this is the converse of what was done in Theorem 3.6. We are given a group H_i, with $i = 1, \ldots, 5$, and try to extend it to a group G in such a way that H_i is exactly the subgroup G^+. This extension problem will turn out to be much simpler for the groups H_1 and H_2 than it is for H_3, H_4, and H_5. When studying types 2 and 5 we showed in detail that tilings with isomorphic symmetry groups have the same type. We will not repeat these types of arguments here.

Before we begin, let us give a small extension of the notion of order; see Definition 3.2. We say that an element g in a group H with unit element e has *order* n if n is the smallest positive integer such that $g^n = e$; if for some g there is no n with this property, we say that the *order of g is infinite*.

Let us first study what happens when we add a reflection to a given symmetry group. The following theorem will be of great help.

Theorem 3.30. *Suppose that the symmetry group $\mathcal{I}(D)$ of a periodic tiling D contains a reflection $\mathcal{S}$. Let P be a point on the reflection axis of $\mathcal{S}$. Then $\mathcal{S}$ maps the lattice of P onto itself. Moreover, every lattice point lies on a reflection axis.*

Proof. This theorem is very similar to Theorem 3.28. The same holds for the proof. We fix a coordinate system at the point P. Let $\mathbf{q}$ be an arbitrary point of the lattice of $P = \mathbf{o}$. There is a translation $\mathcal{T}$ in $\mathcal{I}(D)$ with $\mathcal{T}(\mathbf{o}) = \mathbf{q}$. The map $\mathcal{S} \circ \mathcal{T} \circ \mathcal{S}$ is also a translation; we can see this by writing $\mathcal{T}$ as the product of two reflections. Hence $\mathcal{S} \circ \mathcal{T} \circ \mathcal{S} \in \mathcal{I}(D)$. Consequently, $\mathcal{S} \circ \mathcal{T} \circ \mathcal{S}(\mathbf{o})$ is an

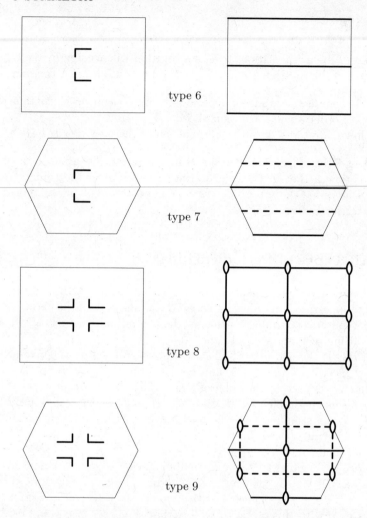

Fig. 3.38. Periodic tilings of types 6 through 9

element of the lattice. Since P lies on the reflection axis, we have $\mathcal{S} \circ \mathcal{T} \circ \mathcal{S}(\mathbf{o}) = \mathcal{S} \circ \mathcal{T}(\mathbf{o}) = \mathcal{S}(\mathbf{q})$. It follows that $\mathcal{S}(\mathbf{q})$ is an element of the lattice; hence $\mathcal{S}$ maps the lattice onto itself. Since $\mathcal{S}$ is its own inverse, it is bijective. For the second result of the theorem, see Exercise 3.23.

How can we extend the group H_1 by adding a reflection? We are dealing with the symmetry group of a periodic tiling without rotational symmetry (type 1), to which we want to add a reflection. For a closer analysis we assume that D is a periodic tiling without rotational symmetry, but with reflection symmetry.

Let $\mathcal{S}$ be a reflection in $\mathcal{I}(D)$. Choose a point P on the reflection axis of $\mathcal{S}$ and consider the lattice of $P = \mathbf{o}$. By the theorem above, the reflection $\mathcal{S}$ maps the lattice of P onto itself. Since the lattice is point-symmetric with respect to $\mathbf{o}$, there is another reflection $\mathcal{S}^*$ that maps the lattice onto itself, with axis perpendicular to that of $\mathcal{S}$. In general, $\mathcal{S}^* \notin \mathcal{I}(D)$. The Voronoi cell must show these symmetries. The lattice is therefore rectangular or rhomboid; see the overview following Figure 3.28 and Exercise 3.27. Choose a lattice point $\mathbf{r} \neq \mathbf{o}$ as close as possible to $\mathbf{o}$, followed by a lattice point $\mathbf{s}$ outside the span of $\mathbf{r}$ that also lies as close as possible to $\mathbf{o}$. If $\mathbf{r}$ and $\mathbf{s}$ lie on the reflection axes mentioned above, the lattice is rectangular; otherwise, it is rhomboid.

Type 6. Let us first consider the case that the lattice is rectangular. In the top row of Fig. 3.38, the figure on the left shows a rectangular Voronoi cell with a motif that has reflection symmetry but no point-symmetry.

The symmetry group of this tiling contains translations and reflections and compositions of these two. These compositions include glide reflections; each glide reflection axis coincides with the axis of a reflection. In the figure on the right the reflection axis is indicated in the usual way with a solid line.

All reflection axes are parallel to each other. There is no choice; if there were reflections with intersecting axes, the intersection point would be the center of a rotation. As we saw, the lattice also has point-symmetry, but this is destroyed by the asymmetry of the motif.

Type 7. Let us now consider the case that the lattice is rhomboid. In the second row of Fig. 3.38, the figure on the left shows a hexagonal Voronoi cell with a motif with reflection symmetry but no point-symmetry. The symmetry group of this tiling contains translations and reflections and compositions of these two. These compositions include glide reflections, but their axes do not always coincide with the normal reflection axes; in the figure on the right the glide reflection axes are shown using dashed lines.

In addition to the situation sketched in Fig. 3.38, there is also another one that at first looks very different; see Fig. 3.39. Nevertheless, the tilings have

Fig. 3.39. Another representation of type 7

isomorphic symmetry groups. The easiest way to see this is by analyzing the lattices; Fig. 3.40. We choose a reflection in each of the symmetry groups and fix a coordinate system at a point on each of the reflection axes. The lattices of the chosen points are both rhomboid. We can see the lattice points as the vectors of the translations in the symmetry groups. We can now find a map Φ that maps one lattice onto the other in such a way that the axes of the chosen reflections correspond to each other. In general, Φ is not a similarity, but rather a similarity followed by an expansion or a contraction in the direction of the coordinate axes. Since the distortions in the directions of the axes are not necessarily of the same size, the Voronoi cells will in general *not* be mapped onto each other. Note that we can choose generating translations in both symmetry groups in such a way that in each lattice the vector of one of these translations lies along the reflection axis, and that the axes of the generating translations are mapped to each other by Φ. The reflection axes and translation vectors are drawn in Fig. 3.40.

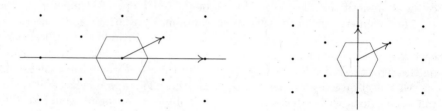

Fig. 3.40. How to construct an isomorphism for type 7

Since the other elements in the symmetry group arise by multiplying the reflection we just mentioned and powers of the generating translations, this fixes the isomorphism between the symmetry groups. We can almost read the isomorphism in the map between the lattices.

It is clear from the above that every tiling whose symmetry group G contains reflections and no rotations has type 6 or 7; we call the symmetry groups H_6 and H_7, respectively.

It may not be clear yet that types 6 and 7 are different. Showing this is not easy. The difference between H_6 and H_7 lies in the indirect isometries. We first note that the different types of indirect isometries, that is, reflections and glide reflections, can be distinguished using their order: a reflection is an element of order 2 in the symmetry group, while a glide reflection has infinite order. In a tiling of type 6, the axis of any glide reflection corresponds to that of a reflection. By contrast, in a tiling of type 7 there is at least one glide reflection whose axis does not coincide with that of a reflection. In terms of properties of the symmetry group we can state this difference as follows. On the one hand, in H_6, every glide reflection $\mathcal{G}$ admits a reflection $\mathcal{S}$ that commutes with it, that is, $\mathcal{G} \circ \mathcal{S} = \mathcal{S} \circ \mathcal{G}$. On the other hand, in H_7, there is a glide reflection $\mathcal{G}$ that no reflection $\mathcal{S}$ commutes with, that is, $\mathcal{G} \circ \mathcal{S} \neq \mathcal{S} \circ \mathcal{G}$. Since the sets consisting of the indirect isometries differ algebraically, types 6 and 7 must be different.

Types 8, 9, and 15. Let us now study how we can extend the group H_2 by adding a reflection. We have the symmetry group $\mathcal{I}(D)$ of a periodic tiling D whose rotations all have order 2. Moreover, $\mathcal{I}(D)$ contains a reflection $\mathcal{S}$. There are two possibilities: either there is a rotation center C on the reflection axis l of $\mathcal{S}$, or there isn't.

In the first case we consider the lattice of $C = \mathbf{o}$. Before beginning our study of type 6, we determined that in this case the lattice is either rectangular or rhomboid. The reflection axis $\mathcal{S}^*$ mentioned there also belongs to $\mathcal{I}(D)$. We are dealing with tilings of types 8 and 9, respectively. These are represented in the usual way in the last two rows of Fig. 3.38. The figure shows the symmetry elements; their presence can be deduced from the fact that we have translations and a rotation with center on the reflection axis.

It is clear that every periodic tiling whose rotations have order 2 and that admits a reflection whose axis contains a rotation center has type 8 or 9. We call the symmetry groups H_8 and H_9, respectively.

The groups H_8 and H_9 can be distinguished as follows. This is similar to the discussion concerning types 6 and 7. In H_8 every glide reflection $\mathcal{G}$ admits a reflection $\mathcal{S}$ that commutes with it, that is, $\mathcal{G} \circ \mathcal{S} = \mathcal{S} \circ \mathcal{G}$. By contrast, in H_9, there is a glide reflection $\mathcal{G}$ that no reflection commutes with.

In the second case there is no rotation center on l. It then follows from Exercise 3.24 that the centers of the rotations in $\mathcal{I}(D)$ are symmetric with respect to l. We then have a tiling of type 15. The existence of further symmetries follows from the fact that we are dealing with a rotation, a reflection, and translations.

Types 10 and 11. How can we extend the group H_3 by adding a reflection? We have a periodic tiling D with rotations of order 3. The group $\mathcal{I}(D)$ also contains a reflection $\mathcal{S}$. Since the rotation centers are symmetric with respect to the reflection axis l of $\mathcal{S}$, we easily see that there must be a rotation center C on l. The lattice of C is hexagonal. Since C is the center of a rotation of order 3, three reflection axes meet at C. It easily follows that there are two possibilities: every center lies on the axis, or at least one doesn't.

In the first case we have a tiling of type 10, and we denote the symmetry group by H_{10}. In the second case we have a tiling of type 11 with symmetry group H_{11}. The first two rows of Fig. 3.41 show the Voronoi cells of tilings of types 10 and 11. As shown in the figure, the symmetry groups also contain glide reflections.

In a tiling of type 10, three reflection axes pass through each center. This means that every subgroup of H_{10} isomorphic to C_3 is also the subgroup of a subgroup of H_{10} that is isomorphic to D_3. In a tiling of type 11, there are centers that do not lie on a reflection axis. This means that H_{11} contains subgroups that are isomorphic to C_3, but are not subgroups of a subgroup of H_{11} itself isomorphic to D_3.

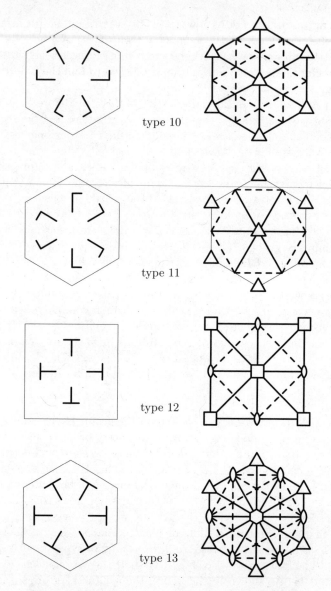

Fig. 3.41. Periodic tilings of types 10 through 13

Types 12 and 17. Next we extend H_4 by a reflection. The corresponding periodic tiling D contains rotations of order 2 and 4 and a reflection. This implies the presence of glide reflections and more rotations of order 2. The reflection can either pass through the rotation centers or not. If a center of a rotation of order 4 lies on a reflection axis, all centers of order 4 do. We then have type 12; we denote the symmetry group of a tiling of type 12 by H_{12}. See the third row of Fig. 3.41.

If no rotation of order 4 has its center on a reflection axis, we have type 17. The corresponding symmetry group is H_{17}; see Fig. 3.43. The groups H_{12} and H_{17} are different: the first contains subgroups isomorphic to D_4; the second does not.

Type 13. Adding a reflection to the symmetry group H_5 leads to the symmetry group H_{13} of the tiling of type 13; see the bottom row of Fig. 3.41.

This concludes our discussion of the effect of adding a reflection to a symmetry group H_i for $1 \leq i \leq 5$. Let us now study the effect of adding a glide reflection. The following theorem will help us.

Theorem 3.31. *Suppose that the symmetry group $\mathcal{I}(D)$ of a periodic tiling D contains a glide reflection $\mathcal{G}$. Let P be a point on the axis l of the glide reflection $\mathcal{G}$. The reflection $\mathcal{S}_l$ in the axis of the glide reflection $\mathcal{G}$ maps the lattice of P onto itself.*

Proof. Fix a coordinate system at the point P. Let $\mathbf{q}$ be an arbitrary lattice point; there is a translation $\mathcal{T}$ in $\mathcal{I}(D)$ with $\mathcal{T}(\mathbf{o}) = \mathbf{q}$. It follows from the definition of a glide reflection that there is a translation $\mathcal{F}$ such that $\mathcal{G} = \mathcal{S}_l \circ \mathcal{F} = \mathcal{F} \circ \mathcal{S}_l$. This translation $\mathcal{F}$ need not be an element of $\mathcal{I}(D)$. Recall that the composition of translations is commutative. We obtain

$$\mathcal{G} \circ \mathcal{T} \circ \mathcal{G}^{-1} = \mathcal{S}_l \circ \mathcal{F} \circ \mathcal{T} \circ \mathcal{F}^{-1} \circ \mathcal{S}_l = \mathcal{S}_l \circ \mathcal{T} \circ \mathcal{S}_l \, .$$

The last map is a translation contained in $\mathcal{I}(D)$; see the proof of Theorem 3.30. Hence $\mathcal{G} \circ \mathcal{T} \circ \mathcal{G}^{-1}(\mathbf{o})$ is an element of the lattice. We have

$$\mathcal{G} \circ \mathcal{T} \circ \mathcal{G}^{-1}(\mathbf{o}) = \mathcal{S}_l \circ \mathcal{T} \circ \mathcal{S}_l(\mathbf{o}) = \mathcal{S}_l \circ \mathcal{T}(\mathbf{o}) = \mathcal{S}_l(\mathbf{q}) \, .$$

Consequently, $\mathcal{S}_l(\mathbf{q})$ is an element of the lattice; $\mathcal{S}_l$ maps the lattice onto itself. Since $\mathcal{S}_l$ is its own inverse, it is also surjective.

We of course need to look at only a limited number of glide reflections. Indeed, if $\mathcal{T}_1$ and $\mathcal{T}_2$ are generating translations of the tiling, we need to consider only glide reflections $\mathcal{G}$ such that $\mathcal{G} \circ \mathcal{G}$ equals $\mathcal{T}_1$ or $\mathcal{T}_2$. This follows from Exercise 3.25.

The following deduction is again similar to the one given in the discussion of types 6 and 7. We study how we can extend the group H_1 by adding a glide reflection. We assume that we have a periodic tiling D without rotational symmetry and that the symmetry group of D contains a glide reflection $\mathcal{G}$ with axis l. We choose a point P on l and consider the lattice of $P = \mathbf{o}$. By the last theorem the lattice is symmetric with respect to l. Since the lattice is also point-symmetric with respect to $\mathbf{o}$, there is a reflection $\mathcal{S}^*$ with axis m such that l and m meet perpendicularly at $\mathbf{o}$ and m is a symmetry axis of the lattice. The lattice is therefore rectangular or rhomboid. Choose a lattice point $\mathbf{r} \neq \mathbf{o}$ as close as possible to $\mathbf{o}$. If $\mathbf{r}$ lies on one of the axes l and m, the lattice is rectangular; otherwise, it is rhomboid.

Type 14. Let us first consider the case that the lattice is rectangular; see the top row of Fig. 3.43. The symmetry group of this tiling contains only translations and glide reflections. The axes of the glide reflections are parallel to each other. This follows from Exercise 3.26: if the axes were not parallel to each other, the symmetry group would contain rotations. We do not find the point-symmetry of the lattice in the symmetry group of the tiling, since it is destroyed by the asymmetry of the motif. Every tiling with translations and glide reflections and no other isometries has type 14. We call the corresponding symmetry group H_{14}.

Type 7, again. We now consider the case that the lattice is rhomboid. In Fig. 3.42, dots indicate the lattice of P.

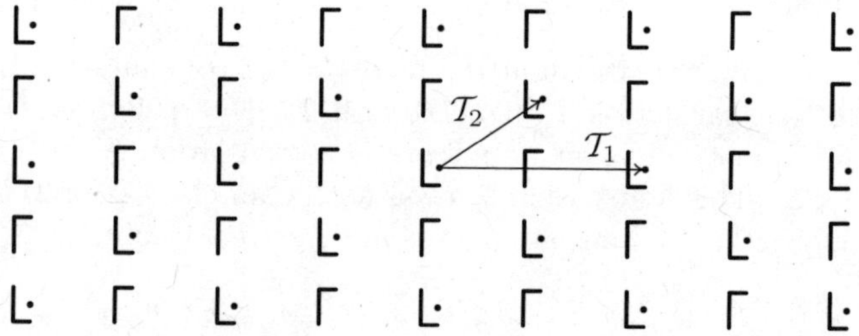

Fig. 3.42. Type 7 arising from a lattice

The motif is the letter L. The first of the generating translations $\mathcal{T}_1$ and $\mathcal{T}_2$ is chosen along the axis of the glide reflection $\mathcal{G}$. Because of the presence of glide reflections we also find upside-down L's in the tiling. The tiling has reflection symmetry; we can check that $\mathcal{T}_2 \circ \mathcal{G}^{-1}$ is a reflection. We have already come across this type when studying the extensions of H_1 by reflections. If we look at the lattice of a point of the reflection axis, we see that this is type 7.

Type 16. The previous section shows that the presence of a glide reflection axis in a rhomboid lattice induces the presence of reflections. But we have already studied all possible extensions of symmetry groups by reflections. When studying possibly new extensions of the group H_2 by glide reflections we can therefore restrict ourselves to rectangular lattices. The axes of the glide reflections either are parallel or intersect perpendicularly; see Exercise 3.26. In this way, we find a tiling of type 16 with symmetry group H_{16}.

It is not a very difficult exercise to check that we have seen all new extensions by glide reflections. By definition, two periodic tilings have the same type if there is an isomorphism between their symmetry groups that maps direct isometries to direct isometries. Now that we have an overview of all

3.6 All Types of Periodic Tilings   141

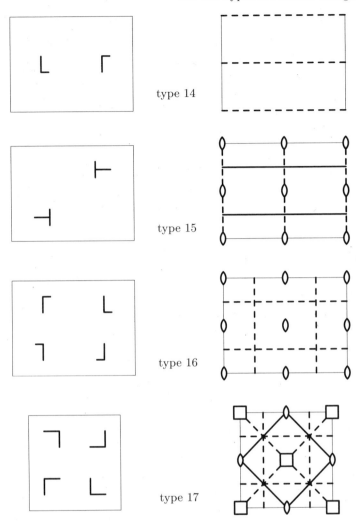

Fig. 3.43. Periodic tilings of types 14 through 17

types of tilings with corresponding symmetry groups, it turns out that the condition concerning the direct isometries can be dropped. That is the gist of the following theorem.

Theorem 3.32. *Two periodic tilings have the same type if and only if their symmetry groups are isomorphic.*

The idea of classifying the tilings according to their symmetry groups originates in [55].

Proof. One direction of the theorem is simple: if the types are the same, the symmetry groups are isomorphic. Let us now prove that if the symmetry

142   3 SYMMETRY

groups are isomorphic, the types are the same. For this it suffices to show that the groups H_i, $i = 1, \ldots, 17$, are all different, which we will do by determining a characteristic algebraic property for each of them. This requires some knowledge of group theory.

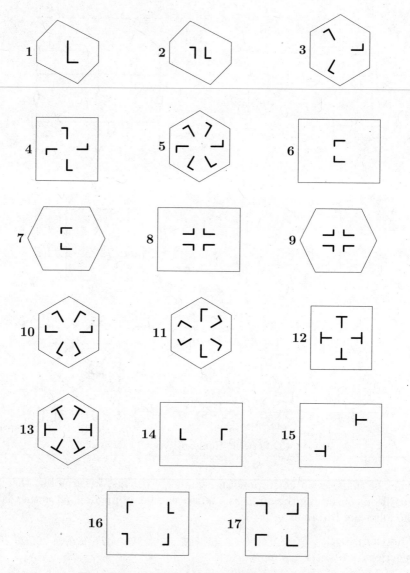

Fig. 3.44. 17 tiles generating periodic tilings

Exactly two of these groups contain an element of order 6, namely H_5 and H_{13}. Of these two, only H_{13} has a subgroup isomorphic to D_6. This characterizes H_5 and H_{13} among the 17 groups.

Exactly three of the groups contain an element of order 4, namely H_4, H_{12}, and H_{17}. Of these three groups, only H_{12} has a subgroup isomorphic to D_4. Of the two groups H_4 and H_{17}, only H_{17} has a subgroup isomorphic to the Klein four-group V_4.

Exactly three of the groups contain an element of order 3 but *no* element of order 6, namely H_3, H_{10}, and H_{11}. Of these three groups, only H_3 has no subgroup isomorphic to D_3. In the group H_{10} every element of order 3 is contained in a subgroup isomorphic to D_3. In the group H_{11} some elements of order 3 have this property, while others do not.

The following groups contain an element of order 2, but no elements of order 4 or 6: H_2, H_6, H_7, H_8, H_9, H_{15}, H_{16}. Of these seven, only H_8 and H_9 have a subgroup isomorphic to D_2. The groups H_8 and H_9 can be distinguished as follows. In H_8 every element $\mathcal{G}$ of infinite order admits an element $\mathcal{S}$ of order 2 that commutes with it, that is, $\mathcal{G} \circ \mathcal{S} = \mathcal{S} \circ \mathcal{G}$. In H_9 there is an element $\mathcal{G}$ of infinite order that no element of order 2 commutes with.

We still have the groups H_2, H_6, H_7, H_{15}, H_{16}. We can characterize H_2 by the property that the product of any element of infinite order (translation) and any element of order 2 (reflection in a point) has order 2.

We still have the groups H_6, H_7, H_{15}, H_{16}. In each of these groups we define the set Q_i ($i = 6, 7, 15, 16$) as follows: Q_i is the set of all elements of infinite order (translations and glide reflections) with the property that every product with an element of order 2 has infinite order. The set Q_{15} is empty, while the sets Q_6, Q_7, and Q_{16} are not. This characterizes Q_{15}.

We are left with the groups H_6, H_7, H_{16}. To every element $\mathcal{F}$ in Q_6 corresponds an element $\mathcal{S}$ of order 2 that commutes with it. The sets Q_7 and Q_{16} do not have this property.

We still need to distinguish H_7 and H_{16}. The group H_7 has an element $\mathcal{F}$ of infinite order, a translation, that admits an element $\mathcal{S}$ of order 2, a reflection, that commutes with it. The group H_{16} does not have this property.

We are now left with the groups H_1 and H_{14}. In H_1, any two elements of infinite order, which are all translations, commute; this is not the case in H_{14}.

The results of this section were first proved by Fedorov (1853–1919) in 1891. There is much literature on tilings and patterns; [42] is an interesting and clear introduction. The standard reference on this subject, [23], is truly fine and richly illustrated. The books [28] and [12] contain a few interesting short texts.

We conclude with a remark. There is no reason to consider only lattices in the plane. The theory of spatial lattices is very important in crystallography. In this setting Voronoi cells are also named after Wigner and Seitz, [37]. See also Sect. 5.5, in particular the section "Crystallography". Astronomers have recently discovered that the distribution of matter in the universe can be modeled using spatial Voronoi diagrams [68]. Most matter is on the boundaries of the diagram, making the universe resemble an enormous bubble bath. In this setting we sometimes talk of the Voronoi foam model.

Exercises

3.23. Let $\mathcal{T}$ be a translation. Let P be a point on the axis of a reflection $\mathcal{S}$. Then $\mathcal{T} \circ \mathcal{S} \circ \mathcal{T}^{-1}$ is a reflection and the point $\mathcal{T}(P)$ lies on its axis.

3.24. Let $\mathcal{R}$ be a rotation about C and let $\mathcal{S}$ be a reflection. Then $\mathcal{S} \circ \mathcal{R} \circ \mathcal{S}$ is a rotation about $\mathcal{S}(C)$. The rotation angles are equal in size after taking the absolute value, but have opposite orientations.

3.25. Let $\mathcal{G}$ be a glide reflection and $\mathcal{T}$ a translation with $\mathcal{T} = \mathcal{G} \circ \mathcal{G}$. If $\mathcal{F}$ is a glide reflection such that $\mathcal{F} \circ \mathcal{F} = \mathcal{T}^k$ for some odd $k > 0$, then $\mathcal{G} = \mathcal{T}^{-(k-1)/2} \circ \mathcal{F}$. Hence if $\mathcal{T}$ and $\mathcal{F}$ belong to a symmetry group, so does $\mathcal{G}$.

3.26. Let $\mathcal{G}_1$ and $\mathcal{G}_2$ be two glide reflections with intersecting axes. Show that $\mathcal{G}_2 \circ \mathcal{G}_1$ is a rotation other than the identity.

3.27. Let O be a lattice point on the axis l of a reflection $\mathcal{S}_l$ that maps the lattice onto itself. Show that the lattice is rectangular or square.
Hint: Let m be the perpendicular from O on l. Then $\mathcal{S}_m$ also maps the lattice onto itself. Choose P and Q on l and m, respectively, as close as possible to O. There is either no lattice point inside the rectangle with vertices P, O, and Q or only one, the intersection point of the diagonals.

4
CURVES

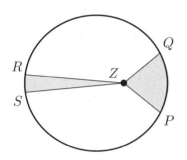

In the third century BC, Apollonius wrote the *Conica*, a detailed study of the conic sections. Almost two thousand years later, Kepler discovered that the motions of the planets around the sun can be described using ellipses with the sun as one of the foci. Newton was later able to prove these results using his general laws of motion.

In the course of time curves have been discovered, or *invented*, and studied for all sorts of reasons and purposes. The circle has been studied the most extensively. To collect all of its properties would require more than one book. In Sect. 4.1 we discuss only the most important properties of the circle in relation to other objects such as triangles and quadrilaterals. In Sect. 4.2 we introduce the trigonometric functions *sine* and *cosine*, which are closely related to the circle. The first application of these functions is the introduction of polar coordinates and the description of rotations of a coordinate system. A very special transformation of the plane, called *inversion*, is studied in Sect. 4.3. The properties of inversion will be of great help in solving some difficult problems related to the circle. Next we introduce the conic sections: the *parabola*, the *ellipse*, and the *hyperbola*. We will present various descriptions of the conics, including the notions of *focus*, *directrix*, and *directrix circle*. In Sect. 4.5 we use homogeneous coordinates to determine the equations of the tangents to a given conic. The discussion will include the definition of the projective plane, which is obtained by adding points at infinity to the plane, and the fact that in projective geometry the conic sections cannot be distinguished from one another. We conclude the chapter with a section on different types of *cycloids*. This includes a discussion of the *dynamics* of the Simson line and its relation with the nine-point circle and deltoid. We will also give a detailed account of the geometric ideas behind the cycloidal pendulum of Christiaan Huygens.

J.M. Aarts, *Plane and Solid Geometry*, DOI: 10.1007/978-0-387-78241-6_4,
© Springer Science+Business Media, LLC 2008

4.1 Circles, Powers, and Cyclic Quadrilaterals

After the straight line, the circle is the most common geometric figure. In this section we will present some of the most important properties of the circle. Let us first recall that for two points M and X, MX can denote any one of the following: the line through M and X, the line segment $[MX]$, and the distance $d(M, X)$ from M to X.

Definition 4.1. *Let M be a point in the plane and let r be a positive real number. The* circle $\odot(M, r)$ *is the set of all points whose distance to M is equal to r,*
$$\odot(M, r) = \{ X : MX = r \},$$
where MX is the shortened notation for $d(M, X)$. We call M the center *of the circle and r its* radius.

We will discuss properties of the circle in relation to points, lines, angles, triangles, and quadrilaterals, in this order. Let us first observe that the circle divides the plane into two parts. To describe this in more detail, we set a coordinate system at the center M of the circle. The circle $C = \odot(M, r)$ has equation
$$\|\mathbf{x}\| = r \quad \text{or} \quad x_1^2 + x_2^2 = r^2.$$
The *interior* $\operatorname{in} C$ and *exterior* $\operatorname{ex} C$ of circle C are given by
$$\operatorname{in} C = \{ \mathbf{x} : \|\mathbf{x}\| < r \} \quad \text{and} \quad \operatorname{ex} C = \{ \mathbf{x} : \|\mathbf{x}\| > r \}.$$
The plane is clearly the disjoint union of the circle C, $\operatorname{in} C$, and $\operatorname{ex} C$; see Fig. 4.1 (a).

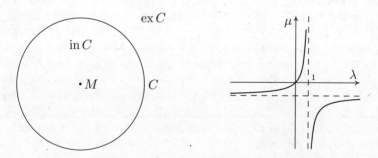

Fig. 4.1. (a) C, $\operatorname{in} C$, $\operatorname{ex} C$; (b) the graph of $\mu = \frac{\lambda}{1-\lambda}$

Let us now turn our attention to the *intersection* of a circle and a straight line. The following theorem gives some properties of such a figure. Though these are intuitively clear, their proofs are not always simple.

Theorem 4.2. *A circle $C = \odot(M, r)$ has the following properties:*

4.1 Circles, Powers, and Cyclic Quadrilaterals

1. The intersection of a line and C consists of zero, one, or two points.
2. The interior of C is convex.
3. If P and Q are distinct points of C, every point of the line segment $[PQ]$ other than P and Q belongs to the interior of C.
4. If the point P lies in the interior of C and the point Q is distinct from P, the intersection of the line PQ and the circle C consists of two points.
5. If P lies in the interior of C, and Q in its exterior, exactly one of the intersection points of the line PQ and the circle C belongs to the line segment $[PQ]$.

If the points P and Q lie on C, we call the line segment $[PQ]$ a *chord* of the circle; we call a chord $[PQ]$ a *diameter* if it passes through the center M of C.

Proof. We set a coordinate system at M. We denote the vectors corresponding to P and Q by $\mathbf{p}$ and $\mathbf{q}$, respectively. The equation of the circle C is $\|\mathbf{x}\| = r$, or $x_1^2 + x_2^2 = r^2$. Suppose that $\mathbf{p}$ and $\mathbf{q}$ are distinct. A point $\mathbf{x}$ on the line $\mathbf{pq}$ can be written as $\mathbf{x} = \lambda \mathbf{p} + (1-\lambda)\mathbf{q}$ for some $\lambda \in \mathbb{R}$. To find the intersection points of $\mathbf{pq}$ and C, we compute $f(\mathbf{x}) = \|\mathbf{x}\|^2 - r^2$ for a point $\mathbf{x}$ on $\mathbf{pq}$. The zeros of $f(\mathbf{x})$ correspond to the intersection points we are looking for. We obtain

$$\begin{aligned} f(\mathbf{x}) &= \|\lambda \mathbf{p} + (1-\lambda)\mathbf{q}\|^2 - r^2 \\ &= \lambda^2 \|\mathbf{p}\|^2 + 2\lambda(1-\lambda)\langle \mathbf{p}, \mathbf{q} \rangle + (1-\lambda)^2 \|\mathbf{q}\|^2 - r^2 \\ &= \lambda^2 f(\mathbf{p}) + 2\lambda(1-\lambda)\left(\langle \mathbf{p}, \mathbf{q} \rangle - r^2\right) + (1-\lambda)^2 f(\mathbf{q}) \,. \end{aligned} \quad (4.1)$$

As a function of λ, $f(x)$ has at most two real zeros; hence the line $\mathbf{pq}$ and the circle C have at most two intersection points.

Let us now show that in C is convex. Let $\mathbf{p}$ and $\mathbf{q}$ be two points in the interior of C, which consequently satisfy $f(\mathbf{p}) < 0$ and $f(\mathbf{q}) < 0$. By the Cauchy–Bunyakovskiĭ–Schwarz inequality, Theorem 1.34, $\langle \mathbf{p}, \mathbf{q} \rangle \leq \|\mathbf{p}\| \|\mathbf{q}\|$, whence $\langle \mathbf{p}, \mathbf{q} \rangle - r^2 < 0$. It then follows from (4.1) that $f(\mathbf{x}) < 0$ for every λ between 0 and 1. Hence $\mathbf{x} \in $ in C for every $\mathbf{x} \in [\mathbf{pq}]$, and we have shown that the interior of C is convex.

Let us now prove the third statement of the theorem. Let P and Q be distinct points on C; then $f(\mathbf{p}) = 0 = f(\mathbf{q})$ and $\langle \mathbf{p}, \mathbf{q} \rangle < r^2$ (Sect. 1.6, Exercise 1.39). It follows from (4.1) that every point between P and Q, that is, every point $\mathbf{x} = \lambda \mathbf{p} + (1-\lambda)\mathbf{q}$ with $0 < \lambda < 1$, lies in the interior of C.

Before proving the fourth statement, we first prove the last one. We divide both sides of (4.1) by $(1-\lambda)^2$ and set $\mu = \lambda/(1-\lambda)$, $\lambda \neq 1$. Figure 4.1 (b) shows the graph of μ as a function of λ. By examining the graph we see that the points for which $\mu > 0$ lie between P and Q, since $\mu > 0$ corresponds to $0 < \lambda < 1$, while the points for which $\mu < 0$ do not lie on $[PQ]$. Equation (4.1) can be rewritten as

$$\frac{f(\mathbf{x})}{(1-\lambda)^2} = \mu^2 f(\mathbf{p}) + 2\mu\left(\langle \mathbf{p}, \mathbf{q} \rangle - r^2\right) + f(\mathbf{q}) \,. \quad (4.2)$$

If $\mathbf{p} \in \operatorname{in} C$ and $\mathbf{q} \in \operatorname{ex} C$, we have $f(\mathbf{p}) < 0$ and $f(\mathbf{q}) > 0$. Since $f(\mathbf{p})$ and $f(\mathbf{q})$ have opposite signs, the discriminant of the right-hand side of (4.2) is positive; it follows that there are two real zeros. Moreover, the product of these two zeros is $f(\mathbf{p})f(\mathbf{q})$, so one must be positive and the other negative. It follows that exactly one of the intersection points lies on $[\mathbf{pq}]$.

Let us finally prove the fourth statement of the theorem. Since $\mathbf{p} \in \operatorname{in} C$, and therefore $f(\mathbf{p}) < 0$, the graph of the right-hand side of (4.2) is a concave parabola. For $\mu = -1$ the right-hand side equals $\|\mathbf{p} - \mathbf{q}\|^2$, which is positive. It follows that the right-hand side has two zeros; the corresponding points of $\mathbf{pq}$ are the intersection points of PQ and C.

Definition 4.3. *Let C be a circle. A line that intersects C at exactly one point is called a* tangent *to C. The intersection point is called the* point of contact.

For any point P on $C = \odot(M, r)$, the line l passing through P and perpendicular to MP is tangent to C; see Fig. 4.2 (a). We can prove this using

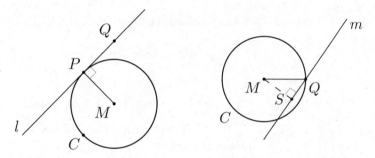

Fig. 4.2. (a) Tangent; (b) or not

the Pythagorean theorem: if Q is a point on l other than P, we have

$$MQ^2 = MP^2 + PQ^2 > r^2.$$

It follows that $Q \notin C$ and that l is a tangent. Conversely, every tangent to C can be obtained in this way. To prove this, suppose that a line m intersects C at Q, and that m is not perpendicular to MQ; see Fig. 4.2 (b). If S is the foot of the perpendicular from M on m, then by the Pythagorean theorem $MS < r$, whence S lies in the interior of C. By the previous theorem, it follows that m has two intersection points with C and is therefore not a tangent.

Circles and Angles

Before we can discuss the properties of circles and angles, we need to extend Basic Assumption 2.12. Consider Fig. 4.3, on the left. By the basic assumption in question, the intersection of angle P and the circle $\odot(P, 1)$ is an arc.

Theorem 4.2 tells us that this arc has exactly one point in common with each leg of the angle; these are the points A and B. We sometimes refer to the resulting arc as arc AB. By definition, the length of arc AB is equal to $\angle P$. Since the circle $\odot(P, R)$ is the image of $\odot(P, 1)$ under the central dilation with center P and scale factor R, it seems logical that the intersection of $\odot(P, R)$ and angle P is an arc of length $R \times \angle P$. At this point we cannot measure

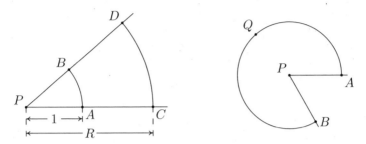

Fig. 4.3. The relation between angles and arcs

arc lengths. Consequently, we cannot prove this last statement, and therefore choose to modify Basic Assumption 2.12 as follows.

Basic Assumption 4.4. *For any angle P, the intersection of angle P and the circle $\odot(P, r)$ is an arc. For $r = 1$ the length of this arc is denoted by $\angle P$; if $r = R$, the length of this arc equals $R \times \angle P$. Furthermore, Properties 1 through 4 of Basic Assumption 2.12 hold.*

The length of arc AB, Fig. 4.3 on the left, is equal to $\angle P$, and that of arc CD is equal to $R \times \angle P$. It is often convenient to use the *angle measure of an arc* rather than the length of the arc. The *angle measure* of arc CD is defined to be $\angle P$, and is denoted by $\widehat{CD}$. The following equivalent equalities hold:

$$\text{length of arc } CD = R \times \widehat{CD} \quad \text{and} \quad \widehat{CD} = \frac{\text{length of arc } CD}{R}.$$

Since P is the center of the circles whose arcs we considered above, we say that P is a *central angle*. The arcs AB and CD are called *intercepted arcs* of angle P. It follows from the definition that *a central angle equals its intercepted arc*, or, more precisely, *the angle measure of a central angle is equal to the angle measure of its intercepted arc*.

Property 4 of Basic Assumption 2.12 concerns the additivity of the angle measure. By the above, this implies the additivity of the angle measures of intercepted arcs. This observation motivates the slight change we want to make to the definition of angle measure. For the moment this measure is defined only for angles that do not exceed π. Since the arc length is defined using the angle measure, this restricts the choice of arcs we can consider. Consider

Fig. 4.3, on the right: arc AQB is the union of the arcs AQ and QB that have only the point Q in common. We would like the length of arc AQB to be the sum of the lengths of the arcs AQ and QB. If the sum of their angle measures is less than π, this follows from property 4 mentioned above. If the sum is greater than π, we define the angle measure of AQB to be the sum of the angle measures of arcs AQ and QB, where we assume that neither of these exceeds π. We can prove that this definition of the angle measure of arc AQB does not depend on the way it is represented as the union of two adjacent arcs, i.e., arcs that have only one point in common.

If the vertex P of an angle lies on a circle and the legs of that angle meet the circle at A and B, we call angle P an *inscribed angle*, and we call the arc from A to B inside this angle the *intercepted arc* of angle P. We use the same terminology if one of the legs of angle P is tangent to the circle; see Fig. 4.5, on the right. There is a simple relation between the angle measure of a central angle and that of an inscribed angle with the same intercepted arc: the central angle measures twice the inscribed angle. This is a corollary of the following theorem.

Theorem 4.5. *The angle measure of an inscribed angle is half the angle measure of its intercepted arc.*

This theorem is sometimes stated less accurately as *an inscribed angle is equal to half its intercepted arc*.

Proof. We need to distinguish several cases. First we assume that the center M of the circle lies on one of the legs of the inscribed angle P; see Fig. 4.4, on the left. We call the intersection points of the legs of angle P and the circle A

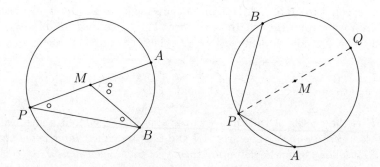

Fig. 4.4. The inscribed angle equals half its intercepted arc

and B. Since $\triangle PMB$ is isosceles, both $\angle BPM$ and $\angle MBP$ are equal to half the exterior angle BMA. The latter equals $\widehat{BA}$, so $\angle BPA = \widehat{BA}/2$.

Next we assume that the center M lies in the interior of the inscribed angle P; see Fig. 4.4, on the right. By what we have just proved, $\angle APQ = \widehat{AQ}/2$ and $\angle QPB = \widehat{QB}/2$. Adding these together gives $\angle APB = \widehat{AB}/2$.

The proofs of the remaining cases are simple. If the center M of the circle lies in the exterior of the inscribed angle P, Fig. 4.5 on the left, we write $\angle BPA$

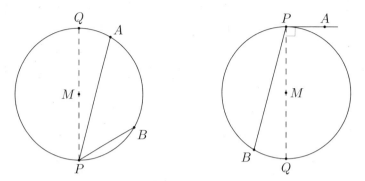

Fig. 4.5. The inscribed angle equals half its intercepted arc

as the difference between $\angle BPQ$ and $\angle APQ$ and complete the proof as in the previous case.

Finally, if one of the legs of the inscribed angle P is tangent to the circle, we prove the statement of the theorem in the same way; see Fig. 4.5, on the right. We only need to remark that the measure of the right angle $\angle APQ$ is half the angle measure of the half circle, arc PQ.

Corollary 4.6. *The angle measure of an inscribed angle is half the angle measure of a central angle on the same arc.*

Let P be a point outside the circle $\odot(M,r)$; see Fig. 4.6, on the left. If the legs of angle P meet $\odot(M,r)$ at A and B, and C and D, respectively, we have

$$\angle P = \tfrac{1}{2}\left(\widehat{BD} - \widehat{AC}\right) . \tag{4.3}$$

To prove this, we first draw the line AD. Since $\angle BAD$ is an exterior angle of $\triangle PAD$, $\angle APD$ is equal to the difference between $\angle BAD$ and $\angle PDA$. Equality (4.3) now follows from Theorem 4.5.

Now suppose that the vertex P lies inside the circle, Fig. 4.6 on the right, and that the legs of angle P meet the circle $\odot(M,r)$ at A and B. Let C and D be the intersection points of the extensions of PA and PB, respectively, and $\odot(M,r)$. In this case we have

$$\angle P = \tfrac{1}{2}\left(\widehat{AB} + \widehat{CD}\right) . \tag{4.4}$$

To prove this we draw the line AD and consider the exterior angle APB of triangle ADP.

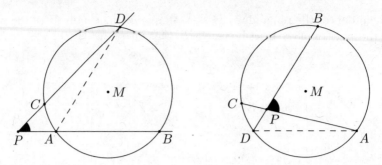

Fig. 4.6. The size of angles with vertex in the exterior or interior of the circle

The Power of a Point and the Radical Axis

In the proof of Theorem 4.2 we introduced a function f; see (4.1). We will now explain its geometric meaning.

Definition 4.7. *The* power *of a point P with respect to the circle $\odot(M,r)$ is the number $PM^2 - r^2$, where $PM = d(P,M)$.*

To every point **x**, the function f defined in (4.1) associates the power of **x** with respect to $\odot(\mathbf{o},r)$. It is clear that the power of P with respect to the circle $C = \odot(M,r)$ is greater than 0, equal to 0, or less than 0 according to whether the point P belongs to ex C, C, or in C, respectively.

Theorem 4.8. *If a line through a point P intersects the circle $\odot(M,r)$ at the points A and B, the product $PA \times PB$ equals the absolute value of the power of P with respect to $\odot(M,r)$.*

Proof. If P lies on the circle, either PA or PB is equal to 0; in this case the theorem holds. Let us assume that P lies outside the circle; see Fig. 4.7, on the left. We will use Theorem 4.5. The angles CBA and CDA have the

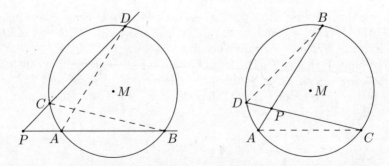

Fig. 4.7. The power of P with respect to $\odot(M,r)$

same measure, because both are inscribed angles with intercepted arc CA. By

Theorem 2.50, we conclude that triangle PBC is similar to triangle PDA. It follows that $PC : PB = PA : PD$, and therefore $PC \times PD = PA \times PB$. Consequently, the product $PA \times PB$ does not depend on the position of the line l through P. Choosing the line l through M, and P, we can easily show that the product is equal to the power of P with respect to the circle.

If the point P lies in the interior of the circle, Fig. 4.7 on the right, we show that the triangles APC and DPB are similar. The proof can then be completed in the same way as above.

Example 4.9 (Radical Axis). The *radical axis* of the circles $\odot(M, R)$ and $\odot(N, r)$, where $M \neq N$, is the set of all points whose powers with respect to both circles are equal. As we will see, the radical axis is a straight line. We choose local coordinates in such a way that $M = \mathbf{m}$ and $N = \mathbf{n}$ are on the x_1-axis; Fig. 4.8. The powers of the point $\mathbf{x}$ with respect to the circles

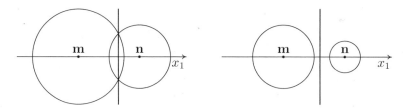

Fig. 4.8. The power of two circles

are equal to each other if and only if $\|\mathbf{x} - \mathbf{m}\|^2 - R^2 = \|\mathbf{x} - \mathbf{n}\|^2 - r^2$. This condition is equivalent to

$$x_1 = \frac{R^2 - r^2 + n_1^2 - m_1^2}{2(n_1 - m_1)}.$$

It follows that the radical axis is a straight line that is perpendicular to the line joining the centers of the circles. If the circles intersect each other in more than one point, the intersection points lie on the radical axis; they are also the intersection points of the radical axis with either circle. It follows that the circles cannot meet in more than two points.

Example 4.10 (Tangent). Let P be a point in the exterior of a circle $C = \odot(M, r)$. Let K be the circle with diameter PM; it meets C at two points, Q and R, as suggested in Fig. 4.9. We can also show that K intersects C in two points through computations using the radical axis. We will show that there are exactly two lines through P that are tangent to C. Since PM is a diameter of K, the angles MQP and MRP are both equal to $\pi/2$. It follows that PQ and PR are tangent to C. Let us show that they are the only lines through P that are tangent to C. Assume that PS is tangent to C for some $S \in C$. By Theorem 4.8, PS^2 is equal to the power of P with respect to C.

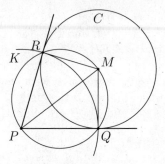

Fig. 4.9. The tangents to C through P

By the Pythagorean theorem we then have $PS = PQ = PR$. Hence, like Q and R, S lies on the circle with center P and radius PQ. This circle is drawn partially in Fig. 4.9. Since two circles have at most two intersection points, S must coincide with either Q or R.

Circles and Triangles

Let A, B, and C be three noncollinear points. There is exactly one point that is equidistant from all three: the intersection point of the perpendicular bisectors of AB, BC, and CA (Sect. 1.4, Exercise 1.20). This implies that for any triangle, exactly one circle passes through all vertices. This circle is called the *circumscribed circle* or *circumcircle* of the triangle ABC; see Fig. 4.10, on the left.

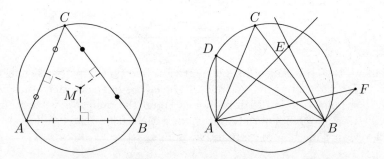

Fig. 4.10. The circumcircle of triangle ABC

For the proof of the next theorem we refer to the figure on the right. We only consider points in the same half-plane of the line AB as C. By Theorem 4.5, a point D on the arc BCA that lies in this half-plane satisfies $\angle ADB = \angle ACB$. For a point E inside the circle, (4.4) implies that $\angle AEB > \angle ACB$. Finally, by (4.3), a point F outside the circle satisfies $\angle AFB < \angle ACB$. This proves the following theorem.

Theorem 4.11. *Let ABC be a triangle. The set of all points X that are on the same side of AB as C and satisfy $\angle AXB = \angle ACB$ is the arc $\overset{\frown}{BCA}$.*

We obtain the following theorem of Thales (ca. 639–ca. 546) as a corollary of this theorem and Theorem 4.5.

Corollary 4.12 (Thales). *For any line segment AB, the set of all points X such that $\angle AXB = \pi/2$ is a circle with diameter AB.*

As an application of the corollary we derive an equation relating the area of a triangle, the radius of its circumcircle, and the lengths of its sides.

Example 4.13. Let $\odot(M, R)$ be the circumcircle of $\triangle ABC$; see Fig. 4.11 (a). Let D be the foot of the altitude from C, and let E be the second endpoint of the diameter through C. The inscribed angles BAC and BEC are equal to each other because they intercept the same arc. Since $[EC]$ is a diameter, $\angle CBE = \pi/2$. The triangles ADC and EBC are therefore similar. It follows that $AC : EC = CD : BC$, whence $ba = 2R \times CD$, where a is the length of the side opposite A, and so on. Denoting the area of triangle ABC by O, we have $CD = 2O/c$, and finally $4RO = abc$.

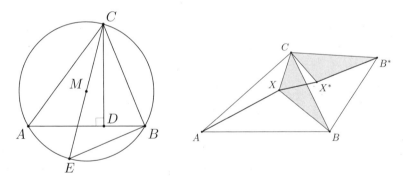

Fig. 4.11. Figures related to the examples: (a) Example 4.13; (b) Example 4.14

Example 4.14. Let ABC be a triangle whose angles are all less than $2\pi/3$; see Fig. 4.11 (b). For a point X of the interior of the triangle we let

$$f(X) = XA + XB + XC.$$

A natural question is whether f has a minimum, and if it does, whether it is unique, and how we can find it. This problem has an elegant and surprisingly simple solution. On the side BC we draw an equilateral triangle BCB^* outward. Let X be an arbitrary point inside ABC. We apply the rotation $\mathcal{R}$ over an angle $\pi/3$ around the point C to the triangle CXB so that B ends up at B^*. We call X^* the image of X; see Fig. 4.11 (b). Since $\angle XCX^* = \pi/3$ and $CX = CX^*$, triangle XCX^* is equilateral, whence $CX = XX^*$. It follows that $f(X)$ gives the length of the *broken line* AXX^*B^*. If the point X can be chosen in such a way that this broken line is a straight line segment, we will have found a minimum for f. The broken line AXX^*B^* can be a line segment only if $\angle CXA = 2\pi/3$, because $\triangle CXX^*$ is equilateral and $\angle CXX^* = \pi/3$. Consequently, the function f has a minimum at the unique point Q on $[AB^*]$ for which $\angle CQA = 2\pi/3$. This is the intersection point of the line segment $[AB^*]$ and the circumcircle of the equilateral triangle drawn outward on CA. We can easily verify that $\mathcal{R}(Q)$ lies on $[AB^*]$, so that AQQ^*B^* is indeed a line segment. It follows that $f(Q)$ is minimal, and also that Q is the only point with this property. The point Q is called the *Torricelli point*.

The uniqueness of Q has an interesting corollary: if CAC^* and ABA^* are the equilateral triangles that were erected outward on CA and AB, respectively, then the segments $[AB^*]$, $[BC^*]$, and $[CA^*]$ have equal lengths, are concurrent at Q, and the angles at Q between any two of them are all $\pi/3$.

In Example 2.22 we showed that the three interior angle bisectors of a triangle ABC are concurrent. The common point I of the bisectors is equidistant from the sides of the triangle; let r denote the distance. The circle $\odot(I, r)$ is tangent to the three sides of the triangle; we call it the *inscribed circle* or *incircle* of triangle ABC; see Fig. 4.12. The figure on the left seems to suggest

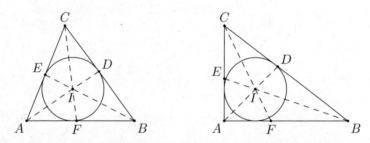

Fig. 4.12. The incircle of triangle ABC

that the incircle touches the sides of the triangle at the feet of the angle bisectors, but the figure on the right shows that this is not the case in general. Let us examine the position of the points of contact more closely; see Fig. 4.13, on the left. The points D, E, and F now denote the points of contact of the triangle and its incircle. Since the power of A with respect to $\odot(I, r)$ equals both AE^2 and AF^2, we have $AE = AF$. Likewise, $CD = CE$ and $BD = BF$. Denoting, as usual, the side opposite a vertex by the corresponding lowercase letter and half the perimeter of $\triangle ABC$ by s, we obtain $AF = s - a$, $BD = s - b$, and $CE = s - c$. In the figure on the right we have drawn the joins of each vertex with the point of contact on the opposite side: AD, BE,

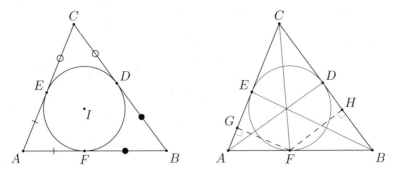

Fig. 4.13. The points of contact of the incircle

and CF. We will show that these lines are concurrent; their intersection point is named after *Gergonne* (1771–1859). We will prove the concurrence using a method based on Ceva's theorem (ca. 1647–ca. 1734); see also [4]. The areas of the triangles AFC and FBC are in the same ratio as the lengths $s-a$ and $s-b$ of their bases. Let FG and FH be the perpendiculars from F on AC and BC, respectively. The area of $\triangle AFC$ equals $(FG/2) \times b$ and that of $\triangle FBC$ equals $(FH/2) \times a$. It follows that $FH/FG = b(s-b)/a(s-a)$. Using similarity, we can show that for every point P of line CF, the ratio of the distances from P to the sides a and b is equal to $b(s-b)/a(s-a)$. This ratio characterizes the points of CF among the points in the interior of $\triangle ABC$. In the same way, we can show that a point Q of triangle ABC lies on AD if and only if the ratio of the distances from Q to the sides b and c is equal to $c(s-c)/b(s-b)$. The ratio of the distances from the intersection point X of CF and AD to the sides a and c consequently equals $c(s-c)/a(s-a)$. In the same way as above, we can show that this ratio characterizes the points of BE inside the triangle. Consequently the three lines AD, BE, and CF concur at X.

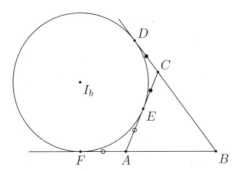

Fig. 4.14. The excircle

The interior bisector of angle B of triangle ABC and the exterior bisectors of angles A and C are concurrent. Their common point is denoted by I_b. It is equidistant from the sides of the triangle; see Fig. 4.14 (see also Sect. 2.3, Exercise 2.20). Let r denote the distance. The circle $\odot(I_b, r_b)$ is tangent to the three sides; we call it the *excircle* of triangle ABC at side b.

If the points of contact of the excircle and the (extended) sides are successively denoted by D, E, and F, we can show in a way similar to the above that $AF = AE$ and $CE = CD$. Since $BD = BF = s$, it follows that $AE = s - c$ and $CE = s - a$.

Cyclic Quadrilaterals

Let us now turn to the relation between circles and quadrilaterals. We consider a quadrilateral $ABCD$ such that the segments $[AC]$ and $[BD]$ have a point in common. We call such quadrilaterals *admissible*. The quadrilateral in Fig. 4.15 (**a**) is not admissible. Figure 4.15 (**b**) shows a cyclic quadrilateral; this quadrilateral is admissible. An admissible quadrilateral $ABCD$ is called *cyclic* if its vertices lie on a circle. A cyclic quadrilateral is sometimes called a *quadrilateral of chords*. The angles BAD and DCB of an admissi-

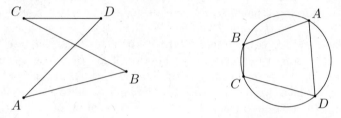

Fig. 4.15. (a) An inadmissible quadrilateral; (b) a cyclic quadrilateral

ble quadrilateral $ABCD$ are called *opposite angles*; CBA and ADC are also opposite angles. Since all angles of a cyclic quadrilateral are inscribed angles, one direction of the following theorem is a consequence of Theorem 4.5.

Theorem 4.15. *An admissible quadrilateral is cyclic if and only if the angle measures of a pair of opposite angles add up to π.*

Proof. We need to prove only the *if* part. Without loss of generality, we may, and do, assume that the angle measures of BAD and DCB add up to π. Let E be a point on the circumcircle of $\triangle ABD$ on the opposite side of BD from A. The sum of the angle measures of DEB and BAD is equal to π. Consequently, the angles DCB and DEB are equal. By Theorem 4.11 it follows that C and E lie on the arc BD, on the same side of the line BD. This shows that the quadrilateral $ABCD$ is cyclic.

The following theorem, named after Miquel, gives a simple, but at the same time surprising, application of cyclic quadrilaterals.

4.1 Circles, Powers, and Cyclic Quadrilaterals   159

Theorem 4.16. *If P_1, P_2, and P_3 are points on the sides BC, CA, and AB, respectively, of a triangle ABC, then the circumcircles of the triangles AP_3P_2, BP_1P_3, and CP_2P_1 are concurrent.*

Proof. Consider Fig. 4.16. The circumcircles of the triangles AP_3P_2 and BP_1P_3 meet at the points P_3 and P. In this proof we assume that P lies in the interior of ABC. The cases that $P = P_3$ and that P lies outside the triangle can be dealt with in a similar way. Since AP_3PP_2 is a cyclic quadrilateral, $\angle PP_3A$ and $\angle AP_2P$ are supplementary angles; hence $\angle PP_3A$ equals $\angle PP_2C$. Since BP_1PP_3 is also a cyclic quadrilateral, a similar argument shows that $\angle BP_3P = \angle CP_1P$. Since angles PP_3A and BP_3P are also supplementary, the same holds for angles PP_2C and CP_1P. It follows that CP_2PP_1 is a cyclic quadrilateral. Hence P lies on the circumcircle of triangle CP_2P_1.

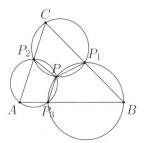

Fig. 4.16. Miquel's theorem

The proofs of the following theorems use cyclic quadrilaterals. We will take a closer look at the result of the second theorem in Sect. 4.6.

Theorem 4.17. *Let X be a point on the circumcircle of a triangle ABC. The feet of the perpendiculars from X on the sides of ABC are collinear.*

The line through the feet is usually named after Simson (1687–1768). However, the line is not even mentioned in Simson's work. The line was discovered by Wallace (1768–1843) [11, p. 41].

Proof. Let J, K, and L be the feet of the perpendiculars from X on the sides BC, CA, and AB, respectively; see Fig. 4.17, on the left. In order to prove that these points are collinear, we will show that $\angle KJC = \angle LJB$. Since the angles BLX and BJX are right angles, the quadrilateral $BXJL$ is cyclic. Consequently, $\angle LJB = \angle LXB$, because these inscribed angles have the same intercepted arc on the circumcircle of $BXJL$. Likewise, since the angles CKX and CJX are right angles, the quadrilateral $JXKC$ is cyclic, whence $\angle KJC = \angle KXC$. To complete the proof it suffices to show that $\angle KXC = \angle LXB$. Without loss of generality, we may, and do, assume that X

is a point of the arc BC. Consequently, the angles $\angle CXB$ and $\angle BAC$ add up to π. It follows that

$$\angle LXB = \angle CXB - \angle CXL = \pi - \angle BAC - \angle CXL.$$

In a similar fashion it follows that the quadrilateral $ALXK$ is cyclic and that the sum of the angle measures of BAC and KXL is π. It follows that

$$\angle KXC = \angle KXL - \angle CXL = \pi - \angle BAC - \angle CXL,$$

which implies the equality $\angle KXC = \angle LXB$.

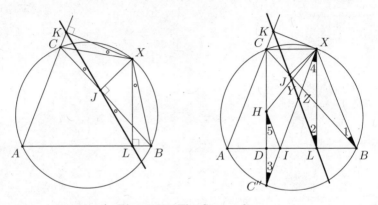

Fig. 4.17. The Simson line

Every point X of the circumcircle of triangle ABC has a Simson line. In two cases we obtain well-known lines: if X coincides with one of the vertices, the Simson line of X is the altitude through that vertex; if X is diametrically opposite one of the vertices, the Simson line of X is the side of the triangle opposite that vertex. In the following theorem we take a closer look at the position of the Simson line.

Theorem 4.18. *Let H denote the orthocenter of triangle ABC; assume that $C \neq H$. Let C'' be the second intersection point of CH and the circumcircle. The Simson line of a point X on the circumcircle bisects the line segment $[HX]$. The line $C''C$ makes the same angle with $C''X$ as it does with the Simson line of X.*

Proof. Figure 4.17, on the right, represents the situation of the theorem. Let us first recall that the nine-point circle is the image of the circumcircle under the central dilation with center H and scaling factor $1/2$; see Example 2.46. Since the foot D of the altitude through C lies on the nine-point circle, we have $DC'' = DH$. Let J, K, and L be the feet of the perpendiculars from X on BC, CA, and AB, respectively. Let Y be the intersection point of $[HX]$

and KL, the Simson line of X. For simplicity we use the numbers introduced in Fig. 4.17 to indicate the angles that play a role in this proof. Since $LBXJ$ is a cyclic quadrilateral, angles 1 and 2 are equal. Angles 1 and 3 are equal because both are inscribed angles with intercepted arc CX. Furthermore, angles 3 and 4 are equal because they are alternate interior angles of the parallel lines CC'' and XL for the transverse line $C''X$. It follows that angles 2 and 4 of triangle XLZ are equal. Since triangle ILX has a right angle at L, the line LZ is a median of this triangle, whence $IZ = ZX$. It follows from the observation at the beginning of the proof that the triangles $DC''I$ and DHI are congruent. Consequently, angles 3 and 5 are also equal. The resulting equality of angles 2 and 5 implies that the lines HI and KL are parallel. Since $IZ = XZ$, we conclude that $HY = YX$; in other words, Y is the midpoint of the segment $[HX]$.

Exercises

4.1. On each side of a triangle ABC, we draw an equilateral triangle outward. The circumcircles of the resulting triangles have a common point.

4.2. Let AD be a median of $\triangle ABC$. If $AD = BC/2$, then $\angle BAC = \pi/2$.

4.3. Let $C = \odot(M, r)$ and let P be a point in ex$C$.

(a) Let Q be point of C. If PQ is tangent to C, then Q lies on the circle with diameter $[PM]$.

(b) The intersection points of C and the circle with diameter $[PM]$ are the points of contact of the tangents of C passing through P.

(c) Let R and S be the intersection points found in (b); then $PR = PS$.

4.4. Circle of Apollonius. Given a line segment $[AB]$ with $A \neq B$ and a positive real number k other than 1, we examine properties of the set

$$K = \left\{ X : \frac{AX}{BX} = k \right\}.$$

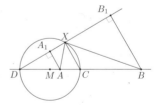

(a) There is exactly one point C of K on the line segment $[AB]$. The line AB also contains a second point D of K.

(b) If $X \in K$, XC is the interior bisector of angle AXB and XD is its exterior bisector.
   Hint: Compare to Sect. 2.3, Exercise 2.3.

(c) Let M be the midpoint of $[DC]$. Any point of K lies on $\odot(M, MC)$.

(d) $K = \odot(M, MC)$. Hint: Let A_1 and B_1 be the feet of the perpendiculars from A and B, respectively, on DX. Show that $\triangle AA_1X$ is similar to $\triangle BB_1X$.

4.5. Irrationality of $\sqrt{2}$. Let $ABCD$ be a square with side of length a. Let b be the length of the diagonal AC. Applying the Pythagorean theorem, we obtain $b = a\sqrt{2}$. The Greeks (the school of Pythagoras) discovered that there is no unit of length that can be used to measure both a and b: there is no segment of length p such that $a = mp$ and $b = np$ for positive integers m and n. In other words, it is impossible to represent b/a as a quotient m/n of positive integers m and n: $\sqrt{2}$ is irrational.

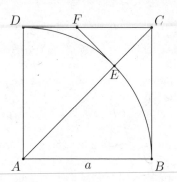

In the figure, EF is a tangent of $\odot(A, a)$.
(a) Prove that $EC = EF = FD$.
(b) Prove that $FC/EC = b/a$.
(c) The irrationality of $\sqrt{2}$ is proved by contradiction. Suppose that there exist positive integers m and n such that $b/a = n/m$. We may, and do, assume that m is minimal. The equality means that there is an integer p such that $a = mp$ and $b = np$. Show that it follows from (a) and (b) that $b/a = (2m - n)/(n - m)$. The denominator of the last fraction is less than m, which gives a contradiction. It follows that $\sqrt{2}$ is irrational.
(d) With hindsight we could prove the irrationality of $\sqrt{2}$ without making use of geometric arguments. We could start as follows: if $n/m = \sqrt{2}$, then $(2m - n)/(n - m) = n/m$, where $n - m < m$.

4.6. Let D, E, and F be the points of contact of the three excircles of a triangle and its sides a, b, and c, respectively. Show that the lines AD, BE, and CF are concurrent. The intersection point is called the *Nagel point*.

4.7. Let $ABCD$ be a cyclic quadrilateral, and let E and F be the orthocenters of $\triangle ABD$ and $\triangle BCD$, respectively. Then $BEDF$ is also a cyclic quadrilateral.

4.8. We call an admissible quadrilateral *tangential* if there exists a circle that is tangent to all four sides. Denoting the successive lengths of the sides of the quadrilateral by a, b, c, and d, we can show that the quadrilateral is tangential if and only if $a + c = b + d$.
Hint: For the *only if* part of the proof see Fig. 4.13, on the left. The *if* part is rather complicated. First we draw a circle that is tangent to three sides of the quadrilateral. We then assume that the fourth side is not tangent to the circle and try to deduce a contradiction.

4.9. Suppose that we are given a point X such that the feet of the perpendiculars from X to the sides of a triangle ABC are collinear. Show that X lies on the circumcircle of ABC.

4.10. Let X be one of the midpoints of the arcs BC on the circumcircle of triangle ABC. Show that the Simson line $\ell(X)$ of X is perpendicular to XA.

4.11. Given a triangle ABC, draw triangles outward on the sides in such a way that $\triangle AFB \sim \triangle DCB \sim \triangle ACE$. In particular, the angle at A in AFB is equal to the angle at D in DCB and to the angle at A in ACE. Show that the lines AD, BE, and CF are concurrent.
Hint: The intersection point is also the intersection point of the circumcircles of the three triangles erected on the sides.

4.12. Let P be a point on the circle $\odot(M, r)$ with diameter $[AB]$. Let Q be the foot of the perpendicular from P on AB.
(a) Show that $PQ^2 = AQ \times QB$.
(b) Let $a = AQ$ and $b = BQ$. We call $(a+b)/2$ the *arithmetic mean* of a and b, and $\sqrt{ab}$ their *geometric mean*. Prove that $\sqrt{ab} \leq (a+b)/2$ and discuss when equality holds.

4.2 Trigonometric Functions and Polar Coordinates

Trigonometric functions are functions from $\mathbb{R}$ to $\mathbb{R}$ whose argument is an angle measure. The most important ones are the *sine*, the *cosine*, and the *tangent*. Quite often these functions can be used for solving geometric problems. In this section we work with oriented angles: an angle has a positive (negative) angle measure if the angle is positively (negatively) oriented.

Definition 4.19. We consider the circle $\odot(\mathbf{o}, 1)$ in $\mathbb{R}^2$. Let $\mathbf{a} = (1, 0)$, and assume that $-\pi < \varphi \leq \pi$. Let $\mathbf{p} = (p_1, p_2)$ be the point on $\odot(\mathbf{o}, 1)$ with $\angle \mathbf{aop} = \varphi$. Then by definition, $\cos \varphi = p_1$ and $\sin \varphi = p_2$. For other values of the argument φ, the values of the sine and cosine functions are defined by periodic extension with period 2π.

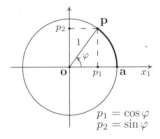

The periodic extension with period 2π means that for an angle ψ we first compute the value $\varphi \in (-\pi, \pi]$ for which $|\varphi - \psi|$ is an integer multiple of 2π. The sine and cosine of ψ are then given by $\sin \psi = \sin \varphi$ and $\cos \psi = \cos \varphi$. Geometrically, it is as if we wrapped the real line around the circle as in Fig. 4.18.

One should keep in mind that in Definition 4.2, $\angle \mathbf{aop}$ is the length of the circular arc between the points $\mathbf{a}$ and $\mathbf{p}$. When we wrap $\mathbb{R}$ around $\odot(\mathbf{o}, 1)$, the point 0 is sent to $\mathbf{a}$, and the points ψ and φ are both sent to the same point $\mathbf{q} = (q_1, q_2)$. We have

$$\cos \psi = \cos \varphi = \cos \angle \mathbf{aoq} = q_1 \quad \text{and} \quad \sin \psi = \sin \varphi = \sin \angle \mathbf{aoq} = q_2.$$

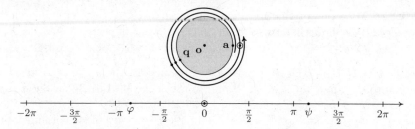

Fig. 4.18. Wrapping the real line $\mathbb{R}$ around $\odot(o, 1)$

Since the radius of the circle is equal to 1, it is obvious that

$$\sin^2 \varphi + \cos^2 \varphi = 1 \text{ for all } \varphi \text{ in } \mathbb{R}.$$

Figure 4.19 gives the graphs of the sine and cosine functions. The following

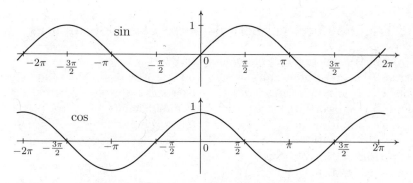

Fig. 4.19. The graphs of the sine and cosine functions

formulas involving the sine and cosine functions can be verified on the graphs: for every φ, we have

$$-1 \leq \sin \varphi \leq 1, \quad \sin(\varphi + \pi) = -\sin \varphi, \quad \sin \varphi = \cos(\tfrac{\pi}{2} - \varphi),$$
$$-1 \leq \cos \varphi \leq 1, \quad \cos(\varphi + \pi) = -\cos \varphi, \quad \cos \varphi = \sin(\tfrac{\pi}{2} - \varphi),$$
$$\sin(-\varphi) = -\sin \varphi, \quad \sin(\pi - \varphi) = \sin \varphi, \quad \sin \varphi = -\cos(\tfrac{\pi}{2} + \varphi),$$
$$\cos(-\varphi) = \cos \varphi, \quad \cos(\pi - \varphi) = -\cos \varphi, \quad \cos \varphi = \sin(\tfrac{\pi}{2} + \varphi).$$

Another common trigonometric function is the *tangent function*; it is defined by the formula $\tan \varphi = \sin \varphi / \cos \varphi$, for all φ outside the set $\{\pi/2 + k\pi : k \in \mathbb{Z}\}$ of zeros of the cosine function. The graph is shown in Fig. 4.20 **(b)**. The tangent function is periodic with period π:

$$\tan(\varphi + \pi) = \tan \varphi$$

for all values of φ for which the function tan is defined. For the applications in geometry, it is important to view the trigonometric functions as ratios of lengths of the sides of a right triangle. This concerns angles φ in the open interval $(0, \pi/2)$. Figure 4.20 (a) shows a triangle MAB with right angle

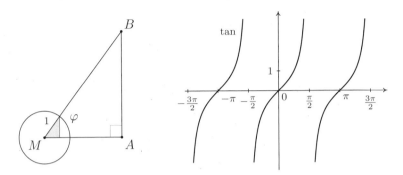

Fig. 4.20. (a) Interpretation of the trigonometric functions; (b) the graph of the tangent function

at A, and a small shaded triangle with hypotenuse of length 1 that is similar to MAB. Triangle MAB is the image of the shaded triangle under a central dilation with scaling factor MB. Setting $\varphi = \angle AMB$, it now follows from Definition 4.2 that

$$\sin\varphi = \frac{AB}{MB}, \quad \cos\varphi = \frac{MA}{MB}, \quad \tan\varphi = \frac{AB}{MA}.$$

Using these formulas we can make the following table:

$\sin 0 = 0$, $\quad \sin\frac{\pi}{6} = \frac{1}{2}$, $\quad \sin\frac{\pi}{4} = \frac{\sqrt{2}}{2}$, $\quad \sin\frac{\pi}{3} = \frac{\sqrt{3}}{2}$, $\quad \sin\frac{\pi}{2} = 1$,

$\cos 0 = 1$, $\quad \cos\frac{\pi}{6} = \frac{\sqrt{3}}{2}$, $\quad \cos\frac{\pi}{4} = \frac{\sqrt{2}}{2}$, $\quad \cos\frac{\pi}{3} = \frac{1}{2}$, $\quad \cos\frac{\pi}{2} = 0$,

$\tan 0 = 0$, $\quad \tan\frac{\pi}{6} = \frac{\sqrt{3}}{3}$, $\quad \tan\frac{\pi}{4} = 1$, $\quad \tan\frac{\pi}{3} = \sqrt{3}$.

Example 4.20 (Law of Sines). Let $\odot(M, R)$ be the circumcircle of triangle ABC; see Fig. 4.21 (a). We denote the lengths of the sides opposite the vertices A, B, and C by a, b, and c, respectively, and the angles at these vertices by α, β, and γ, respectively. We use the same notation for the angle measures. For the sake of simplicity, we assume that all angles are acute. Let F be the second endpoint of the diameter of $\odot(M, R)$ through C. The angles BAC and BFC intercept the same arc and are therefore equal. By Thales' theorem, $\angle CBF = \pi/2$. In triangle CBF, the sine of $\angle BFC$ equals the ratio $a/2R$. It follows that $2R = a/\sin\alpha$. Applying the same reasoning to the other vertices, we obtain the *law of sines*:

$$\frac{a}{\sin\alpha} = \frac{b}{\sin\beta} = \frac{c}{\sin\gamma}.$$

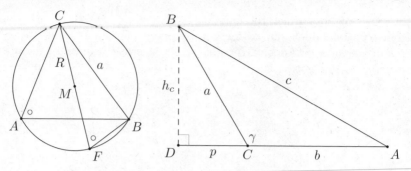

Fig. 4.21. (a) The law of sines; (b) the law of cosines

Example 4.21 (Law of Cosines). Let ABC be a triangle. We use the notation of the previous example. The *law of cosines* states that

$$c^2 = a^2 + b^2 - 2ab\cos\gamma \; .$$

For $\gamma = \pi/2$ this is the Pythagorean theorem. For the proof of the law of cosines, we assume that $\gamma > \pi/2$; see Fig. 4.21 (**b**). The proof for an acute angle γ is similar. By applying the Pythagorean theorem twice, we obtain

$$c^2 = (b+p)^2 + h_c^2 = b^2 + 2bp + \underbrace{p^2 + h_c^2}_{} = b^2 + 2bp + a^2 \; .$$

We conclude by observing that $p = a\cos(\pi - \gamma) = -a\cos\gamma$.

Polar Coordinates

The position of a point X in the coordinate plane is given by its *position vector* $\mathbf{x} = (x_1, x_2)$. The components x_1 and x_2 of $\mathbf{x}$ are sometimes called the *Cartesian coordinates* of X; the name refers to Descartes (1596–1650), who first showed how algebraic methods can be used to solve geometric problems [14]. With his work, Descartes laid the foundations of *analytic geometry*.

For many problems it is more convenient to use *polar coordinates* rather than Cartesian coordinates; see Fig. 4.22, on the left. The polar coordinates of

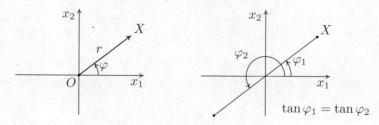

Fig. 4.22. Polar coordinates

a point X other than O with Cartesian coordinates (x_1, x_2) are (r, φ), where r is the distance from X to the origin $O = \mathbf{o}$ of the Cartesian coordinate system, and φ is the positively oriented angle between the positive x_1-axis and OX. In a formula, we have

$$r = \|\mathbf{x}\| = \sqrt{x_1^2 + x_2^2}.$$

The first coordinate r is called the *radial coordinate* of X, the second coordinate φ is called the *angular coordinate* or *polar angle* of X. The polar coordinates of O are $(0, \varphi)$, where φ can be any real number. For any X other than O the relation between the Cartesian coordinates (x_1, x_2) and the polar coordinates (r, φ) is

$$\begin{cases} x_1 = r \cos \varphi, \\ x_2 = r \sin \varphi, \end{cases} \text{and} \quad \begin{cases} r = \sqrt{x_1^2 + x_2^2}, \\ \tan \varphi = x_2/x_1 \text{ and } \cos \varphi = x_1/r. \end{cases}$$

Note that the angle measure of φ is not fully determined by the condition $\tan \varphi = x_2/x_1$; see Fig. 4.22, on the right. For given values of r and $\tan \varphi$ there are still two possibilities for φ; φ is determined uniquely by the extra condition on the cosine.

In the following section we will give new definitions of the conics using polar coordinates. First, however, we discuss a curve that was studied by Nicomedes (ca. 280–ca. 210 BC). The curve is called a conchoid, as its shape is somewhat similar to that of a conch, a large spiral-shelled mollusk.

Example 4.22. The *conchoid* is most easily defined using polar coordinates. We are given a line l, a point O outside the line, and a positive real number c. Let R be an arbitrary point on l. We choose a point X on the line OR such that $RX = c$. There are two such points; see Fig. 4.23, on the left. The conchoid is the set of all points X obtained this way for the different positions of the point R on l.

We choose the origin of the Cartesian coordinate system at O. Moreover, we choose the axes in such a way that the line l is perpendicular to the x_1-axis, and meets it at a positive value of x_1. We let a denote the distance from O to l.

Let us first discuss the case $c \leq a$, represented by the two figures on the right. The polar coordinates of the point X satisfy the equation

$$r = OR \pm c = \frac{a}{\cos \varphi} \pm c \quad \text{for} \quad -\frac{\pi}{2} < \varphi < \frac{\pi}{2}.$$

The plus-or-minus sign appears because we do not know on which side of R the point X is located. Because of the assumption $c \leq a$, the right-hand side of the equation is always positive or zero, regardless of the sign of c. It turns out that the curve has two *branches*: the branch on the right corresponds to the plus sign and the one on the left to the minus sign. From the equation above we deduce the following equalities:

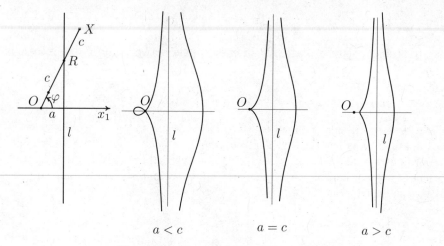

Fig. 4.23. The conchoid of Nicomedes

$$r \cos \varphi = a \pm c \cos \varphi \,,$$
$$x_1 = a \pm c \frac{x_1}{r} \,,$$
$$x_1 - a = \pm c \frac{x_1}{\sqrt{x_1^2 + x_2^2}} \,.$$

Squaring both sides of the last equality and multiplying the result by $x_1^2 + x_2^2$ then gives
$$\left(x_1^2 + x_2^2\right)(x_1 - a)^2 = c^2 x_1^2 \,, \tag{4.5}$$
the equation of a degree-four curve. We can find the intersection points of the curve with the x_1-axis by substituting 0 for x_2. We arrive at the equation
$$x_1^2 \left((x_1 - a)^2 - c^2\right) = 0$$
with roots 0, of multiplicity two, and $a+c$ and $a-c$. If $a = c$, the multiplicity of 0 is three; in this case O, is called a *cusp*. This case is represented in the second figure from the right. If $c < a$, the multiplicity of 0 is two; the point O is called an *isolated point* of the conchoid; see the rightmost figure. We should note that the point O is not a solution of the original equation of the conchoid in polar coordinates; the root 0 is the result of the multiplication by $x_1^2 + x_2^2$ that we used in the derivation of (4.5). Let us finally consider the case $c > a$; see the second figure from the left. This time the point O is called a *node*. There is exactly one number φ_0 between 0 and $\pi/2$ such that $c \cos \varphi_0 = a$.

The above derivation of the formula in Cartesian coordinates remains valid if X and O are on opposite sides of R, or if $\varphi_0 \leq |\varphi| < \pi/2$. The points we have not covered yet are those that lie on the small loop near O. A closer look reveals that for these points $|\varphi| > \pi/2$: the points are located in the second or third quadrant, and

$$\pi - \varphi_0 < \varphi < \pi + \varphi_0 \, .$$

The polar equation for the loop is

$$r = \frac{a}{\cos \varphi} + c \quad \text{and} \quad |\varphi| > \frac{\pi}{2} \, .$$

In Cartesian coordinates, (4.5) always holds.

Digression 7. *The study of curves was an important part of Greek mathematics, because curves could be used as a tool in geometric constructions. According to the ideas of Euclid [18], a geometric figure did not exist until its construction by ruler and compass had been explained. The following rules held. First, given any two distinct points we can construct a line through these points. Second, given any two distinct points, we can construct a circle such that the first point is the center of the circle and the second point lies on the circle. Third, the first and second rules also apply to intersection points of lines and circles that have been constructed using those rules. The book [43] is devoted to these types of constructions and some variations. Greek mathematics knows three famous unsolvable problems:*

1. *The trisection of an angle: find a general method using only ruler and compass to divide any angle into three equal parts.*
2. *The duplication of a cube: given any cube, construct the edge of the cube whose volume is twice that of the given cube, using only ruler and compass.*
3. *The squaring of a circle: given any circle, construct the side of the square whose area is equal to that of the given circle, using only ruler and compass.*

These problems remained unsolved for many centuries. In the nineteenth century, their insolvability was finally shown. L. Wantzel showed this for the first two in 1837, while Lindemann proved the impossibility of the third in 1882.

Figure 4.24 shows the relation between the trisection problem and the conchoid. Suppose that we want to divide an angle β into three equal parts. To obtain Fig. 4.24, we need to introduce some notation. Let l be the extension of one of the legs of β. Let M be the vertex of β and let P be a point on the leg of β that lies on l. Let O be a point on the other leg of β, and let $c = MO$. We now assume that we know how to draw the conchoid of the point O, line l, and number c, and study the intersection points X and Z of the conchoid and $\odot(M, c)$. The line OX meets l at the point R. For this point we have $\angle MRX = \beta/3$. Indeed, $c = RX = XM = MO$, whence $\angle XOM = \angle MXO = 2\angle MRX$. By a similar argument, $\angle POZ = \angle POM/3$. This shows how we can use the conchoid to trisect an angle.

Rotation of the Coordinate Axes

Consider two coordinate systems at the same point C, one of which is a rotation of the other around the origin C. Suppose that a point X has coordinates

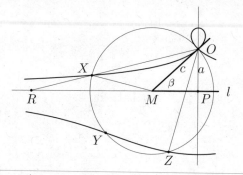

Fig. 4.24. The conchoid and the trisection of angle β

(x_1, x_2) in the first system and coordinates (x'_1, x'_2) in the second system. We call the first the old coordinates and the second the new coordinates. Suppose that a counterclockwise rotation about an angle φ transforms the first coordinate system into the second. We first assume that $0 \leq \varphi < \pi/2$. The

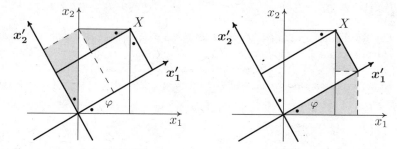

Fig. 4.25. Rotating the coordinate axes: **(a)** from old to new; **(b)** from new to old

following *coordinate transformations* give the new coordinates (x'_1, x'_2) of the point X as a function of the old coordinates (x_1, x_2), and vice versa. Figure 4.25 illustrates the first formula on the left, and the second formula on the right.

$$\begin{cases} x'_1 = x_1 \cos\varphi + x_2 \sin\varphi, \\ x'_2 = -x_1 \sin\varphi + x_2 \cos\varphi. \end{cases} \quad (4.6)$$

$$\begin{cases} x_1 = x'_1 \cos\varphi - x'_2 \sin\varphi, \\ x_2 = x'_1 \sin\varphi + x'_2 \cos\varphi. \end{cases} \quad (4.7)$$

These formulas remain valid for values of φ outside the interval $[0, \pi/2)$, though the corresponding figures are somewhat different.

We can use the coordinate transformations to find summation formulas for the trigonometric functions. Suppose that a point Y has new coordinates $(\cos\alpha, \sin\alpha)$. For convenience we assume that both $0 \le \varphi < \pi/2$ and $0 \le \alpha < \pi/2$. The angle between the x_1'-axis and the line OY is equal to α, and the angle between the x_1-axis and the line OY is equal to $\alpha + \varphi$. The old coordinates of Y are therefore $(\cos(\alpha+\varphi), \sin(\alpha+\varphi))$. Substituting this in (4.7) gives

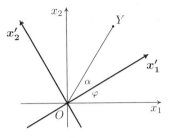

$$\cos(\alpha + \varphi) = \cos\alpha \cos\varphi - \sin\alpha \sin\varphi ,$$
$$\sin(\alpha + \varphi) = \cos\alpha \sin\varphi + \sin\alpha \cos\varphi .$$

Similarly, a point with old coordinates $(\cos\beta, \sin\beta)$, where $0 \le \beta < \pi/2$, has new coordinates $(\cos(\beta-\varphi), \sin(\beta-\varphi))$. Computing the new coordinates from the old using (4.6) leads to the equalities

$$\cos(\beta - \varphi) = \cos\beta \cos\varphi + \sin\beta \sin\varphi,$$
$$\sin(\beta - \varphi) = \sin\beta \cos\varphi - \cos\beta \sin\varphi.$$

These *sum and difference formulas* hold for all values of α, β, and φ; we omit the rather long verification.

Exercises

4.13. Let α, β, and γ be the measures of the angles of a triangle, and let R be the radius of its circumcircle. By Example 4.13 and the law of sines, the area of the triangle is

$$O = 2R^2 \sin\alpha \sin\beta \sin\gamma .$$

4.14. Let **a** and **b** be vectors in the coordinate plane $\mathbb{R}^2$. The spans of these vectors determine four angles; let φ denote the angle whose legs contain the vectors **a** and **b**.

(a) Show that the area O of the parallelogram spanned by **a** and **b** is equal to $O = \|\mathbf{a}\| \|\mathbf{b}\| \sin\varphi$.
(b) Show that $\langle \mathbf{a}, \mathbf{b} \rangle = \|\mathbf{a}\| \|\mathbf{b}\| \cos\varphi$.
Hint: Compute $\|\mathbf{b} - \mathbf{a}\|^2$ using the inner product and the law of cosines.

4.15. Use the law of cosines to derive a formula for the length of the altitude of a triangle. Let ABC be a given triangle. As usual, the letters a, b, and c denote both the sides opposite the vertices A, B, and C, respectively, and their lengths. We want to deduce a formula for the length h_a of the altitude from A.

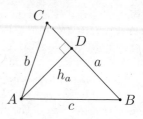

(a) Show that $CD = |b \cos \angle C| = |(b^2 + a^2 - c^2)/2a|$.
(b) Let s denote the semiperimeter of triangle ABC, that is, $2s = a + b + c$. Using (a) and the equality $h_a^2 = b^2 - CD^2$, we can show, through rather involved computations, that $h_a = (2/a)\sqrt{s(s-a)(s-b)(s-c)}$.
(c) For the area O of triangle ABC we obtain *Heron's formula*:
$$O = \sqrt{s(s-a)(s-b)(s-c)}\ .$$
(d) Following the hint given in Exercise 1.2 of Sect. 1.1, we obtain the following formulas for the radius r of the incircle and the radius r_a of the excircle on the side a:
$$r = \sqrt{\frac{(s-a)(s-b)(s-c)}{s}} \quad \text{and} \quad r_a = \sqrt{\frac{s(s-b)(s-c)}{s-a}}\ .$$

4.16. We use the law of cosines to deduce a formula for the length x of the segment $[CD]$ between the top C of triangle ABC and a point D of the base. As usual, the letters a, b, and c denote both the sides opposite A, B, and C respectively and their lengths. We let α denote the angle at A, and we denote both the segments $[AD]$ and $[DB]$ and their lengths by c_1 and c_2, respectively. The result giving the length x of the segment CD is known as Stewart's theorem, after Stewart (1717–1785).

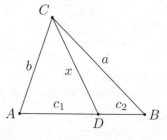

(a) Use the law of cosines to show that $a^2 = b^2 + c^2 - 2bc \cos \alpha$ and $x^2 = b^2 + c_1^2 - 2bc_1 \cos \alpha$.
(b) Eliminate $\cos \alpha$ to obtain the result of Stewart's theorem:
$$cx^2 = c_1 a^2 + c_2 b^2 - cc_1 c_2\ .$$
(c) In particular, the length z_c of the median through C equals
$$z_c^2 = \tfrac{1}{2}\left(a^2 + b^2\right) - \tfrac{1}{4}c^2\ .$$

(d) Use Exercise 2.26 of Sect. 2.3 to show that the length d_c of the angle bisector at C satisfies
$$d_c^2 = \frac{4ab}{(a+b)^2} s(s-c) .$$

4.17. We will now derive an elegant formula for the distance from the orthocenter H of a triangle ABC to the center M of its circumcircle $\odot(M, R)$.

(a) By computing the power of the centroid Z with respect to the circumcircle of triangle ABC we obtain

$$|ZM^2 - R^2| = ZC \times ZE .$$

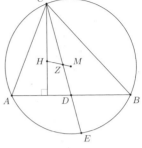

(b) Use Example 2.46, the formula $CD \times DE = AD \times DB$ giving the power of D with respect to the circumcircle, and the formula for the length z_c of the median from C (Stewart's theorem) to derive the following formula:

$$HM^2 = 9R^2 - (a^2 + b^2 + c^2) .$$

4.18. Derive the following formulas from the sum and difference formulas given in this section:

$$\sin \alpha + \sin \beta = 2 \sin\left(\frac{\alpha + \beta}{2}\right) \cos\left(\frac{\alpha - \beta}{2}\right) ,$$

$$\sin \alpha - \sin \beta = 2 \cos\left(\frac{\alpha + \beta}{2}\right) \sin\left(\frac{\alpha - \beta}{2}\right) ,$$

$$\cos \alpha + \cos \beta = 2 \cos\left(\frac{\alpha + \beta}{2}\right) \cos\left(\frac{\alpha - \beta}{2}\right) ,$$

$$\cos \alpha - \cos \beta = -2 \sin\left(\frac{\alpha + \beta}{2}\right) \sin\left(\frac{\alpha - \beta}{2}\right) .$$

4.19. Show that the following equalities hold for all α and β:

$$\sin 2\alpha = 2 \sin \alpha \cos \alpha , \quad \cos 2\alpha = -1 + 2 \cos^2 \alpha ,$$

$$\sin 3\alpha = 3 \sin \alpha - 4 \sin^3 \alpha , \quad \cos 3\alpha = -3 \cos \alpha + 4 \cos^3 \alpha ,$$

$$\tan(\alpha + \beta) = \frac{\tan \alpha + \tan \beta}{1 - \tan \alpha \tan \beta} , \quad \tan(2\alpha) = \frac{2 \tan \alpha}{1 - \tan^2 \alpha} .$$

4.20. The *cissoid* of Diocles (ca. 240–ca. 180 BC) is defined as follows. Given a circle $C = \odot(M, a)$ with diameter OB, we draw the tangent to the circle at B.

Let E be an arbitrary point on this line, and let D be the second intersection point of the line OE and the circle C. Next, let P be the point on the segment OE for which $OP = ED$. The cissoid is the set of all points P obtained in this way as E moves along the tangent. The part of the curve that lies inside the circle looks like an ivy leaf (see insert); this explains the Greek name of the curve.

(a) The equation of the cissoid is $r = (2a/\cos\varphi) - 2a\cos\varphi$ in polar coordinates and $x_1(x_1^2 + x_2^2) = 2ax_2^2$ in Cartesian coordinates.
(b) $\triangle OPQ \cong \triangle DEF$ and the arclength of OR is equal to that of DB.
(c) $\triangle OPQ \sim \triangle BRQ \sim \triangle ROQ$.
(d) $PQ : OQ = OQ : RQ = RQ : QB$.
(e) By multiplying the three ratios in (d) we obtain

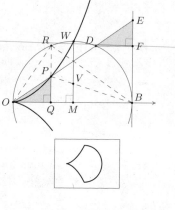

$$\frac{PQ^3}{OQ^3} = \frac{PQ}{QB} = \frac{VM}{MB} = \frac{MV}{MW}$$

and $\dfrac{PQ}{OQ} = \sqrt[3]{\dfrac{MV}{MW}}.$

The cissoid can be used for the duplication of the cube; if V is the midpoint of $[MW]$, the ratio of OQ and PQ is equal to the cube root of 2.

4.21. In this exercise we will compute the values of $\sin(\pi/5)$ and $\cos(\pi/5)$. Consider the isosceles triangle ABC with top angle at C equal to $\pi/5$. Let BE be the bisector of angle B.

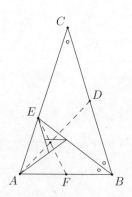

(a) Show that $BC : AB = AB : AE$. This ratio is the golden ratio $1 + \tau = (\sqrt{5} + 1)/2$; see Sect. 2.5, Exercise 2.48.
(b) Setting $AB = 1$, we have $BC = 1 + \tau$ and $AE = \tau$.
(c) Use the law of cosines to show that $\cos(\pi/5) = (\sqrt{5} + 1)/4$.
(d) Show that $\sin(\pi/5) = (\sqrt{10 - 2\sqrt{5}})/4$.
(e) Let D and F be the midpoints of BC and AB, respectively. Let S be the intersection point of AD and EF. Show that $\triangle ABE$ is the image of $\triangle BCA$ under a rotation about S over an angle $-3\pi/5$, followed by a central dilation with center S and scaling factor τ. The figures shows the

successive images of $\triangle BAC$ under iterations of these transformations. Hint: $AS : SD = ES : SF$. The point S is a fixed point of the map that sends $\triangle BCA$ to $\triangle ABE$.

4.22. Let R be a rotation with angle φ and center C. In any coordinate system with origin C the equations for the rotation $\mathbf{y} = \mathcal{R}(\mathbf{x})$ are

$$\begin{cases} y_1 = x_1 \cos\varphi - x_2 \sin\varphi , \\ y_2 = x_1 \sin\varphi + x_2 \cos\varphi . \end{cases}$$

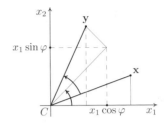

4.3 Turning the Plane Inside Out

To any given circle we are going to associate a map called *inversion*, which turns the plane inside out by interchanging the interior and exterior of the circle. An inversion is a completely different type of transformation from isometries and central dilations. We will see that inversions can be of great help in solving various difficult problems.

Definition 4.23. *Given a circle $C = \odot(O, \rho)$, the inversion ι in C is defined for all points of the plane V with the exception of the center O. For any point P other than O, the image $\iota(P)$ is the unique point of the line OP that lies on the same side of O as P and satisfies the equation*

$$OP \times O\iota(P) = \rho^2 .$$

We call P and $\iota(P)$ each other's inverses with respect to the circle C.

Figure 4.26, on the left, shows a few points and their images under inversion. The point P lies inside the circle C: $OP < \rho$. It follows that $O\iota(P) > \rho$,

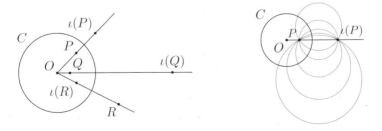

Fig. 4.26. The definition of the inversion ι

so $\iota(P)$ lies outside C. In other words, the inversion $\iota(P)$ indeed interchanges the interior and exterior of the circle C. We note that every point X on the circle is a fixed point of ι, that is, $\iota(X) = X$. Furthermore, we have $\iota(\iota(P)) = P$ for every point P in $V \setminus \{O\}$, so $\iota \circ \iota = $ id on $V \setminus \{O\}$. The following theorem lists the properties we have just mentioned.

Theorem 4.24. *The inversion ι in the circle $C = \odot(O, \rho)$ has the following properties:*

1. *$\iota(P) = P$ for every point P on the circle C.*
2. *$\iota(\text{in } C \setminus \{O\}) = \text{ex } C$ and $\iota(\text{ex } C) = \text{in } C \setminus \{O\}$.*
3. *Let P be a point other than O that does not lie on the circle. Every circle through P and $\iota(P)$ meets the circle C perpendicularly.*
4. *If a circle K meets the circle C perpendicularly, then $K = \iota(K)$.*
5. *Let P be a point other than O that does not lie on the circle C. Every circle through P that meets C perpendicularly also passes through $\iota(P)$.*

Recall that a circle through P meets the circle C perpendicularly if it intersects C and the tangents to the two circles at the intersection points are perpendicular to each other; see Sect. 1.1, Exercise 1.4.

Proof. We already proved the first two properties before stating the theorem. For the third property we refer to Fig. 4.26, on the right. The point O has the same power $OP \times O\iota(P)$ with respect to all circles through both P and $\iota(P)$. Let K be such a circle (not shown in the figure). Let R be the point of contact of the tangent to K that passes through O. According to Theorem 4.8, we have $OR^2 = OP \times O\iota(P)$. It now follows from the definition of the inversion ι that $OR^2 = \rho^2$. Consequently, R lies on C, and the circles K and C meet perpendicularly, since OR is a tangent.

Let us now prove the fourth property. If the circles K and C meet perpendicularly at R, then OR is tangent to K. Suppose $P \in K$. Let Q be the second intersection point of OP and K. By Theorem 4.8 we have $\rho^2 = OR^2 = OP \times OQ$. It now follows from the definition of inversion that $OQ = O\iota(P)$, whence $Q = \iota(P)$ and $\iota(P) \in K$.

Let us finally prove the fifth property. If a circle K passes through P and meets C perpendicularly, it immediately follows from the fourth property that $\iota(P) \in K$.

We will use the following theorem to determine the images of lines and circles under inversion.

Theorem 4.25. *Let ι be the inversion in the circle $C = (O, \rho)$. Let P, Q, and O be noncollinear points. Then P, Q, $\iota(P)$, and $\iota(Q)$ lie on the same circle and $\triangle OPQ \sim \triangle O\iota(Q)\iota(P)$.*

The last result of the theorem implies that the lines PQ and $\iota(P)\iota(Q)$ are antiparallel with respect to angle O; see Sect. 2.5, Exercise 2.41(c).

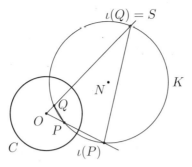

Fig. 4.27. The points P, Q, $\iota(P)$, and $\iota(Q)$ lie on the same circle

Proof. If P or Q lies on the circle C, there is nothing to prove. The circle through P, Q, and $\iota(P)$ is the circumcircle of triangle $PQ\iota(P)$. We call this circle $K = \odot(N, R)$; see Fig. 4.27. The product $OP \times O\iota(P)$ is not only equal to ρ^2 by the definition of inversion, but also equal to m, the power of O with respect to the circle K. Hence $m = \rho^2$. Note that O belongs to the exterior of K, since the points P and $\iota(P)$ of the line OP lie on the same side of O. Let S be the second intersection point of OQ and the circle K. We have $OQ \times OS = m = \rho^2$. Since O belongs to the exterior of K, the points Q and S of the line OQ lie on the same side of O. Since $OQ \times O\iota(Q) = \rho^2$ by the definition of inversion, we have $S = \iota(Q)$. It follows that P, $\iota(P)$, $\iota(Q)$, and Q all lie on K.
Since $OP \times O\iota(P) = OQ \times O\iota(Q) = \rho^2$, we have $\triangle OPQ \sim \triangle O\iota(Q)\iota(P)$ by Theorem 2.52.

In the following examples we look at the images of lines and circles under inversion. But first a remark on terminology: The inversion ι in the circle $C = \odot(O, \rho)$ is not defined in O. Now, suppose that O is a point of the line l. By abuse of language, the expression *the image of l under ι* denotes the set $\iota(l \setminus \{O\})$. The analogue holds for circles through O.

Example 4.26. Let $C = \odot(O, \rho)$. We are going to study the image of straight lines under the inversion ι in C. If a line l passes through O, its image $\iota(l)$ coincides with l as a set, but not pointwise. If l does not pass through O, we first determine the foot P of the perpendicular from O on l. The image $\iota(P)$ of P also lies on this perpendicular; see Fig. 4.28 (a). Next, let Q be a point on l other than P. By Theorem 4.25, $\triangle OPQ \sim \triangle O\iota(Q)\iota(P)$. Consequently, $\angle OPQ = \angle O\iota(Q)\iota(P)$, and $O\iota(Q)\iota(P)$ is a right angle. According to Thales' theorem, $\iota(Q)$ lies on the circle with diameter $[O\iota(P)]$. Every point of this circle other than O is the image $\iota(S)$ of some point S on l. The circle is therefore the image of l under the inversion ι. In the figure, l and C meet; the intersection points are fixed points of ι. We can easily verify that the same proof holds for a line l that lies entirely in the exterior of the circle C; the image $\iota(l)$ then lies in in C.

178   4 CURVES

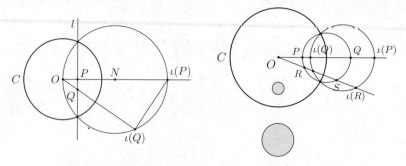

Fig. 4.28. Images under inversion: (a) lines; (b) circles

Example 4.27. Let us now determine the image of circles under the inversion ι in $C = \odot(O, \rho)$. If the circle K passes through O, $\iota(K)$ is the line l through the intersection points of C and K. We can prove this as follows. By the example above, $\iota(l)$ is a circle through O. The intersection points of C and l also lie on this circle. It follows that $\iota(l)$ must coincide with K. Consequently $\iota(K) = \iota \circ \iota(l) = l$. If the circle K does not pass through O, its image $\iota(K)$ is a circle; see Fig. 4.28 (b). The proof of this statement is rather subtle. The special case that K and C meet perpendicularly was already discussed in Theorem 4.24. Let P and Q be the intersection points of K and the line through O and the center of K. Then $OP \times OQ = m$, the power of O with respect to K. By the definition of ι, $OP \times O\iota(P) = \rho^2$. Combining the last two formulas gives $O\iota(P) = (\rho^2/m)OQ$. We will show that a similar formula holds for every point R of K. Let S be the second intersection point of OR and K. If OR is a tangent, this point coincides with R. The same reasoning as above leads to $O\iota(R) = (\rho^2/m)OS$. Furthermore, we can easily show that R lies on K if and only if S lies on K. It follows from these observations that $\iota(K)$ is the image of K under the central dilation with center O and scaling factor ρ^2/m.

Using these examples, we can also determine the image under inversion of more complicated figures. Figure 4.29 shows the image of triangle PQR under the inversion ι in the circle C. The extended sides of the triangle are mapped to circles through O. The image of the triangle under ι consists of three circular arcs. Although the sides of the triangle are deformed into circular arcs, the measures of the angles of the triangle are invariant; the measure of the angle between two circles is the measure of one of the angles formed by the tangents at the intersection point. Because of this property we say that inversion *preserves angles*. What does change is the orientation: it is reversed by inversion. The motion from P to R through Q is counterclockwise, while the motion from $\iota(P)$ to $\iota(R)$ through $\iota(Q)$ is clockwise. Because of this we say that inversion *reverses orientation*. It is not easy to explain this in full detail, but the following example may help clarify the situation.

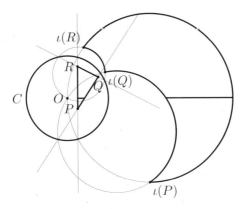

Fig. 4.29. Triangle PQR and its image under inversion

Example 4.28. Figure 4.30 shows the image of angle P under the inversion in the circle $C = \odot(O, \rho)$. The sides of the angle are mapped onto arcs of the circles K_1 and K_2 through O. The centers of K_1 and K_2 are located on the perpendiculars OQ and OR from O on the lines l and m, respectively. It follows that the tangents to K_1 and K_2 at O are perpendicular to OQ

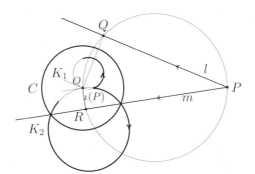

Fig. 4.30. The image of angle P under inversion

and OR, respectively. This in turn means that they are parallel to l and m, respectively. It follows that the measure of the angle between the tangents to K_1 and K_2 at O is equal to $\angle P$. The same holds for the tangents at the other intersection point of K_1 and K_2.

Using inversion we can give a relatively simple proof of a theorem of Ptolemy (ca. 85–ca. 165). A proof without inversion would be much more complicated. The theorem presents a criterion for an admissible quadrilateral to be cyclic; see the discussion on cyclic quadrilaterals above Theorem 4.15.

Theorem 4.29 (Ptolemy). *Let A, B, C, and D be four points such that the line segments $[AC]$ and $[BD]$ intersect. These points lie on a circle if and* *only if*

$$AC \times BD = AB \times CD + BC \times AD .\tag{4.8}$$

Proof. Consider Fig. 4.31. Starting at AB, we successively denote the sides of the quadrilateral $ABCD$ and their respective lengths by a, b, c, and d. Let p and q be the diagonals AC and BD, respectively, and let $\odot(K_1, \rho_1)$ denote the circumcircle of $\triangle ABD$. We apply the inversion ι in a circle $K_2 = \odot(A, \rho_2)$

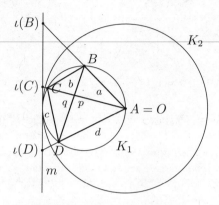

Fig. 4.31. Ptolemy's theorem

to the figure, where ρ_2 is chosen in such a way that the circles K_1 and K_2 are tangent to each other: $\rho_2 = 2\rho_1$. Since the center $A = O$ of K_2 lies on the circle K_1, the image $\iota(K_1)$ of K_1 is a straight line; we will call this line m. The points $\iota(B)$ and $\iota(D)$ lie on m. By Theorem 4.25, we know that $\triangle OBC \sim \triangle O\iota(C)\iota(B)$. Since $O\iota(B) = \rho^2/a$ by the definition of inversion, it follows that $\iota(B)\iota(C) = (\rho^2/ap)b$. In the same way we obtain $\iota(C)\iota(D) = (\rho^2/dp)c$ and $\iota(B)\iota(D) = (\rho^2/ad)q$. If the point C lies on K_1, which implies that A, B, C, and D lie on the same circle, then $\iota(C)$ lies on m. It follows that $\iota(B)\iota(C) + \iota(C)\iota(D) = \iota(B)\iota(D)$, and consequently

$$\frac{\rho^2}{ad}q = \frac{\rho^2}{ap}b + \frac{\rho^2}{dp}c, \text{ or equivalently, } pq = bd + ac ,$$

which is (4.8). If $\iota(C)$ is not a point of m, that is, if C does not lie on K_1, then $\iota(B)\iota(C) + \iota(C)\iota(D) > \iota(B)\iota(D)$, and consequently

$$\frac{\rho^2}{ad}q < \frac{\rho^2}{ap}b + \frac{\rho^2}{dp}c \quad \text{or} \quad pq < bd + ac .$$

In this case (4.8) does not hold. This completes the proof.

In Chap. 2, Example 2.46, we began studying the nine-point circle. Let us return to this figure. This circle passes through nine special points of the triangle. We will show that the nine-point circle touches both the incircle of a triangle and its three excircles. Since inversion has the nice property that it transforms certain circles into straight lines, it will be of great help in studying properties of tangents to circles.

Theorem 4.30 (Feuerbach). *The nine-point circle of a triangle is tangent to both the incircle and the excircles of the triangle.*

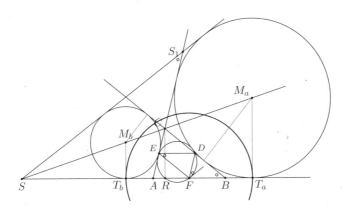

Fig. 4.32. Feuerbach's theorem

Proof. We will prove the tangency to the excircles. The tangency to the incircle can be proved in an analogous way. Consider Fig. 4.32. Let D, E, and F be the midpoints of the sides of $\triangle ABC$. The nine-point circle of $\triangle ABC$ is the circumcircle of $\triangle DEF$. Let $\odot(M_a, r_a)$ and $\odot(M_b, r_b)$, respectively, be the excircles tangent to the sides a and b. The line AB is a common tangent of both circles; the other common tangent is the image of AB under the reflection in the line $M_a M_b$. Let S be the intersection point of these common tangents. Let T_a and T_b, respectively, be the feet of the perpendiculars from M_a and M_b on AB. We can easily see that $\triangle ST_b M_b \sim \triangle ST_a M_a$. It follows that $ST_b/ST_a = r_b/r_a$. In the same way, by considering the perpendiculars from M_a and M_b on side BC, we find that $CM_b/CM_a = r_b/r_a$. Let R be the foot of the perpendicular from C on AB. Since the lines $M_b T_b$, CR, and $M_a T_a$ are parallel to each other, the last equality implies that $RT_b/RT_a = r_b/r_a$. Combining this equality and the first one leads to

$$\frac{RT_b}{RT_a} = \frac{ST_b}{ST_a}. \tag{4.9}$$

Note that $T_b A = BT_a = s - c$, where s denotes the semiperimeter of triangle ABC; see the discussion following Fig. 4.14. Consequently, F is not only the midpoint of AB, but also the midpoint of $T_a T_b$. If we let $p = FT_a = FT_b$, then (4.9) becomes

$$\frac{p-RF}{p+RF} = \frac{SF-p}{SF+p}.$$

By cross-multiplication we obtain $RF \times SF = p^2$. The derivation of this formula is quite tricky; a more insightful way of obtaining this formula uses *cross-ratios*.

Let us study the effect of the inversion ι in $\odot(F,p)$ on the figure. The last equality shows that $\iota(R) = S$. It follows from Theorem 4.24, property 4, that ι maps each of the excircles onto itself. What happens to the nine-point circle? The angle between the tangent to the nine-point circle at F and the side DF is equal to $\angle DEF$, hence also to $\angle FBD$ and $\angle SS_1A$. Consequently, the tangent at F is parallel to SS_1. Since R also lies on the nine-point circle and $\iota(R) = S$, it follows that the line SS_1 is the image of the nine-point circle under ι. This line touches the excircles of triangle ABC. Since $\iota^2 = \mathrm{id}$ and inversion preserves angles, Example 4.28, we see that the nine-point circle is tangent to the excircles.

We conclude this section by showing how inversion can be used to find a circle that is tangent to three given circles. This problem was first studied by Apollonius (ca. 262–ca. 200 BC).

Example 4.31 (Apollonius's problem). Suppose that we are given distinct circles $\odot(M_1, r_1)$, $\odot(M_2, r_2)$, and $\odot(M_3, r_3)$ with the property that each is contained in the exterior of the other two. We are going to discuss how to draw a circle that is tangent to each of the three given circles. See Fig. 4.33; the three given circles are shaded. We begin by enlarging the three circles by adding

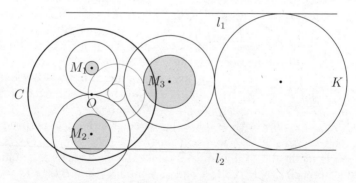

Fig. 4.33. A circle that is tangent to three given circles

the same length δ to each radius, where δ is chosen such that at least two of the enlarged circles touch each other. In the figure, the circles $\odot(M_1, r_1 + \delta)$ and $\odot(M_2, r_2 + \delta)$ are tangent; we call the point of contact O. We will now apply an inversion ι in a suitably chosen circle C with center O. We choose the radius ρ of C such that $\odot(M_3, r_3 + \delta)$ is mapped onto itself by the inversion in C; see Exercise 4.24. Since O is a contact point, the inversion maps the circles $\odot(M_1, r_1 + \delta)$ and $\odot(M_2, r_2 + \delta)$, respectively, onto straight lines l_1 and l_2 that are both parallel to the common tangent to the circles at O. At the same time, the inversion maps the circle we are looking for onto a circle K that is tangent to $\odot(M_3, r_3 + \delta)$, l_1, and l_2. We can easily draw the circle K. Since it is tangent to both l_1 and l_2, its center must lie halfway between l_1 and l_2. This implies that the radius R of K is half the distance between l_1 and l_2. Furthermore, the distance from the center of K to M_3 is equal to $r_3 + \delta + R$. There are two positions of K that meet these requirements. In both cases, the circle $\iota(K)$ is tangent to the three enlarged circles. By adding δ to its radius, we obtain a circle that touches the three original circles, i.e., a solution to the original problem. Can you guess how many solutions there are?

Digression 8. *In the preceding example we showed how to draw a circle that touches three given circles. A quick look in the book [43], for example, reveals how to construct the circle with ruler and compass. Constructions with ruler and compass is a topic that is discussed in many books on geometry; its educational value is undisputed. Finding constructions oneself enhances one's insight into the inner structure of geometry. However, nowadays, geometrical figures are mostly drawn by computers; some of the programs produce better figures than those constructed using ruler and compass. That is one of the reasons why we do not discuss constructions with ruler and compass in this book.*

Exercises

4.23. Let ι be the inversion in a circle $C = (O, \rho)$. Show that

(a) if P and Q either both lie in the interior of C or both lie in the exterior of C, then $P\iota(P)\iota(Q)Q$ is a cyclic quadrilateral;
(b) if P lies in the interior of C and Q in its exterior, or vice versa, then $P\iota(P)Q\iota(Q)$ is a cyclic quadrilateral.

4.24. Given a circle $K = \odot(M, r)$ and a point O in the exterior of K, we want to find a circle $C = (O, \rho)$ such that the inversion in C maps K onto itself. What is the value of ρ?
Hint: What does the inversion do to the line OM?

4.25. Let ι denote the inversion in the circle $C = (O, \rho)$. In Example 4.27 we saw that the image under ι of a circle K that does not pass through O is again a circle. Here is another way to prove this. Consider Fig. 4.28 (b); note that the center O of C lies on PQ. The line OR defines two straight angles at $\iota(R)$. One of these angles is divided into three parts by the lines $\iota(Q)\iota(R)$ and $\iota(P)\iota(R)$.

(a) One of the angles at $\iota(R)$ is equal to $\angle RQP$, and another of these angles is equal to $\angle RPQ$.

(b) Show that $\iota(P)\iota(R)\iota(Q)$ is a right angle.
(c) The image $\iota(K)$ of K is a circle with diameter $[\iota(P)\iota(Q)]$.

4.26. Consider the circles $C = \odot(M, R)$ and $K = \odot(N, r)$ with $M \neq N$ and $R > r$. Let l be the line through M and N. In parts (a) through (c), we assume that $MN > R + r$, in other words, that each circle is contained in the exterior of the other.

(a) There is a unique point S on l such that M and N lie on the same side of S and $SM/SN = R/r$.
(b) The circle C is the image of K under the central dilation with center S and scaling factor R/r. Conclude that there are two common tangents to C and K through S.
(c) There is a unique point T on l lying between M and N such that $TM/TN = R/r$. Show that K is the image of C under the central dilation with center T and scaling factor $-r/R$, and that there are two common tangents to C and K passing through T.
(d) If $MN < R - r$, the circles C and K have no common tangent.
(e) If $R - r < MN < R + r$, the circles intersect in two points. They have two common tangents.
(f) When do C and K have three common tangents? When do they have only one?

4.27. Let l be the line through the centers of the circles $\odot(M, R)$ and $\odot(N, r)$, where $R \geq r$. Suppose that $R - r < MN < R + r$. Let A and B be the intersection points of $\odot(M, R)$ and l, and let C and D be the intersection points of $\odot(N, r)$ and l. We label the points in such a way that C lies between A and B, and B lies between C and D. Prove that the circles $\odot(M, R)$ and $\odot(N, r)$ are perpendicular to each other if and only if $CA/CB = DA/DB$. Hint: Make the following substitutions in the last formula: $CA = MC + R$, $DB = MD - R$, $CB = R - MC$, $DA = R + MD$.

4.28. Show that the nine-point circle of a triangle is tangent to its incircle. Hint: Consider the incircle and one of the excircles, and follow the proof of Theorem 4.30.

4.29. We can derive the law of cosines from Ptolemy's theorem. Let $\odot(M, R)$ be the circumcircle of triangle ABC, and consider the reflection of the triangle in a suitably chosen perpendicular bisector of the triangle.

4.30. Let l be a line perpendicular to the x_1-axis, intersecting it at $A = (a, 0)$ with $a > 0$. Given a point Y on l, let X_1 and X_2 be points on the line OY such that
$$X_1 Y = AY = Y X_2.$$
The set of all points X_i obtained this way as Y varies over l is called the *strophoid*. This name is derived from the Greek word for twisted belt.

(a) The polar coordinates (r, φ) of a point on the strophoid satisfy

$$r = \frac{a}{\cos \varphi} \pm a \tan \varphi \,.$$

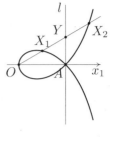

(b) The points X_1 and X_2 in the figure are mapped to each other under the inversion in $\odot(O, a)$.

(c) The Cartesian coordinates (x_1, x_2) of a point on the strophoid satisfy

$$x_1 - a = \pm \frac{a\,x_2}{\sqrt{x_1^2 + x_2^2}} \,.$$

(d) The equation of the strophoid is

$$x_2^2(x_1 - 2a) + (x_1 - a)^2 x_1 = 0 \,.$$

4.4 Conics

In Sect. 3.3, we discussed the conic sections as Voronoi diagrams. We will now view them in a completely different way, defining them using foci and directrices.

Definition 4.32. *Let l be a line, F a point outside l, and let $e > 0$ be a real number. The* conic section K *with* focus F, *directrix* l, *and* eccentricity e *is the set of all points X such that*

$$FX = e\,XQ \,, \tag{4.10}$$

where Q is the foot of the perpendicular XQ from X on l.

Consider Fig. 4.34. By the definition of Q, XQ is the distance from X to the line l. Thus the conic section K is the set of all points for which the ratio of the distances to the focus and directrix is e. Let us first derive the equation of the conic section K in polar coordinates. We introduce a coordinate system at the focus F, in such a way that the x_1-axis is perpendicular to the directrix l and the x_2 axis is parallel to it. Let D be the intersection point of l and the x_1-axis, and let S be the intersection point of the conic section K and the x_2-axis. Finally, let q be the distance from the focus to the directrix, that is, $q = FD$, and let $p = FS$. Since S lies on the conic section K, we have $p = eq$. A point X on K has polar coordinates (r, φ), where $\varphi = \angle DFX$. The distance XQ is equal to q plus the distance from X to the x_2-axis:

$$XQ = q - r \cos \varphi \,.$$

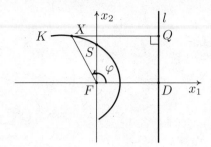

Fig. 4.34. Focal definition of a conic section

Together with (4.10), this gives

$$r = e(q - r\cos\varphi), \text{ or equivalently, } r(1 + e\cos\varphi) = eq\,.$$

This results in the equation of K in polar coordinates:

$$r = \frac{p}{1 + e\cos\varphi}\,. \tag{4.11}$$

This equation also makes sense for $e = 0$. In that case, (4.11) represents a circle with radius p and center F. Let us, for the moment, include the case $e = 0$. We rewrite (4.11) as $r + er\cos\varphi = p$ and turn to Cartesian coordinates. The point X has coordinates (x_1, x_2); see Fig. 4.34. We obtain the equation

$$\sqrt{x_1^2 + x_2^2} + ex_1 = p\,,$$

which leads to

$$\left(1 - e^2\right)x_1^2 + x_2^2 + 2pex_1 - p^2 = 0\,. \tag{4.12}$$

We distinguish the following cases:

1. $e = 1$: a parabola
2. $0 \le e < 1$: an ellipse
3. $e > 1$: a hyperbola

See Fig. 4.35. We will discuss each case in detail.

The Parabola

For a parabola we have $e = 1$: the parabola is therefore the Voronoi diagram of the directrix l and the focus F. We refer to Sect. 3.3, where we also tell how to draw the tangent lines to the parabola using this property. For a parabola, (4.12) has the form

$$x_2^2 + 2px_1 - p^2 = 0\,.$$

We apply the following coordinate transformation:

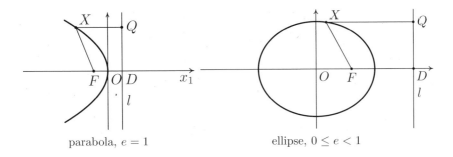

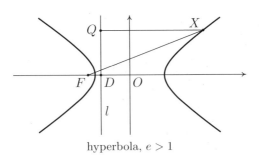

Fig. 4.35. Conic sections

$$\begin{cases} x'_1 = x_1 - p/2, \\ x'_2 = x_2. \end{cases}$$

The effect of this transformation is a shift of the x_2-axis by the vector $(p/2, 0)$. The new origin has old coordinates $(p/2, 0)$. In the case we are considering, we have $p = q$, so O is the midpoint of FD. In the new coordinates, the equation of the parabola becomes

$$\boxed{x_2^2 = -2px_1,} \quad (4.13)$$

where we omit the primes for notational convenience. The focus is $(-p/2, 0)$, and the equation of the directrix is $x = p/2$. The origin is called the *vertex* of the parabola, and the x_1-axis is called the *axis*, or *axis of symmetry*, of the parabola; see Fig. 4.35. The parabola has many applications. For example, the first approximation of the trajectory of a bullet is a parabola. In Sect. 3.3 we mentioned already parabolic mirrors.

Construction 1 *We will now show how to draw, point by point for a given p, the parabola K with equation $x_2^2 = 2px_1$. The equation*

$$x_2^2 = 2px_1$$

is called the vertex equation *of the parabola. The parabola lies in the right half-plane and has its vertex at the origin. Consider Fig. 4.36. To begin with,*

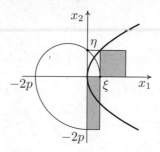

Fig. 4.36. Construction of the parabola

we mark off the point $-2p$ on the x_1-axis. For a given point ξ on the x_1-axis we will construct the point η on the x_2-axis for which (ξ, η) is a point of K. For this we draw the semicircle on the line segment from $(-2p, 0)$ to $(\xi, 0)$. This intersects the x_2-axis at η. We easily see that $\eta^2 = 2p\xi$ (Sect. 4.1, Exercise 4.12), and consequently that (ξ, η) lies on K. As suggested by the shading in Fig. 4.36, the rectangle with area $2p\xi$ is transformed by the parabola K into a square with side η whose area is equal to that of the rectangle. The name of the parabola is related to this construction: the word parabola *is derived from the Greek* parabolē, *which means* comparison *(of square and rectangle)*.

Archimedes Triangles

There is an elementary way to compute the area of a parabolic segment (the definition is given below); the computation goes back to Archimedes (287–212 BC). An *Archimedes triangle* of a parabola is a triangle formed by two tangents to the parabola and the line segment joining the two points of contact; see Fig. 4.37, on the left. The following theorem gives the most important

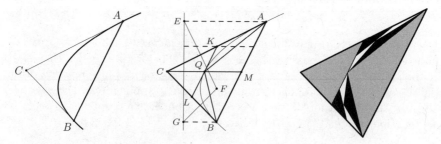

Fig. 4.37. Archimedes triangle ACB

properties of Archimedes triangles. Let us first remark that a parabola divides the plane into two parts: the interior of the parabola and its exterior. The proof is analogous to that for a circle. We can also prove that a chord divides the interior of the parabola into two parts, of which one is bounded. We call this bounded part a *segment* of the parabola.

Theorem 4.33 (Archimedes). *Given a parabola, let ACB be the Archimedes triangle formed by tangents CA and CB to the parabola and the chord AB. Let M be the midpoint of AB. The following statements hold:*

1. *The median CM of triangle ACB is parallel to the axis of the parabola.*
2. *The intersection point Q of the median CM and the parabola is the midpoint of the line segment [CM].*
3. *The tangent to the parabola at Q is a mid-parallel of the Archimedes triangle ACB.*
4. *The area of the segment cut off by chord AB is 2/3 the area of the Archimedes triangle ACB.*

Proof. Consider Fig. 4.37, in the middle. Let F be the focus of the parabola, and let EG be its directrix. The points E and G are the feet of the perpendiculars from A and B, respectively, on the directrix. The tangents CA and CB are the perpendicular bisectors of $[EF]$ and $[GF]$, respectively. The third perpendicular bisector of triangle EFG is the line CM, which is therefore parallel to the axis of the parabola. The line CM also bisects the line segment $[EG]$. It follows that M is the midpoint of the segment $[AB]$. This completes the proof of the first statement.

Let Q be the intersection point of CM and the parabola, and let KL be the tangent to the parabola at Q. Then AKQ is also an Archimedes triangle of the parabola. Consequently, the median through K of triangle AKQ is parallel to the axis of the parabola. Since this median passes through both K and the midpoint of $[AQ]$, it is a mid-parallel of triangle ACM; in particular, K is the midpoint of AC. In the same way it follows that L is the midpoint of $[BC]$. This proves the third statement. It is now also evident that the second statement holds.

Let R be the area of $\triangle ACB$. The area of $\triangle AQB$, which lies inside the parabola, is $R/2$. The area of $\triangle KCL$, which lies outside the parabola, is $R/4$. The area of the Archimedes triangle ACB is the sum of these two areas and the areas of the Archimedes triangles AKQ and BLQ. Repeating this process, we partition each of the triangles AKQ and BLQ into four smaller triangles, one inside the parabola, one outside, and two Archimedes triangles; see Fig. 4.37, on the right. For the sum of the areas of the triangles inside the parabola, we obtain

$$\tfrac{1}{2}R + \tfrac{1}{8}R + \tfrac{1}{32}R + \cdots = \tfrac{2}{3}R,$$

while the areas of the triangles outside the parabola add up to

$$\tfrac{1}{4}R + \tfrac{1}{16}R + \tfrac{1}{64}R + \cdots = \tfrac{1}{3}R.$$

The area of triangle ACB is the sum of the areas of all the small triangles. This proves the fourth statement of the theorem.

Digression 9. *Up to now, we have not made any serious attempts to define the notion of area. Giving such a definition would be quite difficult. For most purposes a couple of simple rules suffice. For a precise definition of the notion of area, see [48, Chap. 21]. By definition, the area of a rectangle is its base times its height; the area of a triangle is half its base times its height. Furthermore, we know that the area is* finitely additive: *if a figure F is the union of finitely many figures $F_1, \ldots, F_n$, where any two have only boundary points in common, we have the following equation for the area of F:*

$$\text{Area}\, F = \text{Area}\, F_1 + \cdots + \text{Area}\, F_n\,.$$

The area function is monotonic: *if F is a subset of G then $\text{Area}\, F \leq \text{Area}\, G$. Finally, the area function is* invariant under isometries.

Let us compute the area of the segment S of the parabola cut off by chord AB. In the proof given above, we showed that we can cover S with small triangles that have either a side, a vertex, or nothing in common. The sum of the areas of these triangles is approximately $(2/3)R$. Because the area function is finitely additive and monotonic, $\text{Area}\, S \geq (2/3)R$. We can also cover the complement of S inside triangle ACB with small triangles. The sum of the areas of these triangles is $(1/3)R$. Consequently, $\text{Area}\, S = (2/3)R$. This completes the computation of the area of segment S. In the time of Archimedes this was a very important result. This simple expression for the area of a parabolic segment provides a solution to the problem of squaring a parabolic segment.

The Ellipse

A conic section with $0 \leq e < 1$ is an ellipse. Dividing (4.12) by $1 - e^2$ and completing the square gives

$$\left(x_1 + \frac{ep}{1-e^2}\right)^2 + \frac{x_2^2}{1-e^2} - \frac{p^2}{(1-e^2)^2} = 0\,. \tag{4.14}$$

Remark 4.34. For $e = 0$, this is the equation of a circle with center O and radius p. In the rest of this chapter we assume that $e > 0$.

Let $c = ep/(1-e^2)$; note that $c > 0$. We apply the following coordinate transformation:
$$\begin{cases} x_1' = x_1 + c\,, \\ x_2' = x_2\,. \end{cases}$$

The effect of this transformation is a shift of the x_2-axis by the vector $(-c, 0)$. The new origin has old coordinates $(-c, 0)$, and the new coordinates of the focus F are $(c, 0)$. In the new coordinates, the equation of the ellipse becomes

$$x_1^2 + \frac{x_2^2}{1-e^2} - \frac{p^2}{(1-e^2)^2} = 0\,,$$

where we again omit the primes. Substituting $a = p/(1-e^2)$ and $b = p/(\sqrt{1-e^2})$, we obtain the equation

$$\boxed{\frac{x_1^2}{a^2} + \frac{x_2^2}{b^2} = 1\,.} \tag{4.15}$$

Figure 4.35, on the top right, shows an ellipse. The ellipse K meets the x_1-axis at $(a,0)$ and the x_2-axis at $(0,b)$. We call the line segment from $(-a,0)$ to $(a,0)$ the *major axis* of the ellipse, and the line segment from $(0,-b)$ to $(0,b)$ its *minor axis*. Note that the ellipse is symmetric with respect to both axes. We call the point O the *center* of the ellipse, and the points $(a,0)$, $(-a,0)$, $(b,0)$, and $(-b,0)$ its *vertices*. The ellipse has two *directrices*, l and l', and two foci, F and F'; see Fig. 4.38. The following properties hold:

1. $a^2 - b^2 = c^2$.
2. $b = a\sqrt{1-e^2}$ and $c = ea$.
3. For all nonzero a and b, (4.15) is the equation of an ellipse.
4. For any point X of the ellipse, $XF + XF' = 2a$.

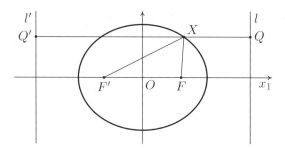

Fig. 4.38. The two directrices of the ellipse

Properties (1) and (2) are easily verified. For the proof of (3) we may, and do, assume that a and b are positive, since $(-x)^2 = x^2$ for all x. We first suppose that $a > b$. The equations in (1) and (2) have unique positive solutions for c and e, with $e < 1$. To these c and e corresponds a unique p, namely $p = (1-e^2)a$. We choose a line l and a point F that is at a distance $q = p/e$ from this line. If we now apply the previous results of this chapter to the conic section with focus F, directrix l, and eccentricity e, we obtain the equation (4.15) for the ellipse. If $a = b$, we are in the case $c = 0$, which was excluded by Remark 4.34. Finally, if $b > a$, we interchange the roles of x_1 and x_2 and apply the same arguments. Finally, we can show that

$$XF + XF' = e\,XQ + e\,XQ' = e\,QQ'\,.$$

By substituting $(a,0)$ for X, we obtain $e\,QQ' = 2a$. We can use (4) to construct the ellipse: on a piece of cardboard, position two tacks in the foci F and F', and place a loop of string of length $2a(1+e)$ around them. Now place a pencil in the loop, pulling it outward until the string is taut. The pencil still has one degree of freedom for its movement: it can trace the ellipse.

Tangent. How does the definition of the ellipse using the focus and directrix compare to the definition of the ellipse as Voronoi diagram discussed in Sect. 3.3? Consider Fig. 4.39. We construct a circle C with center at one

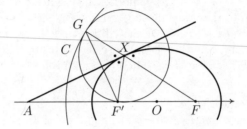

Fig. 4.39. Tangent to the ellipse

of the foci, F in the figure, and with radius $2a$, the length of the major axis. Let G be the intersection point of the line FX and the circle C. Since

$$FG = 2a = XF + XF',$$

we have $XF' = XG$ for every point X of the ellipse K. Moreover, every point X with this property lies on the ellipse. It follows that the ellipse is the Voronoi diagram of the circle C and the point F'. In other words, the circle C is the directrix circle of the ellipse, as defined in Sect. 3.3. Note that an ellipse has two directrix circles, one with center F, the other with center F'. These results can be used to construct the tangents to the ellipse K. Since $XF' = XG$, the triangle $F'XG$ is isosceles. Consequently, XA, the angle bisector of GXF', is also the perpendicular bisector of $F'G$. For every point Y on the line XA other than X, we have

$$YF + YF' = YF + YG > FG = 2a.$$

It follows that such a point Y does not lie on the ellipse. The line XA therefore meets K at exactly one point. Every other line through X intersects K in two points. It follows that XA is tangent to the ellipse. Thus we have shown that for every point G of the directrix circle with center F, the perpendicular bisector of the segment $[GF']$ is tangent to the ellipse; the point of contact is the intersection point with the line GF. Figure 4.39 shows that rays from one of the foci are reflected toward the other focus; see also Fig. 4.40.

Construction 2 *Let us now show how to construct the ellipse point by point given the lengths of the major and minor axes, respectively $2a$ and $2b$. See*

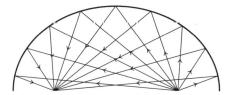

Fig. 4.40. Reflection by the ellipse

Fig. 4.41. To begin with, we can find, and construct, the value of p from the relation $b^2 = pa$ (Sect. 4.1, Exercise 4.12). Given a point ξ on the x_1-axis, we construct a point η on the x_2-axis such that ξ, η lies on the ellipse as follows. We begin by cutting off a strip from the bottom of the rectangle with vertices $(0,0)$, $(0,-2p)$, $(\xi,-2p)$, and $(\xi,0)$ to obtain the shaded rectangle with vertices $(0,0)$, $(0,-2p+\delta_\xi)$, $(\xi,-2p+\delta_\xi)$, and $(\xi,0)$. To construct these new vertices, we draw the line segment from $(0,-2p)$ to $(2a,0)$ and the line $x = \xi$; their intersection point is $(\xi,-2p+\delta_\xi)$. We mark the point $(-2p+\delta_\xi,0)$ and draw a semicircle on the line segment from this point to $(\xi,0)$. This semi-

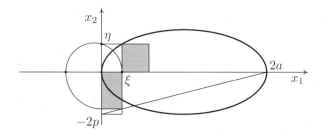

Fig. 4.41. Point-by-point construction of the ellipse

circle intersects the x_2-axis at η. We can easily check that $\delta_\xi : \xi = p : a$ and $\eta^2 = \xi(2p - \delta_\xi)$. The point (ξ, η) therefore lies on the curve with equation

$$x_2^2 + \frac{p}{a}(x_1 - a)^2 - pa = 0 \ .$$

This curve is an ellipse whose major and minor axes have the given lengths. The equation can be rewritten in the following form, which is called the vertex equation *of the ellipse:*

$$x_2^2 = 2px_1 - \frac{p}{a}x_1^2 \ .$$

Let us compare this construction of points of the ellipse to the construction of the parabola. In the case of the parabola, the whole rectangle with area $2p\xi$ was transformed into a square with side η, while in the case of the ellipse it is a reduced *rectangle that is transformed into a square. The name of the ellipse is related to this property: the word ellipse is derived from the Greek word eclipsis, which means leaving out.*

The Hyperbola

Let us now go on to hyperbolas. A hyperbola is a conic section with eccentricity greater than 1.

In order to find the equation of the hyperbola in Cartesian coordinates, we start with (4.14). Since there is a strong similarity between the properties of the ellipse and those of the hyperbola, several proofs will be left to the reader. Let $c = ep/(e^2 - 1)$; note that $c > 0$. We apply the following coordinate transformation:
$$\begin{cases} x_1' = x_1 - c\,, \\ x_2' = x_2\,. \end{cases}$$

The effect of this transformation is a shift of the x_2-axis by the vector $(c, 0)$. The new origin has old coordinates $(c, 0)$, and the new coordinates of the focus F are $(-c, 0)$. In the new coordinates, after substituting $a = p/(e^2 - 1)$ and $b = p/(\sqrt{e^2 - 1})$, (4.14) becomes the equation

$$\boxed{\frac{x_1^2}{a^2} - \frac{x_2^2}{b^2} = 1} \qquad (4.16)$$

of the hyperbola, where we again omit the primes. Figure 4.35, on the bottom, shows the hyperbola K. Note that the hyperbola is symmetric with respect to both axes. It meets the x_1-axis at $(a, 0)$. We call the point O the *center* of the hyperbola, and the points $(a, 0)$ and $(-a, 0)$ its *vertices*. The line segment joining the vertices is called the *axis*. The hyperbola consists of two *branches*, which do not touch and lie in different open half planes of the x_2-axis. The

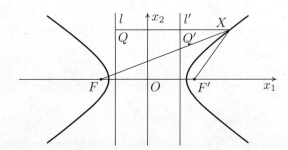

Fig. 4.42. The two directrices of the hyperbola

hyperbola has two foci, F and F', and two directrices, l and l'; see Fig. 4.42. The following properties hold:

1. $a^2 + b^2 = c^2$.
2. $b = a\sqrt{e^2 - 1}$ and $c = ea$.
3. For all a and b, (4.16) is the equation of a hyperbola.
4. For every point X of the hyperbola, $|XF - XF'| = 2a$.

Tangent. How does the definition of the hyperbola using the focus and directrix compare to the definition of the hyperbola as Voronoi diagram discussed in Sect. 3.3? Consider Fig. 4.43. We draw a circle with center at the focus F and radius $2a$. The other focus F' lies outside the circle. Recall that the hyperbola has two branches that lie in different open half-planes of the x_2-axis. The points on one branch satisfy $XF - XF' > 0$, while the points on the other branch satisfy $XF - XF' < 0$. Consider a point X on the first branch, drawn on the right in the figure. Let G be the intersection point of the line segment FX and the circle. Since

$$FG = 2a = XF - XF',$$

we have $XF' = XG$ for every point X on the right branch of the hyperbola. Conversely, every point X that is equidistant from F' and the circle C lies

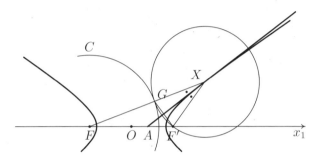

Fig. 4.43. Tangent to the hyperbola

on the right branch of the hyperbola. It follows that the right branch of the hyperbola is the Voronoi diagram of the circle C and the point F'. Thus the circle C is the directrix circle of the hyperbola as defined in Sect. 3.3. The hyperbola has two directrix circles; the directrix circle with center F' is used to define the left branch of the hyperbola. These results can be used to construct the tangents to the hyperbola K. Since $XF' = XG$, the triangle $F'XG$ is isosceles. It follows that XA, the angle bisector of GXF', is also the perpendicular bisector of $[F'G]$. For every point Y on the line XA other than X, we have

$$|YF - YF'| = |YF - YG| < FG = 2a.$$

It follows that such a point Y does not lie on the hyperbola. The line XA therefore meets K at exactly one point. With only two exceptions, every other line intersects K in two points; the two exceptions are the asymptotes through X. It follows that the line XA is a tangent of the hyperbola.

Construction 3 *We will now show how to draw, point by point, the hyperbola given a and b; see Fig. 4.44. To begin with, we can find, and construct, the*

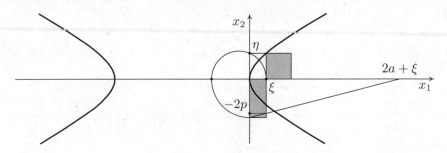

Fig. 4.44. Construction of the hyperbola

value of p from the relation $b^2 = pa$ (Sect. 4.1, Exercise 4.12). Given a point ξ on the x_1-axis, we construct a point η on the x_2-axis such that (ξ, η) lies on the hyperbola as follows. We begin by adding a strip to the bottom of the rectangle with vertices $(0,0)$, $(0,-2p)$, $(\xi,-2p)$, and $(\xi,0)$ to obtain the shaded rectangle with vertices $(0,0)$, $(0,-2p-\delta_\xi)$, $(\xi,-2p-\delta_\xi)$, and $(\xi,0)$. To construct the new vertices we draw the line segment through $(\xi,-2p)$ and $(2a+\xi,0)$. It intersects the x_2-axis at $(0,-2p-\delta_\xi)$. We mark the point $(-2p-\delta_\xi,0)$ and draw a semicircle on the line segment from this point to to $(\xi,0)$. This semicircle intersects the x_2-axis at η. We can easily check that $\delta_\xi : \xi = p : a$ and $\eta^2 = \xi(2p+\delta_\xi)$. The point (ξ,η) therefore lies on the curve with equation

$$x_2^2 - \frac{p}{a}(x_1+a)^2 + pa = 0 \, .$$

This curve is the hyperbola we were looking for. The equation can be rewritten in the following form, which is called the **vertex equation** of the hyperbola:

$$x_2^2 = 2px_1 + \frac{p}{a}x_1^2 \, .$$

Let us compare this construction of points of the hyperbola to the constructions of the parabola and ellipse. In the case of the hyperbola, it is not the rectangle with area $2p\xi$ (parabola) or a reduced rectangle (ellipse) that is transformed into a square with side η, but an extended rectangle. The name of the hyperbola is related to this property: the word hyperbola is derived from the Greek word huperbolē, which means excess.

We recommend reading [21]. It presents a lecture by the famous physicist Feynman in which he explains Kepler's laws concerning the motion of the planets in a geometric fashion, using properties explained in this section.

Classification

The equations of the parabola (4.13), ellipse (4.15), and hyperbola (4.16) are quadratic. Let us show that there are no other essentially different quadratic curves. The general equation of a quadratic curve K is

$$K:\ a_{11}x_1^2 + 2a_{12}x_1x_2 + a_{22}x_2^2 + 2a_{01}x_1 + 2a_{02}x_2 + a_{00} = 0\ . \tag{4.17}$$

We may, and do, assume that not all three coefficients a_{11}, a_{12}, and a_{22} are zero. The number $D = a_{11}a_{22} - a_{12}^2$ is called the *discriminant* of the curve K. This number plays an important role in the classification we are about to give. In the equations referred to above there is no mixed term x_1x_2, that is, the coefficient of x_1x_2 is zero. Therefore we will first try to remove the mixed term in the equation of K. We do this through a coordinate transformation, in this case a counterclockwise rotation over an angle φ:

$$\begin{cases} x_1 = x_1' \cos\varphi - x_2' \sin\varphi\ , \\ x_2 = x_1' \sin\varphi + x_2' \cos\varphi\ . \end{cases} \tag{4.18}$$

In these equations, (x_1, x_2) are the old coordinates and (x_1', x_2') the new ones. We substitute the right-hand side of (4.18) for x_1 and x_2 in (4.17). Omitting the primes and writing down only the quadratic part, we obtain

$$K:\ b_{11}x_1^2 + 2b_{12}x_1x_2 + b_{22}x_2^2 + \cdots = 0 \tag{4.19}$$

with

$$b_{11} = a_{11}\cos^2\varphi + 2a_{12}\cos\varphi\sin\varphi + a_{22}\sin^2\varphi,$$
$$b_{12} = -a_{11}\cos\varphi\sin\varphi + a_{12}\cos^2\varphi - a_{12}\sin^2\varphi + a_{22}\sin\varphi\cos\varphi,$$
$$b_{22} = a_{11}\sin^2\varphi - 2a_{12}\sin\varphi\cos\varphi + a_{22}\cos^2\varphi\ .$$

A rather involved computation shows that

$$b_{11}b_{22} - b_{12}^2 = a_{11}a_{22} - a_{12}^2 = D\ .$$

It follows that the discriminant is invariant under rotation of the coordinate system. We can write the coefficient b_{12} of the mixed term as

$$b_{12} = -\tfrac{1}{2}a_{11}\sin 2\varphi + a_{12}\cos 2\varphi + \tfrac{1}{2}a_{22}\sin 2\varphi\ .$$

If $a_{11} \neq a_{22}$, this coefficient vanishes if

$$\tan 2\varphi = \frac{2a_{12}}{a_{11} - a_{22}}\ . \tag{4.20}$$

If $a_{11} = a_{22}$, it vanishes if $a_{12}\cos(2\varphi) = 0$, or equivalently, when $a_{12} \neq 0$ if $\varphi = \tfrac{\pi}{4}$. We distinguish the cases $D \neq 0$ and $D = 0$.

Let us first consider the case $D \neq 0$. After a counterclockwise rotation of the coordinate system over the angle φ from (4.20), we obtain

$$K:\ b_{11}x_1^2 + b_{22}x_2^2 + 2b_{01}x_1 + 2b_{02}x_2 + b_{00} = 0\ .$$

Since D is invariant, we have $b_{11}b_{22} \neq 0$. By applying the translation of the coordinate system

$$\begin{cases} x_1 = x_1' - \dfrac{b_{01}}{b_{11}}, \\ x_2 = x_2' - \dfrac{b_{02}}{b_{22}}, \end{cases}$$

we can also remove the linear terms from the equation. Omitting the primes, we obtain

$$K: c_{11} x_1^2 + c_{22} x_2^2 + c_{00} = 0.$$

By interchanging the axes, if necessary, we find one of the forms given in the first five rows of Table 4.1.

Table 4.1. Classification of the quadratic curves

		QUADRATIC CURVES	
$D < 0$	$\dfrac{x_1^2}{a^2} - \dfrac{x_2^2}{b^2} = 1$		**hyperbola**
	$\dfrac{x_1^2}{a^2} - \dfrac{x_2^2}{b^2} = 0$		pair of intersecting lines
$D > 0$	$\dfrac{x_1^2}{a^2} + \dfrac{x_2^2}{b^2} = 1$		**ellipse**
	$\dfrac{x_1^2}{a^2} + \dfrac{x_2^2}{b^2} = 0$		a single point
	$\dfrac{x_1^2}{a^2} + \dfrac{x_2^2}{b^2} = -1$		empty set
$D = 0$	$x_1^2 = 2px_2,\ p \neq 0$		**parabola**
	$x_1^2 = k$	$k < 0$	empty set
		$k = 0$	pair of coinciding lines
		$k > 0$	pair of parallel lines

Let us now consider the case $D = 0$. We have $a_{11} a_{22} = a_{12}^2$. Let us first assume that $a_{12} \neq 0$. In this case a_{11} and a_{22} are both nonzero and have the same sign. After substituting (4.18) in (4.17), we obtain (4.19), where

$$b_{11} = a_{11}\left(\cos\varphi + \frac{a_{12}}{a_{11}}\sin\varphi\right)^2,$$

$$b_{22} = a_{11}\left(\sin\varphi - \frac{a_{12}}{a_{11}}\cos\varphi\right)^2.$$

Now, if φ is chosen such that $b_{11} = 0$, then after some computations it follows from the angle-doubling formula applied to $\tan(2\varphi)$ that φ also satisfies (4.20); hence $b_{12} = 0$. Similarly, if we choose φ such that $b_{22} = 0$, then φ satisfies (4.20). In follows that in the case $D = 0$, the equation of the quadratic curve becomes

$$K: b_{11}x_1^2 + 2b_{01}x_1 + 2b_{02}x_2 + b_{00} = 0.$$

After a translation of the coordinate system, this leads to the last four rows of Table 4.1. This includes the case $a_{12} = 0$.

Example 4.35. Let us determine the form of the curve

$$K: x^2 + xy + y^2 - 3x - 3y + 1 = 0$$

and its position in the coordinate plane. In this example we use x and y instead of x_1 and x_2, respectively. The discriminant is $3/4$, so we are dealing with an ellipse. Let us follow the analysis used for the classification given above. Formula (4.20) shows that we can remove the mixed term by rotating the coordinate system over $\pi/4$. After applying the corresponding coordinate transformation

$$\begin{cases} x = \frac{1}{2}\sqrt{2}\,x' - \frac{1}{2}\sqrt{2}\,y', \\ y = \frac{1}{2}\sqrt{2}\,x' + \frac{1}{2}\sqrt{2}\,y', \end{cases}$$

(4.18), the equation of our curve becomes

$$\tfrac{3}{2}x'^2 + \tfrac{1}{2}y'^2 - 3\sqrt{2}\,x' + 1 = 0.$$

Next, we translate the coordinate system by setting $x'' = x' - \sqrt{2}$ and $y'' = y$, which gives

$$\frac{x''^2}{\left(\frac{2}{3}\sqrt{3}\right)^2} + \frac{y''^2}{2^2} = 1.$$

This curve is an ellipse; see Fig. 4.45 (a).

In the $x''y''$-coordinate system, the center of the ellipse is at $(0,0)$, in the $x'y'$-coordinate system it is at $(\sqrt{2}, 0)$, and in the original coordinate system the center is at $(1,1)$. In the original coordinate system, the axes of the ellipse lie on the lines $x = y$ and $x + y = \sqrt{2}$.

Example 4.36. Let us now determine the form of the curve

$$K: 16x^2 + 24xy + 9y^2 - 125y = 0$$

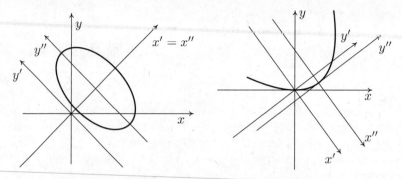

Fig. 4.45. Figures for (**a**) Example 4.35; (**b**) Example 4.36

and its position in the coordinate plane. In this example we will again use x and y instead of x_1 and x_2, respectively. The discriminant is 0, so we are dealing with a, possibly degenerate, parabola. Formula (4.20) shows that we can remove the mixed term by rotating the coordinate system over an angle φ that satisfies the condition $\tan \varphi = -a_{11}/a_{12} = -4/3$. After applying the corresponding coordinate transformation

$$\begin{cases} x = \tfrac{3}{5}x' + \tfrac{4}{5}y', \\ y = -\tfrac{4}{5}x' + \tfrac{3}{5}y', \end{cases}$$

the equation of our curve becomes

$$K: \; 25y'^2 + 100x' - 75y' = 0 \quad \text{or} \quad y'^2 + 4x' - 3y' = 0.$$

After completing the square, we have $(y' - (3/2))^2 = -4(x' - (9/16))$. Next, we translate the coordinate system by setting

$$\begin{cases} x'' = x' - \tfrac{9}{16}, \\ y'' = y' - \tfrac{3}{2}, \end{cases}$$

giving the equation $y''^2 = -4x''$. This curve is a parabola; see Fig. 4.45 (**b**).

Most linear algebra courses provide methods (eigenvalues and eigenvectors) that make it possible to discuss the examples given above without having to compute the coordinate transformations and their effects explicitly. This greatly simplifies the classification of quadratic curves.

Exercises

4.31. The conic sections can be characterized as follows:

(a) A parabola is the set of centers of all circles that are tangent to a given line and go through a given point outside the line.
(b) An ellipse is the set of centers of all circles that touch a given circle and go through a given point inside the circle.
(c) A hyperbola is the set of centers of all circles that touch a given circle and go through a given point outside the circle.

4.32. The eccentricity of a conic section determines its shape.

(a) Let K be a parabola with equation $x_2^2 = 2px_1$. Let $\mathcal{V}$ be the central dilation with center O and scaling factor α. What is the equation of the image $\mathcal{V}(K)$? Conclude that all parabolas are similar to each other.
(b) Two ellipses with the same eccentricity are similar.
(c) Two hyperbolas with the same eccentricity are similar.

4.33. This exercise is about the construction of tangents to a hyperbola. Let F_1 and F_2 be the foci of the hyperbola K and let $2a$ be the length of its axis. Let C be the directix circle with center F_1 and radius $2a$. Finally, let G_1 and G_2 be the two points of contact of the tangents to C from F_2.

(a) The perpendicular bisector of $[G_i F_2]$ is parallel to $F_1 G_i$ for $i = 1, 2$.
(b) Let G be a point on C other than G_1 or G_2. The perpendicular bisector of $[GF_2]$ is tangent to K. The point of contact is the intersection point of this bisector and the line $F_1 G$.
(c) Let $R(G)$ be the intersection point found in (b). The points G_1 and G_2 divide the circle into two arcs. Two points X and Y on C lie on the same arc if and only if $R(X)$ and $R(Y)$ lie on the same branch of the hyperbola.

4.34. In a hyperbolic mirror, the reflected light rays from a focus seem to emanate from the other focus.

4.35. We intersect the hyperbola with equation

$$K: \frac{x_1^2}{a^2} - \frac{x_2^2}{b^2} = 1$$

with the line $x_2 = mx_1$. There are two intersection points if and only if $|m| < |b/a|$. The lines with equations $x_2 = \pm(b/a)x_1$ are called the *asymptotes* of the hyperbola.

4.36. We call the point P the *center* of a set W if the point reflection $\dot{\mathcal{S}}_P$ maps the set W onto itself: $\dot{\mathcal{S}}_P(W) = W$.

(a) Let $f(x_1, x_2) = 0$ be the equation of a plane curve. The origin O is the center of this curve if $f(x_1, x_2) = f(-x_1, -x_2)$ for all x_1, x_2.
(b) We are given a curve K with equation (4.17). If the coefficients a_{01} and a_{02} are both 0, the origin is the center of K. Conversely, if the origin is the center of a nonempty quadratic curve K, the coefficients a_{01} and a_{02} in (4.17) are both 0.
Hint: Use the classification given in this section.

(c) We are again given a curve K with equation (4.17). The equations

$$\begin{cases} a_{11}x_1 + a_{12}x_2 + a_{01} = 0, \\ a_{12}x_1 + a_{22}x_2 + a_{02} = 0, \end{cases} \quad (4.21)$$

have a unique solution if and only if the discriminant of K is nonzero.

(d) Assume that (4.21) has a unique solution $x_1 = m_1$, $x_2 = m_2$. The coordinate transformation

$$x'_1 = x_1 - m_1,$$
$$x'_2 = x_2 - m_2,$$

eliminates the linear terms in the equation of K. Consequently, (m_1, m_2) is the center of K in the old coordinates.

4.37. We are given a linear transformation with equations

$$\begin{cases} x' = x + \tfrac{1}{4}\sqrt{2}\,y, \\ y' = \tfrac{1}{4}\sqrt{2}\,y. \end{cases}$$

Figure 4.46, on the left, shows the square with vertices $(-1, -1)$ and $(1, 1)$ and the circle $C: y^2 + x^2 = 1$. The image of the circle under the linear transformation is an ellipse. Find the equation of this ellipse and verify that the angle between the axes of the ellipse and the coordinate axes is approximately $7°$.

The figure on the right can be considered as a first attempt to draw the

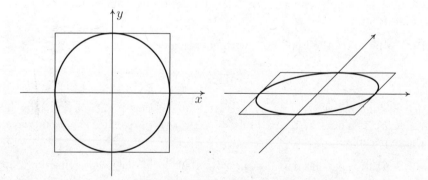

Fig. 4.46. A circle in perspective

picture on the left in perspective. Imagine that the picture on the left is a top view of a figure in a horizontal plane in space. On the right we have tried to draw it in perspective. This picture is not quite right: the major axis of the ellipse is not horizontal. To remedy this problem we apply another rotation over $7°$; see Fig. 5.20.

4.38. Let F be the focus of a parabola. If ACB is an Archimedes triangle on the chord AB of the parabola, then $\triangle FAC \sim \triangle FCB$.
Hint: In Fig. 4.37, in the middle, $\angle EAC = \angle FEG = \angle FCB$.

4.39. Theorem (Apollonius). *Let ACB be an Archimedes triangle on a chord AB of a parabola. Let Q be a point on the parabola that lies inside the triangle. Finally, let K and L be the intersection points of the tangent to the parabola at Q and sides CA and CB, respectively. Then*

$$\frac{KQ}{QL} = \frac{AK}{KC} = \frac{CL}{LB}.$$

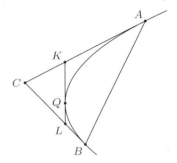

Hint: If X' denotes the projection of X on the directrix of the parabola, then $K'Q' = K'A'$, $L'Q' = L'B'$, $K'Q' + L'Q' = K'C' + L'C'$, and therefore $L'B' = K'C'$.

4.40. Show that the set of all points $C = (c_1, c_2)$ with the property that the Euler line (Example 2.46) of the triangle with top vertex (c_1, c_2), and base vertices $(-1, 0)$ and $(0, 1)$ is parallel to the base is an ellipse.

4.5 Homogeneous Coordinates and Polarity

A useful tool for studying the tangents to conic sections is polarity, which we will consider in this section. Though not essential, some knowledge of the vector space $\mathbb{R}^3$ (Sect. 5.2) can be of help in understanding this subject. We begin by discussing homogeneous coordinates.

Homogeneous Coordinates

We will work in the vector space $\mathbb{R}^2$; a point $\mathbf{x}$ in $\mathbb{R}^2$ is an ordered pair (x_1, x_2) of real numbers. We know that the lines $l : \mathbf{x} = \mathbf{p} + \lambda \mathbf{a}$ and $m : \mathbf{x} = \mathbf{q} + \mu \mathbf{b}$ are parallel if and only if $\mathbf{a}$ and $\mathbf{b}$ have the same span (Sect. 1.5, Exercise 1.23). We also know that the latter holds if and only if $\det(\mathbf{ab}) = 0$ (Sect. 1.6, Exercise 1.37). By definition, two distinct parallel lines have no point in common. Sometimes, however, it is convenient to assign to every pair of parallel lines an intersection point, which will be a point at "infinity". This point can also be visualized as the direction of the lines. For example, the lines $l : \mathbf{x} = \mathbf{p} + \lambda \mathbf{a}$ and $n : \mathbf{x} = \mathbf{q} + \nu \mathbf{a}$ have the direction $\mathbf{a}$ in common. In order to discuss both ordinary points and points at infinity, or directions, we introduce homogeneous coordinates.

Definition 4.37. *The point* $\mathbf{x} = (x_1, x_2)$ *of the vector space* $\mathbb{R}^2$ *has homogeneous coordinates* (ξ_0, ξ_1, ξ_2) *if* $\xi_0 \in \mathbb{R} \setminus \{0\}$, $\xi_1 \in \mathbb{R}$, $\xi_2 \in \mathbb{R}$, *and* $x_1 = \xi_1/\xi_0$ *and* $x_2 = \xi_2/\xi_0$.

A given point has many different triples of homogeneous coordinates. We can easily see that the triples (ξ_0, ξ_1, ξ_2) and (η_0, η_1, η_2) are the homogeneous coordinates of the same point if and only if $\xi_i = (\xi_0/\eta_0)\eta_i$ for $i = 1, 2$. In other words, in homogeneous coordinates the ratios of the coordinates are important, rather than their individual values. For example, the point $(-5, 4)$ has homogeneous coordinates $(1, -5, 4)$, as well as $(2, -10, 8)$ and $(1/5, -1, 4/5)$. The point of the coordinate plane with homogeneous coordinates $(3, 4, 5)$ is $(4/3, 5/3)$.

To a direction vector $\mathbf{a} = (a_1, a_2)$ with $\mathbf{a} \neq \mathbf{o}$ we associate the *point at infinity* $(0, a_1, a_2)$. We sometimes say that $(0, a_1, a_2)$ are the homogeneous coordinates of the direction $\mathbf{a}$. We will say that the directions $\mathbf{a} = (a_1, a_2)$ and $\mathbf{b} = (b_1, b_2)$ determine the same point at infinity if $\mathbf{a}$ and $\mathbf{b}$ have the same span, that is, if $\det(\mathbf{ab}) = 0$. In other words, $(0, a_1, a_2) = (0, b_1, b_2)$ if and only if $a_1 b_2 - a_2 b_1 = 0$. It follows that an point at infinity is also essentially a set of ratios of coordinates. Note that there is no point with homogeneous coordinates $(0, 0, 0)$, either ordinary or at infinity.

In the rest of this section we tacitly assume that for all triples of real numbers at least one of the numbers is nonzero. We call the ordered triples (ξ_0, ξ_1, ξ_2) and (η_0, η_1, η_2) *equivalent* if there exists a real number $\lambda \neq 0$ such that $\xi_i = \lambda \eta_i$ for $i = 0, 1, 2$. We will also say that equivalent triples have the *same ratio*.

Definition 4.38. *A point of the projective plane* $\mathbb{P}^2$ *is a ratio of three real numbers that are not all zero. In other words, it is an equivalence class of the equivalence relation defined above.*

We can visualize the projective plane as follows; see Fig. 4.47. Let β be the horizontal plane at height 1, that is, the x_0-coordinate of each of its points is 1. A point $\mathbf{x} = (x_1, x_2)$ of the coordinate plane $\mathbb{R}^2$ is represented by the point X in β whose spatial coordinates are $(1, x_1, x_2)$. An arbitrary point on the line through X and $(0, 0, 0)$ other than $(0, 0, 0)$ has spatial coordinates (ξ_0, ξ_1, ξ_2) satisfying $x_1 = \xi_1/\xi_0$ and $x_2 = \xi_2/\xi_0$. In other words, the spatial coordinates (ξ_0, ξ_1, ξ_2) are also the homogeneous coordinates of the point $\mathbf{x}$. In this way we assign to every point $\mathbf{x}$ of the coordinate plane a point (ξ_0, ξ_1, ξ_2) of the projective plane $\mathbb{P}^2$. A direction vector $\mathbf{y} = (y_1, y_2)$ is represented by a point Y in the $x_1 x_2$-plane. An arbitrary point on the line through Y and $(0, 0, 0)$ other than $(0, 0, 0)$ has spatial coordinates (η_0, η_1, η_2) satisfying $y_1 \eta_2 = y_2 \eta_1$ and $\eta_0 = 0$. In other words, the spatial coordinates (η_0, η_1, η_2) are also the homogeneous coordinates of the direction vector $\mathbf{y}$. In this way we assign to every direction $\mathbf{y}$ on the coordinate plane a point (η_0, η_1, η_2) of the projective plane $\mathbb{P}^2$.

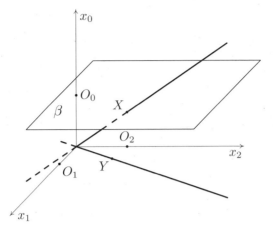

Fig. 4.47. The projective plane

We see that there are two disjoint sets of points in the projective plane: the first consists of the triples $(\zeta_0, \zeta_1, \zeta_2)$ with $\zeta_0 \neq 0$, the second consists of the triples with $\zeta_0 = 0$. Above we gave the correspondence between the first set and the coordinate plane. The points of this set are sometimes called the *finite points* of the projective plane. The second set corresponds to all possible directions in the plane. The projective plane $\mathbb{P}^2$ is an extension of the coordinate plane $\mathbb{R}^2$. The additional points are the *points at infinity*, or directions.

The points $O_0 = (1,0,0)$, $O_1 = (0,1,0)$, $O_2 = (0,0,1)$ and the unit point $E = (1,1,1)$ are called the *reference points* of the projective plane.

The Projective Plane

In this subsection we will give a glimpse of the rich geometric theory of the projective plane. We will discuss only the properties we need to explain the theory of polarity. We must emphasize, though, that projective geometry is a truly beautiful and exciting geometric theory. A study of this theory is a must for everyone who is interested in geometry; for a clear introduction, see [36] or [53]. We begin by defining the lines in the projective plane. Recall that the equation of a line in the coordinate plane $\mathbb{R}^2$ is $a_0 + a_1 x_1 + a_2 x_2 = 0$, where a_0, a_1, and a_2 are real numbers such that $a_1 a_2 \neq 0$; see (1.8).

Definition 4.39. *A line in the projective plane is the set of all triples* (ξ_0, ξ_1, ξ_2) *that satisfy the equation*

$$a_0 \xi_0 + a_1 \xi_1 + a_2 \xi_2 = 0, \qquad (4.22)$$

where the a_i *are real numbers that are not all zero.*

Note that the point at infinity $(0, a_2, -a_1)$ lies on the line with equation (4.22). The line with equation $\xi_0 = 0$ consists of all points at infinity, and is therefore called the *line at infinity*. As usual, a common point of two lines is called an *intersection point*.

Theorem 4.40. *Two distinct lines in the projective plane have a unique intersection point.*

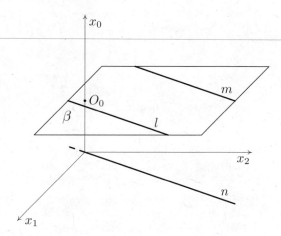

Fig. 4.48. The intersection point of two lines

A consequence of this theorem is that in the projective plane, two lines are not parallel unless they are identical. Figure 4.48 shows two "parallel lines" l and m in the plane β, which is a copy of the coordinate plane $\mathbb{R}^2$. In the projective plane these lines meet at the point at infinity that corresponds to the "direction" of l and m. This is drawn in the $x_1 x_2$-plane as the line n. The property stated in the theorem makes the projective plane completely different from the "usual" plane: Basic Assumption 1.20 states that for every point and every line of the plane, there exists a *unique line* through the point and parallel to the given line. This basic assumption is known as the *parallel postulate*. For centuries, mathematicians tried to derive the parallel postulate from Euclid's other postulates; see [22], [56]. The discoveries of hyperbolic geometry and elliptic geometry showed that such a proof cannot exist. In hyperbolic geometry, for any line and any point outside the line, there exist *infinitely many* lines through the point and "parallel" to the given line, where "parallel" now means that the lines do not intersect. In elliptic geometry, for any line and any point outside the line, there is *no* line through the point that is parallel to the given line; every two lines intersect. The geometry of the projective plane is an elliptic geometry.

Before we begin proving the theorem, let us introduce some notation. In Example 1.36, we introduced the determinant; we will use the following notation for a matrix and its determinant:

$$\begin{vmatrix} a_1 & a_2 \\ b_1 & b_2 \end{vmatrix} = a_1 b_2 - a_2 b_1 \text{ is the determinant of the matrix } \begin{pmatrix} a_1 & a_2 \\ b_1 & b_2 \end{pmatrix}.$$

The matrix is a rectangular array of four numbers arranged in two rows and two columns.

Proof. Let l and m be distinct lines with equations

$$l : a_0 \xi_0 + a_1 \xi_1 + a_2 \xi_2 = 0, \qquad (4.23)$$
$$m : b_0 \xi_0 + b_1 \xi_1 + b_2 \xi_2 = 0. \qquad (4.24)$$

What does it mean that the lines l and m are distinct? If there exists a real number λ such that $b_i = \lambda a_i$ for $i = 0, 1, 2$, we will say that the coefficients of (4.23) and (4.24) have the *same ratio*. If the coefficients of the equations have the same ratio, every solution (ξ_0, ξ_1, ξ_2) of (4.23) is also a solution of (4.24), and vice versa if we interchange (4.24) and (4.23). Conversely, if the equations (4.24) and (4.23) have the same solutions, their coefficients must have the same ratio. We can see this as follows. We assume that $a_0 \neq 0$; the cases $a_1 \neq 0$ and $a_2 \neq 0$ are similar. This implies that $b_0 \neq 0$. Indeed, if $b_0 = 0$, there exist ξ_1 and ξ_2 such that $(0, \xi_1, \xi_2)$ and $(1, \xi_1, \xi_2)$ are both solutions of (4.24) and, consequently, of (4.23); it follows that $a_0 = 0$, which gives a contradiction. Consequently, the solutions of (4.23) and (4.24) are given by $\xi_0 = -(a_1 \xi_1 + a_2 \xi_2)/a_0$ and $\xi_0 = -(b_1 \xi_1 + b_2 \xi_2)/b_0$, respectively, where ξ_1 and ξ_2 are arbitrary real numbers. Substituting $\xi_1 = -1, \xi_2 = 0$ and then $\xi_1 = 0, \xi_2 = -1$, we find that the coefficients of the equations have the same ratio.

Let us determine the common solution of (4.23) and (4.24). Multiplying the first equation by b_0, the second by $-a_0$, and adding the results gives

$$\begin{vmatrix} a_1 & a_0 \\ b_1 & b_0 \end{vmatrix} \xi_1 + \begin{vmatrix} a_2 & a_0 \\ b_2 & b_0 \end{vmatrix} \xi_2 = 0. \qquad (4.25)$$

Next, we multiply the first equation by b_2, the second by $-a_2$, and add the results. This gives

$$\begin{vmatrix} a_0 & a_2 \\ b_0 & b_2 \end{vmatrix} \xi_0 + \begin{vmatrix} a_1 & a_2 \\ b_1 & b_2 \end{vmatrix} \xi_1 = 0. \qquad (4.26)$$

The common solution of (4.25) and (4.26) is therefore the ratio of the numbers

$$\xi_0 = \begin{vmatrix} a_1 & a_2 \\ b_1 & b_2 \end{vmatrix}, \quad \xi_1 = -\begin{vmatrix} a_0 & a_2 \\ b_0 & b_2 \end{vmatrix}, \quad \xi_2 = \begin{vmatrix} a_0 & a_1 \\ b_0 & b_1 \end{vmatrix}.$$

This ratio is a point of the projective plane because the ξ_i are not all zero. If that were the case, the coefficients of (4.23) and (4.24) would have the same ratio.

The Parametric Equation of a Line

By definition, (4.22) is the equation of a line in the projective plane. Recall that at least one of the coefficients a_i is nonzero. We will now derive the parametric equation of the line. We fix distinct points $C = (\gamma_0, \gamma_1, \gamma_2)$ and $D = (\delta_0, \delta_1, \delta_2)$ on the line, and let $X = (\xi_0, \xi_1, \xi_2)$ be an arbitrary point on the line. We then have

$$\begin{cases} a_0 \gamma_0 + a_1 \gamma_1 + a_2 \gamma_2 = 0 \,, \\ a_0 \delta_0 + a_1 \delta_1 + a_2 \delta_2 = 0 \,, \\ a_0 \xi_0 + a_1 \xi_1 + a_2 \xi_2 = 0 \,. \end{cases} \quad (4.27)$$

This system of equations can also be considered from a different point of view: we can see the γ_i, δ_i, and ξ_i as the coefficients of the system and the a_i as the unknowns. By definition, at least one of the a_i is nonzero, so this system has a nonzero solution. As we will see in the digression below, this means that there exist λ and μ such that $\xi_i = \lambda \gamma_i + \mu \delta_i$ for $i = 0, 1, 2$. We can write this result as

$$(\xi_0, \xi_1, \xi_2) = \lambda(\gamma_0, \gamma_1, \gamma_2) + \mu(\delta_0, \delta_1, \delta_2), \text{ or, } X = \lambda C + \mu D \,.$$

Since a point in projective space is determined by the ratios of its coordinates, we may introduce the extra condition $\lambda + \mu = 1$. The parametric equation of the line is then $X = \lambda C + (1 - \lambda) D$.

Digression 10 (Determinant). *Let us define the determinant of a 3×3 matrix. A 3×3 matrix is a rectangular array of nine numbers arranged in three rows and three columns. The* determinant $\det B$ *of the matrix*

$$B = \begin{pmatrix} a_0 & a_1 & a_2 \\ b_0 & b_1 & b_2 \\ c_0 & c_1 & c_2 \end{pmatrix}$$

is given by

$$\det B = \begin{vmatrix} a_0 & a_1 & a_2 \\ b_0 & b_1 & b_2 \\ c_0 & c_1 & c_2 \end{vmatrix}$$
$$= a_0 b_1 c_2 + a_1 b_2 c_0 + a_2 b_0 c_1 - a_0 b_2 c_1 - a_2 b_1 c_0 - a_1 b_0 c_2 \,.$$

The determinant of B is the sum of six terms that are each the product of three numbers, one from each row and one from each column. The signs are determined by the following rule. A product has a plus sign if the numbers are arranged from upper left to lower right (with jumping backward if necessary), and a product has a minus sign if the numbers are arranged from upper right to lower left. The following properties of the determinant are easily verified using the definition. The first property is

4.5 Homogeneous Coordinates and Polarity

$$\begin{vmatrix} a_0 & a_1 & a_2 \\ b_0 & b_1 & b_2 \\ c_0 & c_1 & c_2 \end{vmatrix} = \begin{vmatrix} a_0 & b_0 & c_0 \\ a_1 & b_1 & c_1 \\ a_2 & b_2 & c_2 \end{vmatrix}.$$

In other words, interchanging the rows and columns of the matrix does not change the value of the determinant. In particular, every property of the determinant that holds for the rows also holds for the columns, and vice versa. The second property is

$$\begin{vmatrix} a_0 & a_1 & a_2 \\ \lambda b_0 & \lambda b_1 & \lambda b_2 \\ c_0 & c_1 & c_2 \end{vmatrix} = \lambda \begin{vmatrix} a_0 & a_1 & a_2 \\ b_0 & b_1 & b_2 \\ c_0 & c_1 & c_2 \end{vmatrix}.$$

More generally, if we multiply a row or a column of a matrix by a number λ, the determinant of the new matrix is λ times the determinant of the original matrix. The third property is

$$\begin{vmatrix} a_0 & a_1 & a_2 \\ b_0 & b_1 & b_2 \\ c_0 & c_1 & c_2 \end{vmatrix} = - \begin{vmatrix} b_0 & b_1 & b_2 \\ a_0 & a_1 & a_2 \\ c_0 & c_1 & c_2 \end{vmatrix}.$$

More generally, if we interchange two rows or two columns in a matrix, the sign of the determinant changes. In particular, the determinant of a matrix with two equal rows or columns is 0. The fourth property is

$$\begin{vmatrix} a_0+d_0 & a_1+d_1 & a_2+d_2 \\ b_0 & b_1 & b_2 \\ c_0 & c_1 & c_2 \end{vmatrix} = \begin{vmatrix} a_0 & a_1 & a_2 \\ b_0 & b_1 & b_2 \\ c_0 & c_1 & c_2 \end{vmatrix} + \begin{vmatrix} d_0 & d_1 & d_2 \\ b_0 & b_1 & b_2 \\ c_0 & c_1 & c_2 \end{vmatrix}.$$

More generally, if a row of a given matrix can be written as the sum of two rows, the determinant of the matrix is the sum of the determinants of the two matrices obtained by successively replacing that row by each of the new rows. Finally, the following property is a consequence of the four we have just given:

$$\begin{vmatrix} a_0+\lambda b_0 & a_1+\lambda b_1 & a_2+\lambda b_2 \\ b_0 & b_1 & b_2 \\ c_0 & c_1 & c_2 \end{vmatrix} = \begin{vmatrix} a_0 & a_1 & a_2 \\ b_0 & b_1 & b_2 \\ c_0 & c_1 & c_2 \end{vmatrix}.$$

More generally, if we add a multiple of a row of a matrix to another row, the determinant does not change.

Let us now consider the following system of three equations in three unknowns:
$$\begin{cases} a_0 x_0 + a_1 x_1 + a_2 x_2 = 0, \\ b_0 x_0 + b_1 x_1 + b_2 x_2 = 0, \\ c_0 x_0 + c_1 x_1 + c_2 x_2 = 0. \end{cases}$$

The matrix

210   4 CURVES

$$B = \begin{pmatrix} a_0 & a_1 & a_2 \\ b_0 & b_1 & b_2 \\ c_0 & c_1 & c_2 \end{pmatrix}$$

is called the coefficient matrix of the system. One solution of the system is the *trivial solution* $x_0 = x_1 = x_2 = 0$. Are there any other solutions? We can solve the first two equations for x_0, x_1, and x_2:

$$x_0 = \begin{vmatrix} a_1 & a_2 \\ b_1 & b_2 \end{vmatrix}, \quad x_1 = -\begin{vmatrix} a_0 & a_2 \\ b_0 & b_2 \end{vmatrix}, \quad x_2 = \begin{vmatrix} a_0 & a_1 \\ b_0 & b_1 \end{vmatrix}.$$

This solution also satisfies the third equation, that is, is a solution of the whole system, if and only if

$$c_0 \begin{vmatrix} a_1 & a_2 \\ b_1 & b_2 \end{vmatrix} - c_1 \begin{vmatrix} a_0 & a_2 \\ b_0 & b_2 \end{vmatrix} + c_2 \begin{vmatrix} a_0 & a_1 \\ b_0 & b_1 \end{vmatrix} = \begin{vmatrix} a_0 & a_1 & a_2 \\ b_0 & b_1 & b_2 \\ c_0 & c_1 & c_2 \end{vmatrix} = 0. \qquad (4.28)$$

In other words, the system with coefficient matrix B has a nontrivial solution if and only if $\det B = 0$.

Example 4.41. Let us once more consider system (4.27), in which the a_i are the unknowns and the γ_i, δ_i, and ξ_i are the coefficients. We know that this system has a nontrivial solution, so the determinant of the coefficient matrix must be 0. The first property of the determinant implies that the coefficient matrix of the following system, in the unknowns λ, μ, and ν, is also 0:

$$\begin{cases} \lambda \gamma_0 + \mu \delta_0 + \nu \xi_0 = 0, \\ \lambda \gamma_1 + \mu \delta_1 + \nu \xi_1 = 0, \\ \lambda \gamma_2 + \mu \delta_2 + \nu \xi_2 = 0. \end{cases}$$

It follows that this system has a nontrivial solution $(\lambda^*, \mu^*, \nu^*)$. Since $C = (\gamma_0, \gamma_1, \gamma_2)$ and $D = (\delta_1, \delta_2, \delta_3)$ are distinct points, so that the respective ratios of their coordinates are distinct, ν^* must be nonzero. The nontrivial solution is $X = (\lambda^*/\nu^*)C + (\mu^*/\nu^*)D$.

In passing let us briefly discuss another aspect in which the coordinate plane and the projective plane differ. In Sect. 2.2 we showed that a line divides the plane into two parts. This property is related to the fact that in the usual geometry of the plane, a line is split into two distinct parts if we leave out one point. In the projective plane the two parts of a line minus an ordinary point still meet at infinity. Consequently, the projective plane is not partitioned by any of its lines. Let us illustrate this further; see Fig. 4.49. The figure on the left shows the set of finite points of the projective plane. The figure on the right shows the whole projective plane as a shaded disk. The finite points lie in the interior of the circle; the circle itself represents the line at infinity. Since each line of the projective plane with the exception of the line at infinity contains exactly one point at infinity, each point at infinity, in particular $O_1 = (0, 1, 0)$

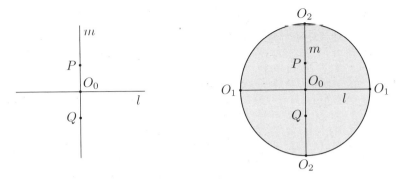

Fig. 4.49. The line l does not partition the projective plane

and $O_2 = (0, 0, 1)$, is represented twice. Every two diametrically opposite points of the circle represent the same projective point. Recall that a point of the projective plane is determined by the ratios of its coordinates. For that reason the point $(0, 0, 1)$, for example, is the same as the point $(0, 0, -1)$. We consider the lines $l\colon \xi_2 = 0$ and $m\colon \xi_1 = 0$. These lines meet at $O_0 = (1, 0, 0)$. We choose the points $P = (1, 0, 1)$ and $Q = (1, 0, -1)$ on m. If the figure on the left represented the coordinate plane, the line l would divide it into two parts, with the points P and Q lying on different sides of l. In the projective plane, things are different. We define three line segments on m:

$$[PQ] = \{\, X : X = \lambda(1, 0, 1) + (1 - \lambda)(1, 0, -1),\ \lambda \in [0, 1]\,\},$$
$$[PO_2] = \{\, X : X = \lambda(1, 0, 1) + (1 - \lambda)(0, 0, 1),\ \lambda \in [0, 1]\,\},$$
$$[QO_2] = \{\, X : X = \lambda(1, 0, -1) + (1 - \lambda)(0, 0, -1),\ \lambda \in [0, 1]\,\}.$$

Note that in the coordinate plane, line segments are defined differently. This is because we cannot use the metric of the coordinate plane in the projective plane. Taking a closer look at the formula, we obtain

$$[QO_2] = \{\, X : X = (\lambda, 0, -1),\ \lambda \in [0, 1]\,\}.$$

For $\lambda = 0$ we obtain the point at infinity O_2; for $\lambda \neq 0$ we obtain the ordinary point with homogeneous coordinates $(1, 0, -1/\lambda)$. In the figure on the right this point lies on the line segment $[QO_2]$. In the same way, we can verify that $[PQ]$ and $[PO_2]$ are the segments indicated in the figure to the right. The line l is disjoint from the line segments $[PO_2]$ and $[QO_2]$. The points P and Q are joined by the union of the segments $[PO_2]$ and $[QO_2]$. This shows that the line l does *not* partition the projective plane.

Conic Sections in $\mathbb{P}^2$

We will use the projective representation of the conic sections to study the tangents to the conics. We begin with the general equation of a conic section K in the coordinate plane, (4.17):

$$a_{11}x_1^2 + 2a_{12}x_1x_2 + a_{22}x_2^2 + 2a_{01}x_1 + 2a_{02}x_2 + a_{00} = 0 \, .$$

In homogeneous coordinates we can rewrite the equation of K in such a way that every term has degree 2, giving the so-called homogeneous quadratic equation

$$F(X) = a_{11}\xi_1^2 + 2a_{12}\xi_1\xi_2 + a_{22}\xi_2^2 + 2a_{01}\xi_0\xi_1 + 2a_{02}\xi_0\xi_2 + a_{00}\xi_0^2 = 0 \, . \quad (4.29)$$

We usually write this equation as

$$F(X) = \sum_{i,k=0}^{2} a_{ik}\xi_i\xi_k = 0, \text{ where } a_{ik} = a_{ki} \, . \quad (4.30)$$

In this equation X is the point in $\mathbb{P}^2$ with coordinates (ξ_0, ξ_1, ξ_2). Note that for $i \neq k$ we have $2a_{ik}\xi_i\xi_k = a_{ik}\xi_i\xi_k + a_{ki}\xi_k\xi_i$.

We consider the intersection of K and the line l through the points $C = (\gamma_0, \gamma_1, \gamma_2)$ and $D = (\delta_0, \delta_1, \delta_2)$. The parametric equation of l is

$$(\xi_0, \xi_1, \xi_2) = \lambda(\gamma_0, \gamma_1, \gamma_2) + \mu(\delta_0, \delta_1, \delta_2), \text{ or, } X = \lambda C + \mu D \, .$$

We can find the intersection points of l and K by substituting $X = \lambda C + \mu D$ in (4.30):

$$F(\lambda C + \mu D) = \sum_{i,k=0}^{2} a_{ik}(\lambda\gamma_i + \mu\delta_i)(\lambda\gamma_k + \mu\delta_k) = 0 \, .$$

By rearranging the terms we obtain

$$\lambda^2 \sum_{i,k=0}^{2} a_{ik}\gamma_i\gamma_k + 2\lambda\mu \sum_{i,k=0}^{2} a_{ik}\gamma_i\delta_k + \mu^2 \sum_{i,k=0}^{2} a_{ik}\delta_i\delta_k = 0 \, .$$

We introduce the following notation:

$$f(C, D) = \sum_{i,k=0}^{2} a_{ik}\gamma_i\delta_k \, . \quad (4.31)$$

Note that $f(C, D) = f(D, C)$ for all C and D. This property will play a prominent role later on. We also have $f(C, C) = F(C)$ for all C. With this notation the last equation becomes

$$F(\lambda C + \mu D) = \lambda^2 F(C) + 2\lambda\mu f(C, D) + \mu^2 F(D) = 0 \, . \quad (4.32)$$

This is a quadratic equation in the ratio λ/μ or μ/λ. The line l and the conic section K therefore have 0, 1, or 2 points in common.

We call the point $D = (\delta_0, \delta_1, \delta_2)$ a *double point* of the conic section K with equation (4.30) if D is a solution of the system of equations

4.5 Homogeneous Coordinates and Polarity

$$a_{00}\delta_0 + a_{10}\delta_1 + a_{20}\delta_2 = 0 \,,$$
$$a_{01}\delta_0 + a_{11}\delta_1 + a_{21}\delta_2 = 0 \,,$$
$$a_{02}\delta_0 + a_{12}\delta_1 + a_{22}\delta_2 = 0 \,.$$

From the digression on determinants, we know that K has a double point if and only if
$$\begin{vmatrix} a_{00} & a_{10} & a_{20} \\ a_{01} & a_{11} & a_{21} \\ a_{02} & a_{12} & a_{22} \end{vmatrix} = 0 \,.$$

If D is a double point of K, every $X = (\xi_0, \xi_1, \xi_2)$ satisfies

$$f(D, X) = \sum_{i,k=0}^{2} a_{ik}\delta_i\xi_k = \sum_{k=0}^{2}\left(\sum_{i=0}^{2} a_{ik}\delta_i\right)\xi_k = 0 \,.$$

In particular, D is a point of the conic section K. Let us now take an arbitrary point $C = (\gamma_0, \gamma_1, \gamma_2)$ of the projective plane, and consider the intersection of K and the line through C and D. Equation (4.32) now reads

$$F(\lambda C + \mu D) = \lambda^2 F(C) = 0 \,.$$

We see that $\lambda = 0$ is a double root of the equation and that D is an intersection point of order 2; this is why we call D a double point. Conversely, if $\lambda = 0$ is a double root of the equation $F(\lambda D + \mu X) = 0$ for every X, then $f(D, X) = 0$ for every X. It follows that $\sum_{i=0}^{2} a_{ik}\delta_i = 0$ for $k = 0, 1, 2$, so D is a double point of K.

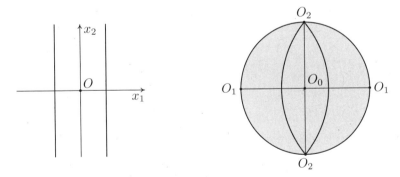

Fig. 4.50. (a) The conic section $x_1^2 = 1$; (b) its projective version

Examples 1 *The conic section $\xi_1^2 - \xi_2^2 = 0$ has a double point at $(1, 0, 0)$. In the Cartesian plane, this conic section is a degenerate conic, namely a pair of lines that meet at $(0, 0)$; this point is a double point.*
Let us now consider the conic section $x_1^2 - 1 = 0$. In the Cartesian plane

this is again a degenerate conic, namely a pair of parallel lines (Sect. 4.4). In homogeneous coordinates, the equation of this conic section is $\xi_1^2 - \xi_0^2 = 0$; see Fig. 4.50. The point at infinity O_2 is a double point.

Polarity

Let us study the conic section K with equation (4.30) in the projective plane. We assume that the conic section has no double points, which excludes only the degenerate conics; see Exercise 4.43.

Definition 4.42. *The points C and D of the projective plane are said to be* conjugate *with respect to the conic section K if $f(C, D) = 0$ (with $f(C, D)$ as defined in (4.31)). For a fixed C, the equation $f(C, X) = 0$ defines a line called the* polar *of C.*

The first property of the polar is obvious.

Theorem 4.43. *The point C belongs to the polar of the point D if and only if D belongs to the polar of C.*

The following property will help us determine the equations of the tangents to the conics.

Theorem 4.44. *If C is a point of the conic section K, the polar of C is tangent to K at C.*

Proof. If C lies on the conic section and D lies on the polar of C, (4.32) becomes
$$F(\lambda C + \mu D) = \mu^2 F(D) = 0 \,.$$
We see that $\mu = 0$ is a double root of the equation. This shows that C is an intersection point of multiplicity 2 of the polar of C and the conic section. The polar has no other intersection points with K. Hence, the polar of C is tangent to K at C.

Example 4.45. Let $C = (c_1, c_2)$ be a point on the parabola $x^2 = 2py$. We are looking for the equation of the tangent to the parabola at C.
To solve this problem, we use homogeneous coordinates. In homogeneous coordinates, the point C is $(1, c_1, c_2)$, and the equation of the parabola is $\xi_1^2 - 2p\xi_0\xi_2 = 0$. The term $2p\xi_0\xi_2$ is equal to $p\xi_0\xi_2 + p\xi_2\xi_0$. Consequently, we can rewrite the equation as
$$F(X) = f(X, X) = \xi_1 \xi_1 - p\xi_0 \xi_2 - p\xi_2 \xi_0 = 0 \,.$$
Replacing the first factor of each product $\xi_i \xi_k$ by the corresponding coordinate of C, we obtain
$$f(C, X) = c_1 \xi_1 - p\xi_2 - pc_2 \xi_0 = 0 \,.$$
Changing back to Cartesian coordinates gives the equation of the tangent:
$$c_1 x = p(y + c_2) \,.$$

Example 4.46. In Example 4.35 we studied the form of the conic section K with equation $x^2 + xy + y^2 - 3x - 3y + 1 = 0$ and its position in the coordinate plane; see Fig. 4.51. The point $(1, 1+\sqrt{2})$ lies on the ellipse. In homogeneous

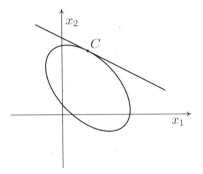

Fig. 4.51. Tangent to the ellipse at C

coordinates, the equation of the ellipse is

$$\xi_1^2 + \xi_1\xi_2 + \xi_2^2 - 3\xi_0\xi_1 - 3\xi_0\xi_2 + \xi_0^2 = 0\ .$$

The coefficient -3 of $\xi_0\xi_1$ comes from a_{01} and a_{10}. Therefore we rewrite the equation in the form

$$f(X, X) = \xi_0^2 + \xi_1^2 + \xi_2^2 - \tfrac{3}{2}\xi_0\xi_1 - \tfrac{3}{2}\xi_1\xi_0 - \tfrac{3}{2}\xi_0\xi_2 - \tfrac{3}{2}\xi_2\xi_0 + \tfrac{1}{2}\xi_1\xi_2 + \tfrac{1}{2}\xi_2\xi_1$$
$$= 0\ .$$

The point C has homogeneous coordinates $(1, 1, 1+\sqrt{2})$. We obtain the equation of the polar of C, which is also the tangent to the ellipse at C, by replacing the first factor of each term $\xi_i\xi_k$ by the corresponding coordinate of C. The result is

$$f(C, X) = \xi_0 + \xi_1 + (1+\sqrt{2})\xi_2 - \tfrac{3}{2}\xi_1 - \tfrac{3}{2}\xi_0 - \tfrac{3}{2}\xi_2 - \tfrac{3}{2}\left(1+\sqrt{2}\right)\xi_0 + \tfrac{1}{2}\xi_2$$
$$+ \tfrac{1}{2}\left(1+\sqrt{2}\right)\xi_1$$
$$= 0\ .$$

Rearranging the terms, we obtain $(-2\sqrt{2}-3)\xi_0 + \xi_1 + 2\xi_2 = 0$. In the coordinate plane, this equation becomes

$$2y + x - 2\sqrt{2} - 3 = 0\ .$$

Theorem 4.47. *Let C be a point not on the conic section K. For any point D of K, the line through C and D is a tangent of K if and only if D lies on the polar of C. The tangents to the conic through C are given by the equation*

$$(f(C, X))^2 - F(C)F(X) = 0\ .$$

Proof. Since C does not lie on the conic, while D does, (4.32) has the form
$$\lambda^2 F(C) + 2\lambda\mu f(C,D) = 0,$$
where $F(C) \neq 0$. Consequently, $\lambda = 0$ is a double root of the equation, whence D is a point of contact if and only if $f(C,D) = 0$. This proves the first statement of the theorem.

Next let D be an arbitrary point of the projective plane. The line through C and D is a tangent of K if and only if (4.32) has a double root. This happens if and only if the discriminant of the equation is 0. In that case the point D satisfies
$$(f(C,D))^2 - F(C)F(D) = 0.$$
This proves the second statement of the theorem.

Example 4.48. What are the equations of the tangents to the ellipse $x^2 + 2y^2 = 4$ through the point $C = (2,3)$? See Fig. 4.52. The homogeneous coordinates

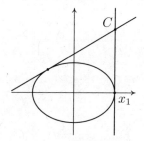

Fig. 4.52. Tangents of the ellipse through C

of C are $(1, 2, 3)$, and the equation of the ellipse in homogeneous coordinates is
$$F(X) = -4\xi_0^2 + \xi_1^2 + 2\xi_2^2 = 0,$$
so $F(C) = 18$. The equation of the polar of C is
$$f(C,X) = -4\xi_0 + 2\xi_1 + 6\xi_2.$$
According to the theorem, the tangents are the solutions of
$$(-4\xi_0 + 2\xi_1 + 6\xi_2)^2 - 18(-4\xi_0^2 + \xi_1^2 + 2\xi_2^2) = 0.$$
This equation can also be written as
$$2(-2\xi_0 + \xi_1)(-22\xi_0 - 7\xi_1 + 12\xi_2) = 0.$$
In Cartesian coordinates, the equations of the tangents through $C = (2,3)$ are therefore $x = 2$ and $12y = 7x + 22$. The first of these equations does not come as a surprise. The points of contact are the common solutions of the equation of the polar of C and the equations of the tangents. These are $(2,0)$ and $(-14/11, 12/11)$.

Example 4.49. We will now look for the equations of the tangents to the hyperbola $b^2x_1^2 - a^2x_2^2 = a^2b^2$ through the point $(0,0)$; see Fig. 4.53. The equation

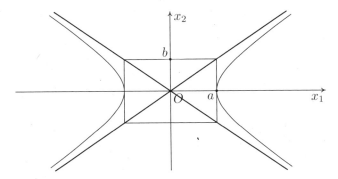

Fig. 4.53. The asymptotes of the hyperbola $b^2x_1^2 + a^2x_2^2 = a^2b^2$

of the hyperbola in homogeneous coordinates is

$$F(X) = a^2b^2\xi_0^2 - b^2\xi_1^2 + a^2\xi_2^2 = 0.$$

The polar of $(1,0,0)$ is $\xi_0 = 0$. The tangents through $(1,0,0)$ are

$$b\xi_1 - a\xi_2 = 0 \quad \text{and} \quad b\xi_1 + a\xi_2 = 0,$$

with the points of contact $(0,a,b)$ and $(0,a,-b)$, respectively. In Cartesian coordinates, the two tangents through $(0,0)$ have the same equation as that through $(1,0,0)$ in homogeneous coordinates. Since the points of contact lie on the line at infinity, we call the tangent lines *asymptotes*. The figure illustrates the geometric meaning of the numbers a and b. For an arbitrary point $P = (1,p_1,p_2)$, the line

$$b\xi_1 - a\xi_2 = (bp_1 - ap_2)\xi_0$$

through P intersects the hyperbola at a point at infinity. Because of this, $(0,a,b)$ is called an *asymptotic direction*.

Example 4.50 (Pedal curve). Let us demonstrate a general procedure for constructing curves on an example, for which we take the *pedal curve*. Consider the parabola with focus F. Let D be the foot of the perpendicular from F on the directrix. As usual, we write $p = FD$. This number is called the *focal parameter* of the parabola. The point O on the line DF is such that $OD = DF$, and O lies on the other side of D from F; see Fig. 4.54. We draw the tangent to the parabola at a point Y. Next, we construct the foot X of the perpendicular from O on the tangent. The curve that consists of all points X obtained in this way for a variable point Y on the parabola is called the *pedal curve of the parabola with respect to the point O*. This curve is also called the

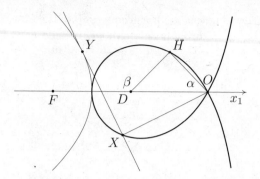

Fig. 4.54. The trisectrix, an example of a pedal curve

Maclaurin *trisectrix*; we will explain this later. Let us first derive the equation of the trisectrix. We choose a coordinate system at O with the line FO as x_1-axis, oriented such that $D = (-p, 0)$ and $F = (-2p, 0)$. The equation of the parabola is
$$x_2^2 + 2px_1 + 3p^2 = 0 \ .$$
Let $Y = (\eta_1, \eta_2)$ be an arbitrary point on the parabola. Using homogeneous coordinates, we obtain the equation of the tangent at Y:
$$\eta_2 x_2 + px_1 + p\eta_1 + 3p^2 = 0 \ .$$
The perpendicular from O on this tangent has equation
$$\eta_2 x_1 - px_2 = 0 \ .$$
Since Y lies on the parabola, we have
$$\eta_2^2 + 2p\eta_1 + 3p^2 = 0 \ .$$
We can now find the equation of the trisectrix by eliminating η_1 and η_2 from the last three equations. The easiest way is first to solve η_1 and η_2 from the equations of the tangent and perpendicular, and then to substitute the resulting values in the third equation. This gives the equation
$$2x_1 \left(x_1^2 + x_2^2\right) = -3px_1^2 + px_2^2 \ . \tag{4.33}$$
The intersection points with the x_1-axis are O, with multiplicity two, and $(-3p/2, 0)$. The point O is a node. For a better representation of this curve we use polar coordinates. Substituting $x_1 = r\cos\varphi$ and $x_2 = r\sin\varphi$, we obtain $r = 0$ and
$$r = \frac{-3p\cos^2\varphi + p\sin^2\varphi}{2\cos\varphi} = \frac{p}{2\cos\varphi}\left(1 - 4\cos^2\varphi\right) \ .$$
Since $r \geq 0$ by definition, there are some restrictions on the values of φ. An analysis of the graph of r as a function of φ gives a good idea of the shape of the trisectrix. Here are the results of the analysis:

1. For $-\pi/3 < \varphi < \pi/3$ there are no points on the trisectrix because r is negative for these values of φ.
2. For $\pi/3 \le \varphi < \pi/2$ we obtain the part of the curve that lies in the first quadrant.
3. For $\pi/2 \le \varphi < 2\pi/3$ there are no points.
4. For $2\pi/3 \le \varphi \le 4\pi/3$ we obtain the part of the curve that lies in the second and third quadrants, that is, the loop.
5. For $4\pi/3 < \varphi < 3\pi/2$ there are no points.
6. For $3\pi/2 < \varphi \le 5\pi/3$ we obtain the part of the curve that lies in the fourth quadrant.

Let us now explain why the curve is called a trisectrix. Consider a point H located on the loop of the curve, above the x_1-axis. The line segments from H to O and D define two angles with the x_1-axis: $\alpha = HOD$ and $\beta = HDF$, respectively. These angles satisfy $\beta = 3\alpha$; the graph of the trisectrix can therefore be used to trisect an angle. We prove the relation between α and β as follows. The second polar coordinate of H is $\varphi = \pi - \alpha$. Note that if H lies on the loop and above the x_1-axis, its angular coordinate satisfies $2\pi/3 \le \phi < \pi$. Consequently, $0 < \alpha \le \pi/3$. Using the formula found above for the radial coordinate, we obtain the Cartesian coordinates of H:

$$x_{1H} = \frac{p}{2}\left(1 - 4\cos^2(\pi - \alpha)\right)$$
$$= -p + \frac{p}{2}\left(3 - 4\cos^2\alpha\right)$$
$$= -p - \frac{p}{2\cos\alpha}\cos(3\alpha)$$

and

$$x_{2H} = \frac{p\sin(\pi - \alpha)}{2\cos(\pi - \alpha)}\left(1 - 4\cos^2(\pi - \alpha)\right)$$
$$= -\frac{p}{2\cos\alpha}\left(-3\sin\alpha + 4\sin^3\alpha\right)$$
$$= \frac{p}{2\cos\alpha}\sin(3\alpha).$$

If we translate the coordinate system to the left and place the origin at D, the polar coordinates of H become $(\rho, \pi - \beta)$. The Cartesian coordinates of H in this system are

$$x^*_{1H} = \rho\cos(\pi - \beta) = -\rho\cos\beta \quad \text{and} \quad x^*_{2H} = \rho\sin(\pi - \beta) = \rho\sin\beta.$$

But then we must have

$$-p - \frac{p}{2\cos\alpha}\cos(3\alpha) = -p - \rho\cos\beta \quad \text{and} \quad \frac{p}{2\cos\alpha}\sin(3\alpha) = \rho\sin\beta.$$

It follows that $\rho = p/(2\cos\alpha)$ and $\beta = 3\alpha$.

Projective Transformations

We will now briefly discuss projective transformations. Keep in mind that a point of $\mathbb{P}^2$ is a triple ratio, in other words, the triples (ξ_0, ξ_1, ξ_2) and $(\lambda\xi_0, \lambda\xi_1, \lambda\xi_2)$ denote the same point for all values of $\lambda \neq 0$.

Definition 4.51. *A* projective transformation *is a function* $\mathbf{S}$ *that to a point* $X = (\xi_0, \xi_1, \xi_2)$ *of* $\mathbb{P}^2$ *assigns the point* $Y = (\eta_0, \eta_1, \eta_2) = \mathbf{S}(X)$ *of* $\mathbb{P}^2$ *defined by the formula*

$$\begin{cases} \eta_0 = s_{00}\xi_0 + s_{01}\xi_1 + s_{02}\xi_2 , \\ \eta_1 = s_{10}\xi_0 + s_{11}\xi_1 + s_{12}\xi_2 , \\ \eta_2 = s_{20}\xi_0 + s_{21}\xi_1 + s_{22}\xi_2 , \end{cases}$$

where

$$\begin{vmatrix} s_{00} & s_{01} & s_{02} \\ s_{10} & s_{11} & s_{12} \\ s_{20} & s_{21} & s_{22} \end{vmatrix} \neq 0 .$$

In *projective geometry*, we study the properties that remain invariant under projective transformations. Using Exercise 4.46, we can write the solution X of $Y = \mathbf{S}X$ as $X = \mathbf{T}Y$ for some projective transformation $\mathbf{T}$. From this it immediately follows that, for example, lines and conic sections are notions from projective geometry.

We will use the reference points O_0, O_1, O_2 defined earlier, as well as the unit point $E = (1,1,1)$. The following theorem shows that there is an abundance of projective maps.

Theorem 4.52. *Let* $A = (\alpha_0, \alpha_1, \alpha_2)$, $B = (\beta_0, \beta_1, \beta_2)$, $C = (\gamma_0, \gamma_1, \gamma_2)$, *and* $D = (\delta_0, \delta_1, \delta_2)$ *be points of which no three are collinear. There is a unique projective map* $\mathbf{S}$ *such that* $\mathbf{S}(O_0) = A$, $\mathbf{S}(O_1) = B$, $\mathbf{S}(O_2) = C$, *and* $\mathbf{S}(E) = D$.

Proof. For all λ, μ, and ν the transformation

$$\begin{aligned} \eta_0 &= \lambda\alpha_0\xi_0 + \mu\beta_0\xi_1 + \nu\gamma_0\xi_2 , \\ \eta_1 &= \lambda\alpha_1\xi_0 + \mu\beta_1\xi_1 + \nu\gamma_1\xi_2 , \\ \eta_2 &= \lambda\alpha_2\xi_0 + \mu\beta_2\xi_1 + \nu\gamma_2\xi_2 , \end{aligned}$$

sends O_0 to A, O_1 to B, and O_2 to C. This easily follows by substitution. We still can use any values for λ, μ, and ν; at this point we make explicit use of the fact that points of $\mathbb{P}^2$ are ratios. The condition $\mathbf{S}(E) = D$ becomes

$$\begin{aligned} \delta_0 &= \lambda\alpha_0 + \mu\beta_0 + \nu\gamma_0 , \\ \delta_1 &= \lambda\alpha_1 + \mu\beta_1 + \nu\gamma_1 , \\ \delta_2 &= \lambda\alpha_2 + \mu\beta_2 + \nu\gamma_2 . \end{aligned}$$

We can see this as a system of equations in the unknowns λ, μ, and ν. Since A, B, and C are noncollinear, the determinant of the coefficient matrix is nonzero; see Exercise 4.45. According to Exercise 4.46, there is a unique solution. We denote this solution by λ, μ, and ν.

4.5 Homogeneous Coordinates and Polarity

The formula we just found for **S** can also be used to describe a coordinate transformation. In the proof of the theorem we used the formula to send O_0, O_1, O_2, and E to A, B, C, and D, respectively. We could say that we have given the points an *alibi* (a new position). In the case of coordinate transformation, on the other hand, we assign a new name to the points, an *alias*. Given four points A, B, C, and D of which no three are collinear, we can choose new reference points and a new unit point at these points: $\widetilde{O}_0$ at A, $\widetilde{O}_1$ at B, $\widetilde{O}_2$ at C, and $\widetilde{E}$ at D. The formula of **S** shows how to compute the old coordinates from the new ones.

Classification

Let us finally show that the distinction between ellipse, parabola, and hyperbola disappears in projective geometry. In introducing the projective plane we make the distinction between ordinary, or finite, points and infinite points. However, we showed above that there is an abundance of coordinate transformations in projective geometry. Consequently, we can designate any line as the line at infinity.

Let K be a conic section given by (4.30). We choose the new reference points $\widetilde{O}_0$, $\widetilde{O}_1$, and $\widetilde{O}_2$ in such a way that $\widetilde{O}_i$ are $\widetilde{O}_j$ are conjugates with respect to K for $i \neq j$. That can easily be arranged; see Fig. 4.55. We assume

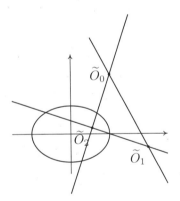

Fig. 4.55. Conjugate basic points

that the conic section is not a double line. We first choose the point $\widetilde{O}_0$ not on the conic section. Next we choose $\widetilde{O}_1$ on the polar of $\widetilde{O}_0$, but outside the conic section. The point $\widetilde{O}_2$ is then the intersection point of the polars of $\widetilde{O}_0$ and $\widetilde{O}_1$. What is the equation of the conic in the new coordinates? The general form of the equation of a conic section is (4.30). For $i \neq j$, the points $\widetilde{O}_i$ and $\widetilde{O}_j$ are conjugate; hence $a_{ij} = 0$. The equation of K with respect to the new basic points is therefore

$$a_{00}x_0^2 + a_{11}x_1^2 + a_{22}x_2^2 = 0 \ .$$

After scaling, where we replace x_i by $x_i/\sqrt{|a_{ii}|}$, and possibly after interchanging the reference points, we obtain one of the following forms for the equation of K:

1. $\xi_0^2 + \xi_1^2 + \xi_2^2 = 0$: no solution in $\mathbb{P}^2$.
2. $\xi_0^2 + \xi_1^2 - \xi_2^2 = 0$: a "hyperbola".
3. $\xi_0^2 + \xi_1^2 = 0$: the point $\tilde{O}_2$.
4. $\xi_0^2 - \xi_1^2 = 0$: two intersecting lines.
5. $\xi_0^2 = 0$: two coinciding lines.

We see that the distinction between ellipse, parabola, and hyperbola has disappeared. The distinction reappears if we switch to Cartesian coordinates. If

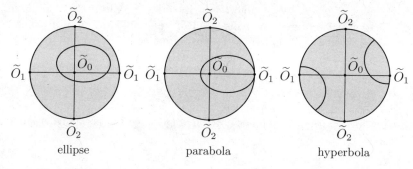

Fig. 4.56. The position of the conic with respect to the line at infinity

the conic section does not meet the line at infinity, we have an ellipse in the coordinate plane. If the conic section touches the line at infinity, we have a parabola; and if the conic section intersects the line at infinity, as in case 2 above, we have a hyperbola. See Fig. 4.56.

Exercises

4.41. The system of equations

$$\begin{cases} a_1 x_1 + a_2 x_2 = a_3 \ , \\ b_1 x_1 + b_2 x_2 = b_3 \ , \end{cases}$$

in the unknowns x_1 and x_2, with $\begin{vmatrix} a_1 & a_2 \\ b_1 & b_2 \end{vmatrix} \neq 0$, has the unique solution

$$x_1 = \frac{\begin{vmatrix} a_3 & a_2 \\ b_3 & b_2 \end{vmatrix}}{\begin{vmatrix} a_1 & a_2 \\ b_1 & b_2 \end{vmatrix}}, \quad x_2 = \frac{\begin{vmatrix} a_1 & a_3 \\ b_1 & b_3 \end{vmatrix}}{\begin{vmatrix} a_1 & a_2 \\ b_1 & b_2 \end{vmatrix}} \ .$$

Hint: See the proof of Theorem 4.40.

4.42. Show that

$$\begin{vmatrix} a_0 & a_1 & a_2 \\ b_0 & b_1 & b_2 \\ c_0 & c_1 & c_2 \end{vmatrix} = a_0 \begin{vmatrix} b_1 & b_2 \\ c_1 & c_2 \end{vmatrix} - a_1 \begin{vmatrix} b_0 & b_2 \\ c_0 & c_2 \end{vmatrix} + a_2 \begin{vmatrix} b_0 & b_1 \\ c_0 & c_1 \end{vmatrix}$$

$$= a_0 \begin{vmatrix} b_1 & b_2 \\ c_1 & c_2 \end{vmatrix} - b_0 \begin{vmatrix} a_1 & a_2 \\ c_1 & c_2 \end{vmatrix} + c_0 \begin{vmatrix} a_1 & a_2 \\ b_1 & b_2 \end{vmatrix}.$$

See (4.28).

4.43. The general equation of a conic section K in the coordinate plane is

$$a_{11}x_1^2 + 2a_{12}x_1x_2 + a_{22}x_2^2 + 2a_{01}x_1 + 2a_{02}x_2 + a_{00} = 0.$$

In homogeneous coordinates it is

$$F(X) = \sum_{i,k=0}^{2} a_{ik}\xi_i\xi_k = 0, \text{ with } a_{ik} = a_{ki}.$$

(a) If the conic section has a finite double point, this point is the center of the conic section; see (4.21).
(b) If K is an ellipse, parabola, or hyperbola, it has no double points.

4.44. Let (c_1, c_2) be a point on the ellipse with equation $x^2/a^2 + y^2/b^2 = 1$. The equation of the tangent to the ellipse at (c_1, c_2) is

$$\frac{c_1 x}{a^2} + \frac{c_2 y}{b^2} = 1.$$

4.45. The points $B = (\beta_0, \beta_1, \beta_2)$, $C = (\gamma_0, \gamma_1, \gamma_2)$, and $D = (\delta_0, \delta_1, \delta_2)$ of the projective plane are collinear if and only if

$$\begin{vmatrix} \beta_0 & \beta_1 & \beta_2 \\ \gamma_0 & \gamma_1 & \gamma_2 \\ \delta_0 & \delta_1 & \delta_2 \end{vmatrix} = 0.$$

Hint: Substituting B, C, and D in the equation $a_0\xi_0 + a_1\xi_1 + a_2\xi_2 = 0$ gives three equations in the unknowns a_0, a_1, a_2.

4.46. In this exercise we discuss a generalization of Exercise 4.41. We are given a system of equations

$$\begin{cases} a_0x_0 + a_1x_1 + a_2x_2 = a_3, \\ b_0x_0 + b_1x_1 + b_2x_2 = b_3, \\ c_0x_0 + c_1x_1 + c_2x_2 = c_3, \end{cases}$$

in the unknowns x_0, x_1, x_2 that satisfies

$$\begin{vmatrix} a_0 & a_1 & a_2 \\ b_0 & b_1 & b_2 \\ c_0 & c_1 & c_2 \end{vmatrix} \neq 0.$$

In this exercise we determine the *unique solution* of this system.

(a) At least one of the determinants $\begin{vmatrix} a_1 & a_2 \\ b_1 & b_2 \end{vmatrix}$, $\begin{vmatrix} a_1 & a_2 \\ c_1 & c_2 \end{vmatrix}$, $\begin{vmatrix} b_1 & b_2 \\ c_1 & c_2 \end{vmatrix}$ is nonzero.
Hint: See Exercise 4.42.
From now on we will assume that $\begin{vmatrix} a_1 & a_2 \\ b_1 & b_2 \end{vmatrix} \neq 0$.

(b) Use Exercise 4.41 to find λ and μ such that
$$\begin{cases} \lambda a_1 + \mu b_1 = c_1, \\ \lambda a_2 + \mu b_2 = c_2. \end{cases}$$

(c) Simultaneously eliminate x_1 and x_2 by multiplying the first equation of the system by $\lambda \begin{vmatrix} a_1 & a_2 \\ b_1 & b_2 \end{vmatrix}$, the second by $\mu \begin{vmatrix} a_1 & a_2 \\ b_1 & b_2 \end{vmatrix}$, and the third by $-\begin{vmatrix} a_1 & a_2 \\ b_1 & b_2 \end{vmatrix}$, and adding the results. This gives the equation

$$x_0 \left(-a_0 \begin{vmatrix} b_1 & b_2 \\ c_1 & c_2 \end{vmatrix} + b_0 \begin{vmatrix} a_1 & a_2 \\ c_1 & c_2 \end{vmatrix} - c_0 \begin{vmatrix} a_1 & a_2 \\ b_1 & b_2 \end{vmatrix} \right)$$
$$= \left(-a_3 \begin{vmatrix} b_1 & b_2 \\ c_1 & c_2 \end{vmatrix} + b_3 \begin{vmatrix} a_1 & a_2 \\ c_1 & c_2 \end{vmatrix} - c_3 \begin{vmatrix} a_1 & a_2 \\ b_1 & b_2 \end{vmatrix} \right).$$

(d) It follows that
$$x_0 \begin{vmatrix} a_0 & a_1 & a_2 \\ b_0 & b_1 & b_2 \\ c_0 & c_1 & c_2 \end{vmatrix} = \begin{vmatrix} a_3 & a_1 & a_2 \\ b_3 & b_1 & b_2 \\ c_3 & c_1 & c_2 \end{vmatrix}.$$

(e) Find x_1 and x_2.

4.47. In this exercise we take a further look at Example 4.50.

(a) Determine the polar of O with respect to the parabola.
(b) Show that the angle between the tangents to the parabola through O is equal to $\pi/3$.
(c) Every point (x_1, x_2) on the trisectrix satisfies $-3p/2 \leq x_1 < p/2$ ($x_1 = p/2$ is a vertical asymptote).

4.48. We consider the parabola with focus $F = (-2a, 0)$, where $a > 0$, and the x_2-axis as directrix.

(a) The pedal curve of the parabola with respect to $O = (0,0)$ has equation
$$x_2^2(a - x_1) - x_1^2(a + x_1) = 0.$$

(b) Show that the pedal curve and the strophoid (Sect. 4.3, Exercise 4.30) are congruent.

4.49. Let K be a parabola that touches the x_1-axis at $(1,0)$ and the line with equation $qx_1 - px_2 = 0$ at (p, q), where $p \neq 0 \neq q$. The equation of K is
$$q^2 x_1^2 - 2(p+1)qx_1 x_2 + (p+1)^2 x_2^2 - 2q^2 x_1 + 2(p-1)qx_2 + q^2 = 0.$$

This formula can be obtained using homogeneous coordinates. We begin with the general equation (4.29) of a conic section.

(a) Since the tangent at $(1,1,0)$ has equation $\xi_2 = 0$, we have $a_{11} = -a_{01} = a_{00}$ and $a_{12} \neq a_{02}$.
(b) Since K is a parabola, the general equation has the form
$$\xi_1^2 + 2b_{12}\xi_1\xi_2 + b_{12}^2\xi_2^2 - 2\xi_0\xi_1 + 2b_{02}\xi_0\xi_2 + \xi_0^2 = 0, \qquad b_{02} \neq b_{12}.$$
(c) Since the parabola passes through $(1, p, q)$ and the tangent at that point has equation $q\xi_1 - p\xi_2 = 0$, we must have
$$p - qb_{02} - 1 = 0 \quad \text{and} \quad (p + b_{12}q)^2 = 1.$$
One of the solutions for b_{12} is not admissible, since it results in a double line.

4.6 The Parametric Equations of a Curve, Cycloids

Browsing through the books [7], [41], and [66], you may be surprised by the enormous variety of curves that have been invented through the centuries. In earlier sections we discussed the most important curves. In this section we introduce several new kinds of curves. Quite frequently it is convenient to represent curves using coordinate functions of the type
$$x_1 = f_1(t), \quad x_2 = f_2(t).$$
Such functions are called *parametric equations* of the curve. They suggest

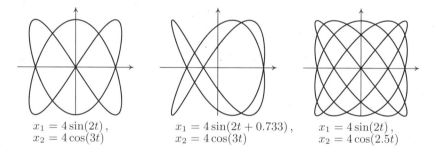

Fig. 4.57. Lissajous curves

that the curve has a dynamical nature. Thinking of t as time, we see a point $(f_1(t), f_2(t))$ whose orbit traces the curve. In most examples the domain of the functions f_1 and f_2 is the real line; in that case we do not mention it explicitly. Our first example consists of the well-known *Lissajous* (1822–1880) curves; see Fig. 4.57.

Let us briefly discuss the parametric representations of the conic sections. It comes as no surprise that the parametric equations of the circle with radius A and center at the origin are

$$x_1 = A\cos t, \qquad x_2 = A\sin t.$$

That is how the trigonometric functions were defined. The following representation of the ellipse is related to the parametric equations of the circle; see Fig. 4.58. The ellipse with equation $b^2 x_1^2 + a^2 x_2^2 = a^2 b^2$ in Cartesian coordi-

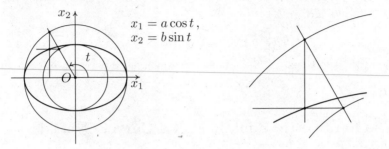

Fig. 4.58. (a) Parametric representation of the ellipse; (b) detail

nates has parametric equations

$$x_1 = a\cos t, \qquad x_2 = b\sin t.$$

For simplicity we assume that $0 < b < a$. Note that the parameter t has a simple geometric meaning. To see this, we first draw the circles with center O and radii a respectively b. Next we draw a half-line through O at an angle t with the x_1-axis. We can find the point of the ellipse corresponding to t as follows. We draw a vertical line through the intersection point of the half-line and the circle $\odot(O, a)$, and a horizontal line through the intersection point of the half-line and the circle $\odot(O, b)$. The intersection point of these lines is the point $(a\cos t, b\sin t)$ on the ellipse.

The parametric representation of the circle and ellipse we have just discussed is not the only representation of its type. There is another parametrization using rational functions, based on the fact that $\sin t$ and $\cos t$ are rational functions of $u = \tan(t/2)$. For the circle we obtain the parametrization

$$x_1 = \cos t = \frac{\cos^2 \tfrac{1}{2}t - \sin^2 \tfrac{1}{2}t}{\cos^2 \tfrac{1}{2}t + \sin^2 \tfrac{1}{2}t} = \frac{1-u^2}{1+u^2},$$

$$x_2 = \sin t = \frac{2\sin \tfrac{1}{2}t \cos \tfrac{1}{2}t}{\cos^2 \tfrac{1}{2}t + \sin^2 \tfrac{1}{2}t} = \frac{2u}{1+u^2}.$$

This representation maps the real line bijectively onto the circle minus the point $(-1, 0)$.

A *spiral* is a curve with parametric equations

$$x_1 = r(\varphi) \cos \varphi ,$$
$$x_2 = r(\varphi) \sin \varphi ,$$

where $r(\varphi)$ is a strictly monotonic function of φ. A well-known example is the *Archimedean spiral* given by $r(\varphi) = a\varphi$, where a is a nonzero real number. Figure 4.59 shows such a spiral with positive a.

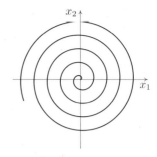

Fig. 4.59. An Archimedean spiral

Example 4.53 (Spirals). Often a curve is defined by a function $r(\varphi)$ that describes the relation between the polar coordinates of the points of the curve. The curve then has parametric equations

$$x_1 = r(\varphi) \cos \varphi , \qquad x_2 = r(\varphi) \sin \varphi .$$

If the function $r(\varphi)$ is strictly monotonic, the curve is called a spiral; see Fig. 4.59.

In our discussion of the golden mean in Sect. 2.5, Exercise 2.49, we defined a similarity transformation: a rotation over $\pi/2$ about S followed by a central dilation with center S and scaling factor τ. Let us denote this transformation by $\mathcal{F}$. Consider Fig. 4.60. The center S is the intersection point of

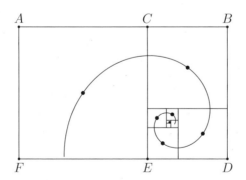

Fig. 4.60. A logarithmic spiral

the diagonal AD and the line EB, and τ is the inverse of the golden mean. If we choose a coordinate system with origin at S, the action of $\mathcal{F}$ is given by $\mathcal{F}((\rho, \varphi)) = (\rho\tau, \varphi - \pi/2)$. Starting with the point (ρ, φ) and repeatedly applying the map $\mathcal{F}$, we obtain the sequence

$$(\rho, \varphi), \quad \left(\tau\rho, \varphi - \tfrac{\pi}{2}\right), \quad (\tau^2\rho, \varphi - \pi), \quad \left(\tau^3\rho, \varphi - \tfrac{3\pi}{2}\right), \quad (\tau^4\rho, \varphi), \quad \ldots$$

The figure shows the beginning of such a sequence. The graph of the function $\mathbf{f}(t) = (\rho\tau^t, \varphi-(\pi/2)t), t \in \mathbb{R}$, is a smooth curve that goes through all points of the sequence, in particular the point (ρ, φ), which is $\mathbf{f}(0)$. This curve is called the *logarithmic spiral*. The parametric equations of the logarithmic spiral are

$$x_1 = \rho\tau^t \cos\left(\varphi - \tfrac{\pi}{2}t\right), \quad x_2 = \rho\tau^t \sin\left(\varphi - \tfrac{\pi}{2}t\right).$$

Note that $\mathcal{F}$ maps the logarithmic spiral onto itself. It follows that the central dilation with center S and scaling factor τ^4 also maps this curve onto itself. Consequently, every line segment SP meets the logarithmic spiral under the same angle.

Cycloids

In the following discussion we will frequently use the relation between angle measures and arcs described in Basic Assumption 4.4. A cycloid is a curve traced by a point on a circle that rolls along a line; see Fig. 4.61. At time 0 the

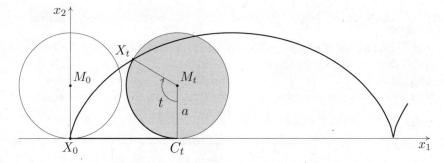

Fig. 4.61. A cycloid

center of the circle with radius a is at M_0. We denote the point of contact with the x_1-axis by X_0. In the figure, X_0 is the origin of the coordinate system. We follow the motion of the point X_0 on the rolling circle. We assume that the circle rolls along the line at a constant speed. Consequently, at time t, the circle touches the x_1-axis at $C_t = (at, 0)$. This assumption obviously does not affect the shape of the curve. At time t, the center of the circle is at $M_t = (at, 1)$ and the special point X_0 is at X_t. As the circle rolls along the x_1-axis, the arc $C_t X_t$ and the segment $[X_0 C_t]$ have equal length, namely at. This implies that $\angle C_t M_t X_t = t$. From the figure, we conclude that the coordinates of $X(t)$ are

$$x_1 = at - a\sin t, \quad x_2 = a - a\cos t.$$

These are the parametric equations of the *cycloid*. The Greek word *kyklos* means *wheel*; you might say that the cycloid is a wheel-curve. The orbit of the tire valve on a bicycle is approximately a cycloid.

4.6 The Parametric Equations of a Curve, Cycloids

There are several variations of the construction of the cycloid [44, volume three]. Instead of rolling along a line, the circle can roll over another circle; see Fig. 4.62. There are then two options: the moving circle can either roll along the inside or along the outside of the fixed circle.

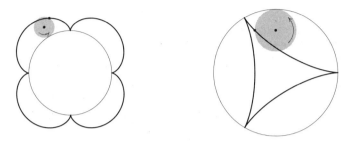

Fig. 4.62. Two variations of the construction of a cycloid

We are given a circle with radius A. We choose a coordinate system with origin at the center of the circle; see Fig. 4.63 (a). Let us determine the parametric equations of the curve described by a point on a circle that rolls over the outside of this given circle. At time 0, the center of the rolling circle, whose

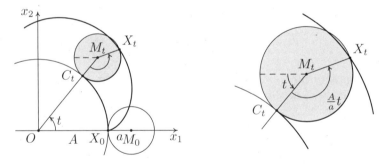

Fig. 4.63. (a) An epicycloid; (b) detail

radius is a, is at M_0. We call the point of contact of the two circles X_0. Suppose that at time t the point of contact of the circles is at C_t and that $\angle X_0 O C_t = t$. We let M_t denote the center of the rolling circle at time t, and X_t the position of the point we are following. We have $M_t = ((A+a)\cos t, (A+a)\sin t)$. As the rolling circle moves over the circle, the length At of arc $X_0 C_t$ of the fixed circle is equal to the length of the arc $C_t X_t$ on the rolling circle, which is $a \times \angle C_t M_t X_t$. It follows that $\angle C_t M_t X_t = (A/a)t$. From the figure, we conclude that the coordinates of $X(t)$ are

$$x_1 = (A+a)\cos t - a\cos\frac{A+a}{a}t,$$
$$x_2 = (A+a)\sin t - a\sin\frac{A+a}{a}t.$$

These are the parametric equations of the *epicycloid*. The Greek word *epi* means *on top of*; it refers to the fact that the moving circle rolls along the outside of the fixed circle, on top of it, so to speak. If A/a is rational, the resulting curve is closed. We will discuss several special cases further on.

Let us now derive the parametric equation of a curve traced by a point on a circle that rolls along the inside of a fixed circle; see Fig. 4.64. At time

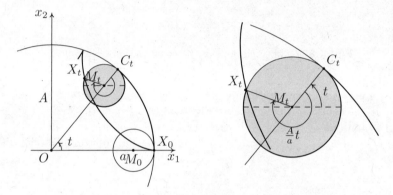

Fig. 4.64. (a) A hypocycloid; (b) detail

$t=0$, the center of the rolling circle, whose radius is again a, is at M_0. We call the point of contact of the two circles X_0. Suppose that at time t the point of contact of the circles is at C_t and that $\angle X_0 O C_t = t$. We let M_t denote the center of the rolling circle at time t, and X_t the position of the point we are following. We have $M_t = ((A-a)\cos t, (A-a)\cos t)$. As the rolling circle moves along the inside of the circle, the length of arc $X_0 C_t$ of the fixed circle is equal to the length of the arc $C_t X_t$ on the rolling circle. Consequently, $\angle C_t M_t X_t = (A/a)t$. From the figure, we conclude that the coordinates of $X(t)$ are

$$x_1 = (A-a)\cos t + a\cos\frac{A-a}{a}t,$$
$$x_2 = (A-a)\sin t - a\sin\frac{A-a}{a}t.$$

These are the parametric equations of the *hypocycloid*. The Greek word *hypo* means *under*; it refers to the fact that the rolling circle moves along the inside of the fixed circle. If A/a is rational, the resulting curve is closed. We will discuss several special cases further on.

4.6 The Parametric Equations of a Curve, Cycloids

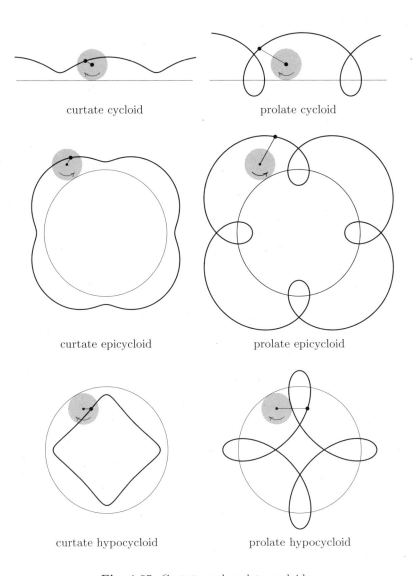

Fig. 4.65. Curtate and prolate cycloids

The three types of cycloids we have just seen each have two more variants, the *curtate* and the *prolate*. In the first case we follow the orbit of a point inside the rolling circle; in the second case we follow a point outside the rolling circle; see Fig. 4.65.

The equations of the curtate and prolate cycloids are easy to find. The parametric equations of the curtate cycloid are

$$x_1 = at - h\sin t,$$
$$x_2 = a - h\cos t, \text{ where } 0 < h < a.$$

Example 4.54 (Cardioid). A *cardioid* is an epicycloid for which the radii of the fixed and rolling circle are equal. There are various descriptions of the cardioid, each of which exhibits a special geometric property of the cardioid. Figure 4.66 shows a cardioid. The parametric equations of the cardioid are

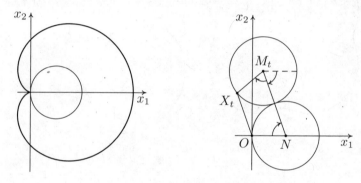

Fig. 4.66. (a) A cardioid; (b) its generation (enlarged)

$$x_1 = a - 2a\cos t + a\cos 2t,$$
$$x_2 = 2a\sin t - a\sin 2t,$$

where a is the radius of the circles. We can derive these equations from the general equations of the epicycloid through a coordinate transformation, or directly using Fig. 4.66 (b). The arrows at the angles each denote an angle of measure t. The figure on the right also shows that

$$X_t O = 2a(1 - \cos t).$$

Through a straightforward computation we derive the following equation of the cardioid in Cartesian coordinates from the parametric equations:

$$(x_1 - a)^2 + x_2^2 = 5a^2 - 4a^2 \cos t.$$

Combining the last two equations gives

$$x_1^2 + x_2^2 - 2ax_1 = 4a^2(1 - \cos t) - 2a\sqrt{x_1^2 + x_2^2}.$$

Squaring both sides and substituting $b = 2a$ gives

$$\left(x_1^2 + x_2^2 - bx_1\right)^2 = b^2\left(x_1^2 + x_2^2\right). \tag{4.34}$$

We will now give three more descriptions of the cardioid. To begin with, the cardioid can be seen as the pedal curve of a circle with respect to a point on that circle; see Fig. 4.67 (a). Let us determine the pedal curve of the circle

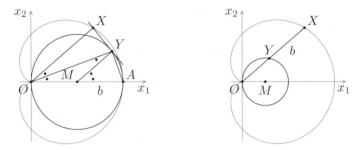

Fig. 4.67. A cardioid: (a) as pedal curve; (b) as limaçon

$\odot(M, b)$ with respect to the origin O. The center M lies on the x_1-axis and the circle meets this axis at O and at A. We draw the tangent to the circle at Y. The point X is the foot of the perpendicular from O on that tangent. Since angle AMY is an exterior angle of the isosceles triangle OMY and $OX \parallel MY$, it follows that OY is the bisector of angle MOX. We find the following equations for the polar coordinates (r, φ) of X:

$$r = OX = OY \cos \tfrac{\varphi}{2} = OA \cos^2 \tfrac{\varphi}{2} = b(\cos \varphi - 1).$$

The equation in Cartesian coordinates easily follows:

$$\left(x_1^2 + x_2^2 - bx_1\right)^2 = b^2\left(x_1^2 + x_2^2\right).$$

This equation is the same as (4.34); hence the pedal curve is a cardioid.
Anyone familiar with complex numbers can interpret the figure of the pedal curve as follows: if in the complex plane, the points O, A, and Y represent the complex numbers 0, 1, and z, respectively, then X represents the complex number z^2, because $\triangle OAY \sim \triangle OYX$ (Theorem 2.50). It follows that the complex map $z \mapsto z^2$ transforms the circle $|z - (1/2)| = 1/2$ into a cardioid. In passing, we note that many geometric properties admit an elegant proof using complex numbers [24].
A second description of the cardioid is related to the *limaçon (snail) of Pascal*; see Fig. 4.67 (b). This Pascal is Etienne Pascal (1588–1651), the father of the famous Blaise Pascal (1623–1662). The construction of the limaçon is a

variant of the construction of the conchoid, Example 4.22. Let O be a point on the circle $\odot(M, b/2)$. To a point Y on the circle we associate one of the two points X with $YX = b$. The limaçon is the set of all X that can be obtained in this way as Y varies over the circle. Choosing the coordinates as indicated in the figure, we obtain the following polar coordinates (r, φ) for the point X:

$$r = b\cos\varphi \pm b.$$

The same method that was used for the conchoid gives the following equation in Cartesian coordinates:

$$\left(x_1^2 + x_2^2 - bx_1\right)^2 = b^2 \left(x_1^2 + x_2^2\right).$$

This equation is the same as (4.34); hence the limaçon is a cardioid.

A third description of the cardioid is as a *pericycloid*. Let $\odot(M, b)$ be a fixed circle. A circle with radius $2b$ swings like a hoop around the fixed circle; the fixed circle remains inside the moving circle, and the circles always have one point of contact. The orbit of a fixed point of the moving circle is called a pericycloid; see Fig. 4.68. More generally, we can consider a rolling circle with

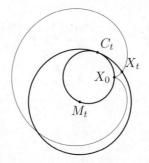

Fig. 4.68. A cardioid as pericycloid

any radius greater than b. The pericycloid is a hypocycloid with the property that the radius of the rolling circle is greater than that of the fixed circle. In the case we were considering, we obtain the following equations:

$$x_1 = -b\cos t + 2b\cos\tfrac{1}{2}t,$$
$$x_2 = -b\sin t + 2b\sin\tfrac{1}{2}t.$$

We can easily see that this pericycloid is a cardioid: substituting $s = t/2$, we obtain

$$x_1 = 2b\cos s - b\cos(2s),$$
$$x_2 = 2b\sin s - b\sin(2s).$$

These are the parametric equations of an epicycloid with fixed and rolling circles of equal radius. Hence this curve is a cardioid.

Tangents

We have already discussed tangents several times. The most precise description of tangents was in Sect. 4.5, in particular in Theorem 4.44. This description was of algebraic nature: a tangent is a line whose intersection with the conic is a point of multiplicity two, called the point of contact. This method cannot be used for cycloids. In order to define tangents to curves of this type, we need to use methods from calculus. We will briefly say how tangents are defined in this setting, and will come back to it later when necessary. We try to stay succinct so as not to interrupt our exposition of the geometric theory.

Let P be a point on a curve, and let $Q_1, Q_2, Q_3, \ldots$ be a sequence of points with limit P; see Fig. 4.69. If the sequence of chords $PQ_1, PQ_2, PQ_3, \ldots$ has

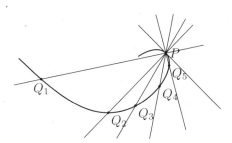

Fig. 4.69. Definition of a tangent

a limit, and this limit is independent of the choice of the sequence $Q_1, Q_2, Q_3, \ldots$, we call the limit the *tangent* at P. We deduce the equation of the tangent from the parametric equations

$$x_1 = f_1(t), \quad x_2 = f_2(t)$$

of the curve. Let $P = (f_1(t_0), f_2(t_0))$ and $Q_n = (f_1(t_n), f_2(t_n))$ for $n = 1, 2, \ldots$. The chord $Q_n P$ has parametric equations $\mathbf{x} = \mathbf{p} + (\lambda/(t_n - t_0))(\mathbf{q}_n - \mathbf{p})$, that is,

$$x_1 = f_1(t_0) + \lambda \frac{f_1(t_n) - f_1(t_0)}{t_n - t_0},$$

$$x_2 = f_2(t_0) + \lambda \frac{f_2(t_n) - f_2(t_0)}{t_n - t_0}.$$

If the functions f_1 and f_2 are differentiable in t_0, the vector equation of the tangent is obtained by taking limits $t_n \to t_0$:

$$x_1 = f_1(t_0) + \lambda f_1'(t_0), \quad x_2 = f_2(t_0) + \lambda f_2'(t_0),$$

that is,

$$(x_1, x_2) = (f_1(t_0), f_2(t_0)) + \lambda (f_1'(t_0), f_2'(t_0)).$$

For example, the tangent to the circle $x_1 = \cos t$, $x_2 = \sin t$ at the point $(\cos t_0, \sin t_0)$ has equation

$$(x_1, x_2) = (\cos t_0, \sin t_0) + \lambda(-\sin t_0, \cos t_0) \, .$$

Another example: we can see the graph of a function $y = f(x)$ from $\mathbb{R}$ to $\mathbb{R}$ as a curve with parametric equations $x_1 = x$ and $x_2 = f(x)$; see Fig. 4.70. The

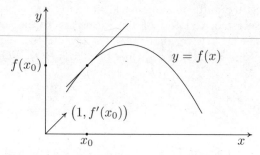

Fig. 4.70. Tangent to f at x_0: $y = f(x_0) + f'(x_0)(x - x_0)$

tangent to this curve at the point $(x_0, f(x_0))$ has parametric equations

$$(x_1, x_2) = (x_0, f(x_0)) + \lambda(1, f'(x_0)) \, ,$$

that is,

$$x_1 = x_0 + \lambda,$$
$$x_2 = f(x_0) + \lambda f'(x_0) \, .$$

Eliminating λ and replacing x_1 by x and x_2 by y, we obtain

$$y = f(x_0) + f'(x_0)(x - x_0) \, ,$$

the equation of the tangent that is used in calculus.

The Triangle and the Deltoid

A *deltoid* is a hypocycloid with three cusps; the ratio of the radius of the fixed circle and that of the rolling circle is 3. We will show that we can associate a deltoid to any triangle using the Simson lines of the points of its circumcircle. To this end we must first define the envelope of a system of lines. We do this through two examples. For a further explanation, see [61]; all topics in this section are discussed extensively in that book.

Example 4.55. Let C be a circle, Q a point in its interior, and P a point on the circle; see Fig. 4.71 (**a**). We draw the perpendicular bisector of the line segment $[PQ]$. As P moves along the circle, these perpendicular bisectors form a system of lines that all touch the same ellipse. This follows from the construction of the tangents to an ellipse in Fig. 4.39. We say that the ellipse is the *envelope* of the system of perpendicular bisectors.

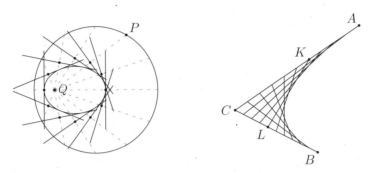

Fig. 4.71. Conics as envelopes: (a) an ellipse; (b) a parabola

Example 4.56. On the sides of an angle C, we set out line segments CA and CB. On CA we choose a variable point K. Let L be the point on CB such that
$$AK : KC = CL : LB \ .$$
As K moves along the line segment CA, we obtain a system of lines KL that all touch (part of) a parabola; see Fig. 4.71 (b). By Sect. 4.5, Exercise 4.49, there is a parabola that is tangent to the lines CA and CB, touching them at A and B, respectively. The point of contact of the line KL and the parabola is the point Q with the property that $KQ : QL = AK : KC$.

We will now explain, through a simple example, how to compute the parametric equations of the envelope of a system of lines. We assume that the reader has some knowledge of calculus.

Example 4.57 (Envelope). Let $p > 0$. Suppose that for every t in $\mathbb{R}$, we are given a line
$$\ell(t) : -\tfrac{1}{2}px_1 + tx_2 = t^2 \ .$$
We will compute the equations of the envelope of the system of lines $\ell(t), t \in \mathbb{R}$; see Fig. 4.72. A close look at the figures reveals that the point of contact of a line $\ell(t)$ and the envelope is quite close to the point of intersection of that line and a "neighboring" line of the system. Our method of computing the envelope is based on this observation. Let Δt be a real number. The point of intersection (p_1, p_2) of the lines $\ell(t)$ and $\ell(t + \Delta t)$ has coordinates
$$p_1 = \frac{2t^2}{p} + \frac{2t\Delta t}{p} \ , \qquad p_2 = 2t + \Delta t \ .$$

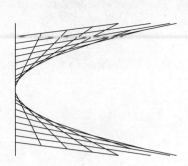

Fig. 4.72. Computing the parametric equations

For small values of Δt this intersection point is close to $(2t^2/p, 2t)$. Thus we obtain the following parametric equations for the envelope:

$$x_1 = \frac{2t^2}{p}, \qquad x_2 = 2t.$$

This is the parabola with equation $x_2^2 = 2px_1$.

In general, we compute the envelope of a system of lines as follows. Suppose that the system of lines is given by the equation

$$\ell(t) : \langle \mathbf{a}(t), \mathbf{x} \rangle = c(t), \text{ or } a_1(t)x_1 + a_2(t)x_2 = c(t), \qquad (4.35)$$

where t ranges over some interval of $\mathbb{R}$ or over $\mathbb{R}$. For each value of t, (4.35) is the equation of a straight line $\ell(t)$ that is perpendicular to the span of $\mathbf{a}(t)$. We also consider the system

$$\ell'(t) : \langle \mathbf{a}'(t), \mathbf{x} \rangle = c'(t), \text{ or } a_1'(t)x_1 + a_2'(t)x_2 = c'(t), \qquad (4.36)$$

where the primes, as usual, denote differentiation with respect to t. We can find the parametric equations of the envelope by solving the equations (4.35) and (4.36) for $\mathbf{x}$. There is a solution if and only if $\det(\mathbf{a}(t)\mathbf{a}'(t)) \neq 0$. For each value of t for which the determinant does not vanish, we obtain a value of $\mathbf{x}$. The result is $\mathbf{x}$ as a function of t. Applying this method to the example given above, we obtain

$$\ell(t) : -\tfrac{1}{2}px_1 + tx_2 = t^2 \quad \text{and} \quad \ell'(t) : x_2 = 2t,$$

whence $x_1 = 2t^2/p$. Why does this method produce the right answer? The solution $\mathbf{x}(t)$ satisfies (4.35). Differentiating this equation gives

$$\langle \mathbf{a}'(t), \mathbf{x}(t) \rangle + \langle \mathbf{a}(t), \mathbf{x}'(t) \rangle = c'(t).$$

Since $\mathbf{x}(t)$ also satisfies (4.36), we obtain

$$\langle \mathbf{a}(t), \mathbf{x}'(t) \rangle = 0.$$

This means that for every t, the span of $\mathbf{x}'(t)$ is parallel to $\ell(t)$, since both lines are perpendicular to $\mathbf{a}(t)$. Since $\mathbf{x}(t)$ lies on $\ell(t)$, the latter is the tangent to the envelope at $\mathbf{x}(t)$.

The following theorem shows that we can associate a deltoid to any triangle in a natural way.

Theorem 4.58. *Consider the circumcircle $\odot(M,R)$ of a triangle ABC, and let $\ell(X)$ be the Simson line of the point X on this circle. The envelope of the system of Simson lines $\{\,\ell(X) : X \text{ on } \odot(M,R)\,\}$ is a deltoid. The radius of the rolling circle that generates the deltoid is equal to the radius of the nine-point circle of triangle ABC, and the nine-point circle is the inscribed circle of the deltoid.*

Figure 4.74 illustrates the theorem. The proof follows from the properties mentioned below. We begin by taking a closer look at the result of Theorem 4.18. Consider Fig. 4.73, on the left. Let X_1 and X_2 be points on the cir-

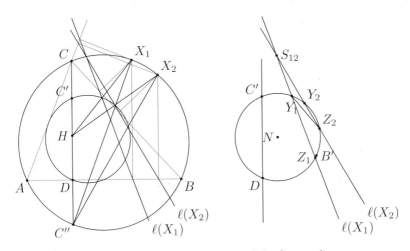

Fig. 4.73. Relative positions of the Simson lines

cumcircle of triangle ABC. The measure of the angle between $C''C$ and $C''X_i$ is equal to that of the angle between $C''C$ and the Simson line $\ell(X_i)$ at X_i, $i = 1, 2$. Since $X_iC''C$ is an inscribed angle for $i = 1, 2$, it follows that $\widehat{X_1X_2}$, the angle measure of arc X_1X_2, is equal to twice the measure of the angle between $\ell(X_1)$ and $\ell(X_2)$. As the point X moves clockwise along the circumcircle, the Simson lines $\ell(X)$ rotates counterclockwise, which implies that the two angle measures we just mentioned have opposite signs. In particular, if X_1 and X_2 are diametrically opposite points of the circumcircle, their Simson lines $\ell(X_1)$ and $\ell(X_2)$ are perpendicular to each other, and if X has made a full turn around the circumcircle, its Simson line will have rotated over π. Let Y_i and Z_i be the intersection points of $\ell(X_i)$ and the nine-point circle, $i = 1, 2$. The point Y_i is the image of X_i under the central dilation with center H, the orthocenter of triangle ABC, and scaling factor $1/2$. The figure on the right shows the relevant parts of the figure on the left. If we set

$\widehat{X_2X_1} = 2\varphi$, we also have $\widehat{Y_2Y_1} = 2\varphi$. Note that $\angle Y_1 S_{12} Y_2$ is equal to φ, but also equal to half the angle measure of $\widehat{Z_1Z_2} - \widehat{Y_2Y_1}$. It follows that $\widehat{Z_1Z_2} = 4\varphi$. The converse also holds: if Y_1Z_1 is the Simson line of X_1, Y_1 is the midpoint of $[HX_1]$, and $\widehat{Z_1Z_2} = 2\widehat{Y_2Y_1}$, then Y_2Z_2 is the Simson line of the point X_2 on the circumcircle, and $\widehat{X_2X_1} = \widehat{Y_2Y_1}$. This completes the proof of the following property.

Property 4.59. *In the situation sketched above, suppose that Y_1Z_1 is the Simson line of the point X_1 and that Y_1 is the midpoint of the segment $[HX_1]$. Then we have both that Y_2Z_2 is the Simson line of the point X_2 and that Y_2 is the midpoint of the segment $[HX_2]$ if and only if $\widehat{Z_1Z_2} = 2\widehat{Y_2Y_1}$.*

Let us continue our analysis of the figure on the right. Note that $\angle Z_1 Y_1 Z_2 = 2\varphi$, whence $\angle Y_2 Z_2 Y_1 = \varphi$. It follows that $\triangle Z_2 S_{12} Y_1$ is isosceles, that is, $S_{12}Y_1 = Y_1Z_2$. This means that if Z_2 is close to Z_1, $S_{12}Y_1$ is approximately equal to Y_1Z_1. In view of the above observations concerning the envelope, these results suggest that the following property holds. We will not give a proof, since it requires results from calculus.

Property 4.60. *Let H be the orthocenter of a triangle ABC. We denote the Simson line of a point X on the circumcircle by $\ell(X)$. Let Y and Z be the intersection points of $\ell(X)$ and the nine-point circle, where Y is the midpoint of $[HX]$. The envelope of the Simson lines touches $\ell(X)$ at the point S for which $SY = YZ$.*

Note that $C'D$, the altitude from C on AB, is also a Simson line. We will use this line as fixed reference line, but we do want to point out that we could have used any other Simson line as reference line. We use the notation $\widehat{Y_1C'} = 2\psi$. It follows from the computations given above that $\widehat{DZ_1} = 4\psi$ and $\widehat{Z_1Y_1} = \widehat{DB'C'} - 6\psi$. The figure on the left shows that

$$\widehat{C''BC} = \widehat{C''B} + \widehat{BC}$$
$$= 2\left(\tfrac{\pi}{2} - \angle CBA\right) + 2\angle BAC$$
$$= \pi - 2(\angle CBA - \angle BAC).$$

Consequently, $\widehat{Z_1Y_1} = \pi - 2(\angle CBA - \angle BAC) - 6\psi$. It is immediately clear that the value of $\widehat{Z_1Y_1}$ does not change if we replace 2ψ by $2\psi + (2/3)\pi$ or by $2\psi + (4/3)\pi$; consequently, the distance from the corresponding Simson line to N does not change either. We have proved the following property.

Property 4.61. *The envelope of the Simson lines has a threefold rotational symmetry with center at the center N of the nine-point circle.*

In the special case that Y_1Z_1 is a diameter of the nine-point circle, the distance from N to the point of contact of the envelope with this diameter is

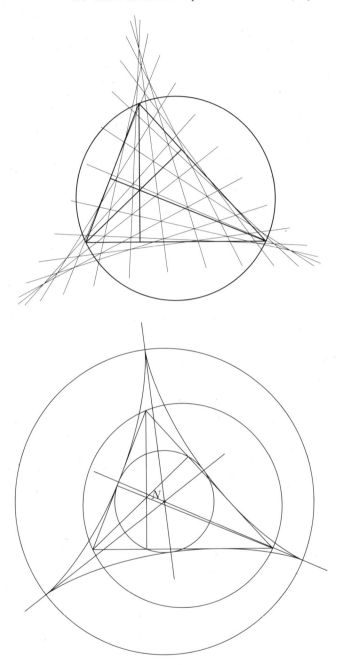

Fig. 4.74. The envelope of the Simson lines of a triangle is a deltoid; see the top figure. Both figures show the triangle and its altitudes. The bottom figure shows the relative positions of the nine-point circle, the circumcircle, and the deltoid. It also shows the tangents to the deltoid at the cusps

equal to $3r$, where $r = R/2$ is the radius of the nine point circle. It follows that the distance from a point on the envelope to N cannot exceed $3r$. We will see that if $Y_1 Z_1$ is a diameter of the nine-point circle, the point of contact with the envelope is a cusp. In order to determine the position of the envelope with respect to the triangle, we have to find the values of ψ for which the point of contact is a cusp. These values follow from the computations given above.

Property 4.62. *The Simson line $\ell(X)$ of X is the tangent at a cusp if and only if $\widehat{XC}$ equals $(2/3)(\angle BAC - \angle CBA)$ plus a multiple of $(2/3)\pi$.*

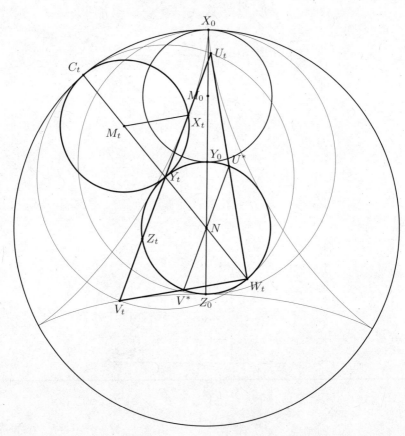

Fig. 4.75. The envelope of the Simson lines is a deltoid. The chord of the deltoid that is also tangent to the deltoid can be viewed as the diameter of a circle that rolls along the inside of the fixed circle and has radius twice that of the nine-point circle

We can now complete the proof of Theorem 4.58. Consider Fig. 4.75. We choose a coordinate system such that the tangent at a cusp, namely $Z_0 X_0$, is the x_2-axis. We have $Z_0Y_0 = Y_0X_0$. Hence X_0 lies on the envelope of the Simson lines. We take the circle $\odot(N, 3r)$, where r is the radius of the nine-point circle, as fixed circle, and take $\odot(M_0, r)$ as rolling circle. We will show that the deltoid traced by X_0 is the envelope of the Simson lines. We let M_t denote the position of the center of the rolling circle at time t, C_t the point of contact of the two circles, and X_t the position of X_0. We use the notation $\angle X_0 N M_t = 2\varphi$. The length of the arc $X_0 C_t$ is $6r\varphi$; this is also the length of the arc $X_t C_t$ of the rolling circle. It follows that $\angle X_t M_t C_t = 6\varphi$. Let Y_t be the midpoint of $[NM_t]$. The angles $X_t Y_t M_t$ and $Y_t X_t M_t$ are both equal to 3φ. Let us consider a second rolling circle, $\odot(Y_0, 2r)$, which we will call the *large* rolling circle to distinguish it from the circle that generates the deltoid. We let Y_t denote the position of the center of the large rolling circle at time t, and U_t the position of X_0, seen as a point on the large rolling circle. Since the length of the arc $U_t C_t$ of the large rolling circle is equal to 6φ, we have $\angle U_t Y_t M_t = 3\varphi$. We see that Y_t, X_t, and U_t are collinear. It follows that $\angle Z_t Y_t N = 3\varphi$ and $\angle W_t N Z_0 = 2\varphi$, and consequently $\widehat{Z_t Z_0} = 4\varphi$. Since $\widehat{Y_0 Y_t} = 2\varphi$, Property 4.59 now implies that $Y_t Z_t$ is a Simson line. Moreover, it is clear that all Simson lines can be obtained in this way. We can easily show that $\widehat{Y_t X_t}$ and $\widehat{Y_t Z_t}$ are both equal to $\pi - 6\varphi$. Hence, $Z_t Y_t = Y_t X_t$ and X_t is a point of the envelope by Property 4.60. This completes the proof of the theorem. There is, however, more beauty to cycloids than meets the eye. The points X_t, Y_t, and Z_t lie on the diameter $U_t V_t$ of the large rolling circle. Let $U^* V^*$ be the diameter of the nine-point circle that is parallel to $U_t V_t$. Note that $[W_t Y_t]$ is a diameter of the nine-point circle. Since $\angle U^* N Y_0$ is equal to the angle between the Simson lines $Y_0 Z_0$ and $Y_t Z_t$, hence equal to φ, we easily deduce that both $W_t U^*$ and $W_t V^*$ are Simson lines. Since U^* is the midpoint of $W_t U_t$, the point U_t lies on the envelope. Likewise, V_t is a point of the envelope. It follows that *the diameter $U_t V_t$ of the large rolling circle is both a chord and a tangent of the deltoid: the diameter has its endpoints U_t and V_t on the deltoid and touches the deltoid at X_t.*

The Cycloidal Pendulum

We conclude this section on cycloids with some remarks concerning applications. Cycloids have an important application in the construction of gear wheels. For optimal performance it is important that gear wheels be always in contact with each other. To achieve this, gear wheels have a special shape that is closely related to cycloids [29]. Cycloids were also used to explain the motion of the planets when it was still generally believed that the earth was the center of the universe. According to Ptolemy, the orbits of the planets and the sun around the earth are epicycloids.

The shape of a cycloid is similar to that of a skateboarding halfpipe. This is no accident. It is related to the following problem. We are given two points A and B on a vertical surface, with A higher up than B, but not directly above it. The question is, what is the shape of the optimal slide connecting A and B, that is, the slide on which an object moves from A to B in the shortest possible time under the sole influence of gravity. If A were to lie directly above B this would be a straight line. The curve of the optimal slide is called the *brachistochrone*, which literally means shortest time. In 1697, Johann Bernoulli (1667–1748) discovered that the brachistochrone is the same curve as the cycloid. To find the brachistochrone, we first draw an arbitrary cycloid, as shown in Fig. 4.76: the thin cycloid through A and D. This cycloid is traced

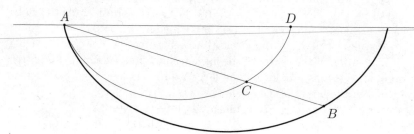

Fig. 4.76. Finding the optimal slide

by a circle with arbitrary radius r that rolls along the line AD, starting at A. We then draw the line through A and B. This meets the cycloid at the point C. The cycloid of the optimal slide from A to B is traced by a point on the circle with radius $(AB/AC)r$ that rolls along the line AD, starting at A.

The cycloid had already made its appearance in a completely different problem related to the research of Christiaan Huygens (1629–1695). Huygens tried to construct a pendulum whose frequency is independent of its amplitude. This independence was considered important for precise measurement of time, which in turn was indispensable for finding the position of a ship in open sea. The latitude of the position of a ship was easily found using the position of the sun and the stars, but for the longitude a precise clockwork was essential. The degree of longitude was derived from the difference between the local time and the time at the home port. The first could be determined using the position of the sun. Huygens discovered that the frequency of a cycloidal pendulum is independent of its amplitude. We will discuss the solution of Huygens's problem in Exercise 4.64. Huygens also provided a design for a cycloidal pendulum. This design was patented in 1656. Clocks that were constructed according to this design were first marketed by Salomen Coster, and after 1660 by Pieter Visbagh. The exhibit at the Huygens museum Hofwijck in Voorburg (the Netherlands) contains parts of the original cycloidal pendulum, as well as an original Visbagh clock.

For a discussion of the cycloidal pendulum we must first introduce a few properties of the *arc length* of the cycloid. We will use results from calculus. Consider Fig. 4.77. We are looking for a formula to compute the length of the arc of the cycloid between the points X_0 and X_t. The cycloid is traced by a point on a circle with radius a that rolls along a line. We assume that the

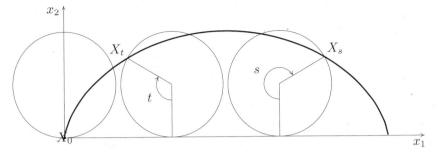

Fig. 4.77. Determining the arc length

circle moves at a constant speed. We follow the motion of a special point on the circle that starts at X_0 at time 0. At time t the special point has arrived at X_t; the coordinates of X_t are

$$x_1 = at - a\sin t, \qquad x_2 = a - a\cos t.$$

The velocity vector at the point X_t is

$$x_1'(t) = a - a\cos t, \qquad x_2'(t) = a\sin t.$$

The value of the velocity $v(t)$, that is, the length of the velocity vector at time t, follows from

$$v(t)^2 = (x_1'(t))^2 + (x_2'(t))^2 = a^2(2 - 2\cos t) = 4a^2 \sin^2(\tfrac{1}{2}t).$$

The result is $v(t) = 2a\sin(t/2)$, for $0 \leq t \leq 2\pi$. To find the length L_t of the arc from X_0 to X_t, we integrate the velocity $v(\tau)$ from 0 to t. When the velocity is constant, this corresponds to taking velocity × time. We obtain

$$L_t = \int_0^t v(\tau)\,d\tau = \int_0^t 2a\sin(\tfrac{1}{2}\tau)\,d\tau = 4a - 4a\cos(\tfrac{1}{2}t). \tag{4.37}$$

It follows that the length of a cycloid from one cusp to the next is equal to $8a$. For $s = \pi + u$, the length of the arc from X_0 to X_s is $4a + 4a\sin(u/2)$.

Let us now study the motion in a vertical plane of a point mass that is accelerated by gravity only; we assume that the point moves along a given curve in the vertical plane. We will need to use differential equations.

Example 4.63 (Mathematical Pendulum). Let us first study the motion of a point mass along a circular arc; see Fig. 4.78 **(a)**. More specifically, our goal is to find equations for the motion of a bead under the influence of gravity along a circular arc, without slipping and ignoring friction. We choose a coordinate system with origin at M, the center of the circle containing the arc. We denote the radius of the circle by r. The motion of the bead is the same as that of a pendulum of length r suspended from M. Let φ be the angle between the

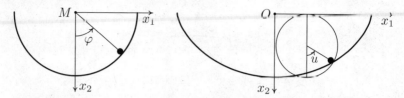

Fig. 4.78. Motion: (a) along a circular arc; (b) along a cycloid

x_2-axis and the position vector of the bead; this angle is a function of time, $\varphi = \varphi(t)$. We aim to derive a differential equation for $\varphi(t)$. At time t, the Cartesian coordinates of the bead are

$$x_1(t) = r \cos(\varphi(t)) , \qquad x_2(t) = r \sin(\varphi(t)) .$$

Differentiating, we get

$$x_1'(t) = -r (\sin \varphi(t)) \varphi'(t) , \qquad x_2'(t) = r (\cos \varphi(t)) \varphi'(t) .$$

The velocity of the bead at time t is therefore

$$\sqrt{(x_1'(t))^2 + (x_2'(t))^2} = r|\varphi'(t)| ,$$

and the kinetic energy of the bead is $T(t) = (1/2)mr^2\varphi'(t)^2$, where m is the mass of the bead. Since, up to an additive constant, the potential energy is equal to

$$V(t) = -mgr \cos \varphi(t) = -mgr + 2mgr \sin^2(\tfrac{1}{2}\varphi(t)) ,$$

and the sum of the kinetic and potential energies is constant, the equation of the motion of the bead is

$$\tfrac{1}{2}mr^2\varphi'(t)^2 - mgr + 2mgr \sin^2(\tfrac{1}{2}\varphi(t)) = \text{constant} .$$

Differentiating this formula and simplifying the result gives

$$r\varphi''(t) + g \sin \varphi(t) = 0 .$$

This is the equation of the *mathematical pendulum*. For small values of φ an approximate solution can be found by replacing the function $\sin \varphi(t)$ by $\varphi(t)$. This leads to the equation

$$r\varphi''(t) + g\varphi(t) = 0 ,$$

the equation of the *harmonic oscillator* with solutions

$$\varphi(t) = C_1 \sin \sqrt{\tfrac{g}{r}} t + C_2 \cos \sqrt{\tfrac{g}{r}} t ,$$

where C_1 and C_2 are real numbers. The frequency of this approximate solution is $2\pi\sqrt{(r/g)}$, which does not depend on the amplitude. It should be emphasized, however, that the frequency of the solutions of the equation of the mathematical pendulum do depend on the amplitude.

4.6 The Parametric Equations of a Curve, Cycloids

Example 4.64 (Cycloidal Pendulum). This time, we study the motion of a bead along a cycloidal arc; see Fig. 4.78 (b). The derivation of the corresponding equation is similar to that for the mathematical pendulum. We want to find equations for the motion of a bead under the influence of gravity along a cycloid, without slipping and ignoring friction. We assume that a circle with radius r rolls along the underside of the x_1-axis; the x_2-axis points downward, and $(0, 2r)$ is the lowest point of the cycloid. The figure shows the circle at the moment that the bead and the special point that traces the cycloid are at the same point. We let u denote the angle between the line joining the center of the rolling circle and the bead on the one hand and the vertical line through the center on the other hand; the angle measure is a function of time, $u = u(t)$. At time t, the Cartesian coordinates of the bead are

$$x_1(t) = ru(t) + r\sin(u(t)),$$
$$x_2(t) = r + r\cos(u(t)).$$

Differentiating, we get

$$x_1'(t) = ru'(t) + r\cos(u(t))\,u'(t),$$
$$x_2'(t) = -r\sin(u(t))\,u'(t).$$

The velocity of the bead at time t is therefore $2r|u'(t)\cos(u(t)/2)|$. By (4.37) and the remarks following it, the length of the arc from the bead to the lowest point of the cycloid is equal to $s(t) = 4r\sin(u(t)/2)$. Let us derive a differential equation for $s(t)$. At this point, the derivation differs from that of the mathematical pendulum. Assuming that the mass of the bead is m and its velocity at time t is equal to $v(t)$, the kinetic energy of the bead is

$$T(t) = \tfrac{1}{2}m\,(v(t))^2 = 2r^2 m\,(u'(t))^2 \cos^2\left(\tfrac{1}{2}u(t)\right) = \tfrac{1}{2}m\,(s'(t))^2.$$

Since, up to an additive constant, the potential energy is equal to

$$V(t) = -mgr\cos(u(t)) = -mgr + \frac{mg}{8r}(s(t))^2,$$

and the sum of the kinetic and potential energies is constant, the equation of the motion of the bead is

$$\tfrac{1}{2}ms'(t)^2 - mgr + \frac{mg}{8r}s(t)^2 = \text{constant}.$$

Differentiating this formula and simplifying the result gives

$$s''(t) + \frac{g}{4r}s(t) = 0,$$

the equation of the harmonic oscillator. The solutions of the equation are

$$s(t) = C_1 \sin\sqrt{\frac{g}{4r}}\,t + C_2 \cos\sqrt{\frac{g}{4r}}\,t,$$

where C_1 and C_2 are real numbers. The frequency $2\pi\sqrt{(4r/g)}$ of these solutions is independent of the amplitude, i.e., of the maximal displacement measured along the cycloidal arc. Because of this property the cycloid is called the *tautochrone* or *isochrone curve*.

When we discussed the mathematical pendulum, we remarked that the motion of a bead along a circular arc is the same as that of a pendulum. Christiaan Huygens discovered that the motion along a cycloid can also be realized as the motion of a pendulum. Figure 4.79 shows how this works. We

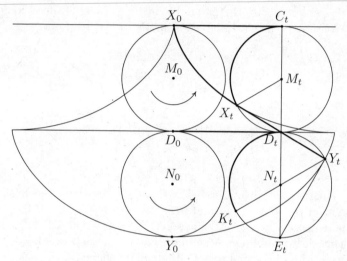

Fig. 4.79. The cycloidal pendulum of Christiaan Huygens

consider two circles with radius r that roll simultaneously along two horizontal lines $X_0 C_t$ and $D_0 D_t$, in counterclockwise direction, as indicated by the arrows. The distance between the horizontal lines is $2r$. At time 0, the top circle touches the top line at X_0 and has its center at M_0. We follow the point X_0. We let C_t denote the point of contact of the top circle and the top line at time t, M_t the center of the top circle, and X_t the position of the point X_0. At time 0, the bottom circle touches the bottom line at D_0 and has its center at N_0. On this circle, we follow the point Y_0. We let D_t denote the point of contact of the bottom circle and the bottom line at time t, N_t the center of the bottom circle, Y_t the position of the point Y_0, and K_t the position of the point D_0. Using the notation $\angle C_t M_t X_t = 2\varphi$, we have $\angle M_t D_t X_t = \varphi$, $\angle D_t N_t K_t = 2\varphi$, and therefore $\angle N_t Y_t D_t = \varphi$ and $\angle E_t D_t Y_t = \varphi$. It follows that X_t, D_t, and Y_t are collinear. Simple computations show that $X_t Y_t = 4r \cos\varphi$. It now follows from (4.37) that the length L_t of the cycloidal arc from X_0 to X_t is equal to $4r - 4r\cos\varphi$. Hence
$$L_t + X_t Y_t = 4r \, .$$

In Exercise 4.54 we will show that the line $X_t Y_t$ touches the top cycloid at the point X_t. Together with the last formula, this implies that if we suspend a pendulum of length $4r$ from X_0 and let it swing between the arcs of the upper cycloid, its lower end moves along the bottom cycloid. Thus we see that we can force the bead of the pendulum to move along a cycloidal orbit. Huygens made his pendulum clock in precisely the same way: at the suspension point of the pendulum he attached cycloidal arcs to restrict the motion of the pendulum. These arcs are traced by a circle with radius one-fourth the length of the pendulum.

Exercises

4.50. We study the logarithmic spiral of Example 4.53. Its parametric equations are

$$x_1(t) = \rho \tau^t \cos(\varphi - \tfrac{\pi}{2} t), \quad x_2(t) = \rho \tau^t \sin(\varphi - \tfrac{\pi}{2} t). \qquad (4.38)$$

The vector $(x_1'(t), x_2'(t))$ is called the *tangent vector* at t.

(a) The parametric equations of the tangent to the spiral at the point associated to the value t are

$$(x_1(\lambda), x_2(\lambda)) = (x_1(t), x_2(t)) + \lambda (x_1'(t), x_2'(t)), \quad \lambda \in \mathbb{R}.$$

(b) By differentiating (4.38) we obtain the tangent vector

$$x_1'(t) = \rho \tau^t \ln(\tau) \cos(\varphi - \tfrac{\pi}{2} t) + \rho \tau^t \tfrac{\pi}{2} \sin(\varphi - \tfrac{\pi}{2} t),$$
$$x_2'(t) = \rho \tau^t \ln(\tau) \sin(\varphi - \tfrac{\pi}{2} t) - \rho \tau^t \tfrac{\pi}{2} \cos(\varphi - \tfrac{\pi}{2} t).$$

(c) For any given t, let $\psi(t)$ denote the angle between the position vector $(x_1(t), x_2(t))$ and the tangent vector $(x_1'(t), x_2'(t))$. We have

$$\cos \psi(t) = \frac{-1}{\sqrt{1 + (\pi/(2 \ln \tau))^2}}.$$

It follows that $\psi(t)$ is a constant function; in other words, the measure of the angle between the position vector and the tangent vector is independent of t.

4.51. The *astroid* is the hypocycloid obtained by rolling a circle with radius a along the inside of a circle with radius $4a$.

(a) Show that the parametric equations of the astroid can be written as

$$x_1 = 4a \cos^3 t,$$
$$x_2 = 4a \sin^3 t.$$

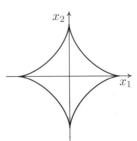

(b) By eliminating t, we obtain the equation

$$x_1^{2/3} + x_2^{2/3} = (4a)^{2/3} .$$

4.52. We can generate the cardioid by letting two circles roll simultaneously along the outside of a fixed circle. The three circles have the same radius, and the rolling circles are always diametrically opposite each other. This way of generating the cardioid shows that every chord through the cusp has length $4r$. Hint: Show that $X_1 X_2$ is parallel to the line joining the centers of the rolling circles, and passes through the cusp.

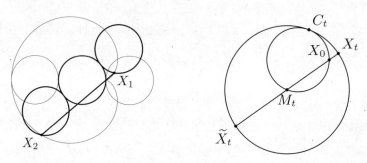

Fig. 4.80. Cycloids: (a) in Exercise 4.52; (b) in Exercise 4.53

4.53. The cardioid can also be described as a pericycloid. At time 0, circles with radii a and $2a$ touch at the point X_0. We let X_t denote the position of the point X_0 at time t, C_t the point of contact of the two circles, and M_t the center of the rolling circle.

(a) The points M_t, X_0, and X_t are collinear.
(b) Let $\widetilde{X}_t$ be the second intersection point of $M_t X_t$ and $\odot(M_t, 2a)$. Use Exercise 4.52 to show that $\widetilde{X}_t$ lies on the cardioid.

4.54. The *folium of Descartes* has parametric equations

$$x_1 = \frac{3at}{1+t^3}, \quad x_2 = \frac{3at^2}{1+t^3},$$

where $a > 0$.

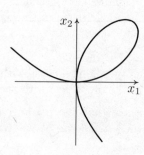

(a) Which part of the curve is determined by the condition $-\infty < t < -1$, and which part by the condition $-1 < t < \infty$?
(b) Determine the equation of the tangent at $(0,0)$.

(c) The equation of the other tangent at $(0,0)$ is obtained by interchanging the equations of x_1 and x_2.

4.55. Consider Fig. 4.61. The tangent to the cycloid at the point X_t meets the line $M_t C_t$ at the point $(at, 2a)$. This tangent is therefore perpendicular to $C_t X_t$.
Hint: The parametric equations of the tangent at X_t are

$$(x_1, x_2) = (at - a\sin t, a - a\cos t) + \lambda(a - a\cos t, a\sin t), \quad \lambda \in \mathbb{R}.$$

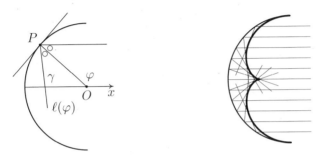

Fig. 4.81. The caustic curve of parallel lines reflected by a circle

4.56. A beam of parallel light rays is reflected by a circular mirror with radius R, as shown in Fig. 4.81. The reflected rays are not concurrent, as in the case of a parabolic mirror, but envelop a so-called *caustic curve*. This curve can be observed in a cup of tea that is placed under a light bulb. Let us derive the equations of the caustic curve. We use the notation of the figure.

(a) $\gamma = 2\varphi - \pi$.
(b) The equation of $\ell(\varphi)$ is

$$x_1 \sin(2\varphi) - x_2 \cos(2\varphi) = R \sin \varphi.$$

(c) Using the method of Example 4.57, we find that $\ell'(\varphi)$ has equation

$$2x_1 \cos(2\varphi) + 2x_2 \sin(2\varphi) = R \cos \varphi$$

with solution

$$x_1 = \tfrac{3}{4} R \cos \varphi - \tfrac{1}{4} R \cos(3\varphi),$$
$$x_2 = \tfrac{3}{4} R \sin \varphi - \tfrac{1}{4} R \sin(3\varphi).$$

Hint: Sect. 4.2, Exercise 4.19.
(d) The caustic curve is part of an epicycloid with two cusps.

4.57. A beam of light is reflected by a circular mirror with radius R, as shown in Fig. 4.82. The light source is on the circle. The reflected rays are not concurrent, but envelop a caustic curve. Let us derive the equations of this curve. We use the notation of the figure.

(a) $\gamma = (3/2)\varphi - \pi/2$.
(b) The equation of $\ell(\varphi)$ is

$$x_1 \cos(\tfrac{3}{2}\varphi) + x_2 \sin(\tfrac{3}{2}\varphi) = R\cos(\tfrac{1}{2}\varphi).$$

(c) Using the method of Example 4.57, we find that $\ell'(\varphi)$ has equation

$$-\tfrac{3}{2}x_1 \sin(\tfrac{3}{2}\varphi) + \tfrac{3}{2}x_2 \cos(\tfrac{3}{2}\varphi) = -\tfrac{1}{2}R\sin(\tfrac{1}{2}\varphi).$$

The equations of the caustic curve are

$$x_1 = \tfrac{2}{3}R\cos\varphi + \tfrac{1}{3}R\cos(2\varphi),$$
$$x_2 = \tfrac{2}{3}R\sin\varphi + \tfrac{1}{3}R\sin(2\varphi).$$

Hint: Use Sect. 4.2, Exercise 4.18.
(d) The caustic curve is part of a cardioid.

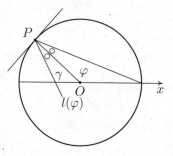
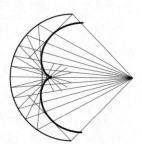

Fig. 4.82. The caustic curve of a light source on a circle

4.58. At each point of a cardioid we draw the *normal*, the line perpendicular to the tangent at that point. The envelope of the system of normals is again a cardioid. It is the image of the original cardioid under a central dilation with scaling factor $-1/3$. We use the following equations for the original cardioid. The point $\mathbf{q}(t)$ of the cardioid with parameter t has coordinates

$$q_1(t) = 2\cos t - \cos 2t,$$
$$q_2(t) = 2\sin t - \sin 2t. \qquad (4.39)$$

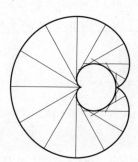

4.6 The Parametric Equations of a Curve, Cycloids

(a) The parametric equations of the tangent at $\mathbf{q}(t)$ are

$$x_1 = q_1(t) + \lambda(-\sin t + \sin 2t),$$
$$x_2 = q_2(t) + \lambda(\cos t - \cos 2t).$$

(b) The equation of the normal $n(t)$ at $\mathbf{q}(t)$ is

$$(-\sin t + \sin 2t)x_1 + (\cos t - \cos 2t)x_2 = \sin t.$$

(c) The equation of the envelope of the system of normals $n(t)$, $t \in [0, 2\pi)$, is

$$x_1(t) = \tfrac{2}{3}\cos t + \tfrac{1}{3}\cos 2t,$$
$$x_2(t) = \tfrac{2}{3}\sin t + \tfrac{1}{3}\sin 2t.$$

(d) To show that the envelope in (c) is the image of the original curve under a central dilation, we substitute $t + \pi$ for t in the equation (4.39) of the original cardioid. After reflecting in O, the curve has equations

$$x_1(t) = 2\cos t + \cos 2t,$$
$$x_2(t) = 2\sin t + \sin 2t.$$

4.59. The tangent of the deltoid at a point that is not a cusp meets the deltoid in two more points. Show that the normals to the deltoid at these intersection points meet at a point on the circumscribed circle of the deltoid.
Hint: Reread the final part of the proof of Theorem 4.58.

4.60. Let us study the pedal curve of the astroid in Exercise 4.51 with respect to the origin O.

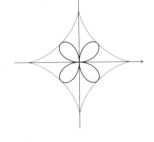

(a) The parametric equations of the tangent $l(t)$ to the astroid at the point $Y(t)$ are

$$x_1(\lambda) = 4a\cos^3 t - \lambda\cos t,$$
$$x_2(\lambda) = 4a\sin^3 t + \lambda\sin t,$$

where λ is the parameter.

(b) The perpendicular $m(t)$ from O on $l(t)$ has equation $-x_1\cos t + x_2\sin t = 0$.
(c) The intersection point $X(t)$ of $l(t)$ and $m(t)$ satisfies $\lambda = 4a\cos 2t$.
(d) The pedal curve has parametric equations

$$x_1 = 2a\sin(2t)\sin t, \quad x_2 = 2a\sin(2t)\cos t.$$

(e) In polar coordinates, the equation is $r(\varphi) = 2a|\sin(2\varphi)|$.

5
SOLID GEOMETRY

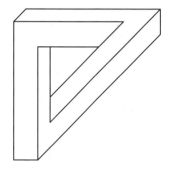

Now that we have completed our study of plane geometry, we are strongly tempted to check whether the methods we used can also be applied to the study of solid geometry. This may give an unbalanced image of solid geometry, but the advantage of this approach is that a number of important subjects from solid geometry can be dealt with efficiently. As the diagram at the beginning of the book shows, our treatment of solid geometry runs parallel to that of plane geometry.

The object we study in solid geometry is 3-space; this must be a model for the space around us. We fix 3-space by giving a list of properties, stated as basic assumptions. The first assumption says that we have a metric space. In this space we define straight lines and planes in a natural way, using the real line $\mathbb{R}$ and the coordinate plane $\mathbb{R}^2$ as models. The most important basic assumption may be that intersecting planes meet in a straight line; in fact, this assumption implies that the dimension of 3-space is at most 3. We define a coordinate space $\mathbb{R}^3$ in analogy to $\mathbb{R}$ and $\mathbb{R}^2$. In this space we define the outer product of vectors, which can be used in defining the orientation of tripods. In Sect. 5.3 we digress and explain how the figures in this chapter were made. In Sect. 5.4 we clarify the structure of the group of all isometries of 3-space; as in dimension two, the reflection plays a key role. Then we analyze the symmetry groups of the regular polyhedra. The main theorem lists all finite subgroups of the group of motions of 3-space. We also discuss applications in crystallography. We conclude this chapter on solid geometry with a discussion of quadratic surfaces, which can be described using a degree-two equation in a suitable coordinate system. There are many similarities to the treatment of conics in Sect. 4.4.

J.M. Aarts, *Plane and Solid Geometry*, DOI: 10.1007/978-0-387-78241-6_5,
© Springer Science+Business Media, LLC 2008

5.1 The Basic Assumptions of Solid Geometry

In this section we fix the basic assumptions for our study of solid geometry. In stating our assumptions we use the results of plane geometry from previous chapters. Our approach is close to that of Sect. 1.4. We try to state and arrange the assumptions in such a way that we can quickly introduce local coordinates.

Basic Assumption 5.1. *Euclidean 3-space is a metric space: a set W with metric d. The set W contains more than one point.*

Given points A and B of W, the line segment $[AB]$ and the straight line AB are defined in exactly the same way as in Definitions 1.12 and 1.13, respectively. The distance between A and B is denoted by $d(A, B)$ or, as in plane geometry, simply by AB.

Basic Assumption 5.2. *For every line l in 3-space there is an isometric surjection from l to $\mathbb{R}$.*

If φ is an isometric surjection from the line l to $\mathbb{R}$, all A and B on l satisfy

$$\varphi([AB]) = [\varphi(A)\varphi(B)] ;$$

compare to Theorem 1.15. Moreover, for every pair of distinct points A and B there is a unique line m passing through A and B; compare to Theorem 1.16.

Basic Assumption 5.3. *W contains three noncollinear points.*

Definition 5.4. *Let A, B, and C be three noncollinear points. The* plane α *through A, B, and C is the subset of W with the following properties:*

1. *A, B, and C lie in α.*
2. *If D and E are distinct points of α, then the line DE lies completely inside α.*
3. *A point X lies in α if and only if there is a series of points $X_1, X_2, \ldots$, $X_n = X$ such that for $1 \leq i \leq n$ the point X_i either is equal to A, B, or C, or lies on the line $X_k X_l$ for some $1 \leq k, l < i$.*

The plane α is sometimes also denoted by ABC; see Theorem 5.8.

Condition (3) of the definition says that conditions (1) and (2) already determine whether a point belongs to α. We sometimes say that the plane through A, B, and C is the smallest subset of W that satisfies (1) and (2).

As with points and lines, the expressions *X lies in α*, *α passes through X*, and *X and α are incident* are equivalent and simply mean $X \in \alpha$. The expressions *a line in a plane* and *a plane through a line* also have the same meaning. The following theorem implies the existence of a plane, which we had not shown yet.

5.1 The Basic Assumptions of Solid Geometry

Theorem 5.5. *Through any three noncollinear points there is a plane.*

Proof. Let A, B, and C be three noncollinear points. We are going to define sets Q_n inductively for $n = 0, 1, 2, \ldots$. As a start, let $Q_0 = \{A, B, C\}$. Then we let $Q_1 = AB \cup BC \cup CA$ and

$$Q_2 = \{X : X \in YZ \text{ with } Y \in Q_1 \text{ and } Z \in Q_1\}.$$

The set Q_2 consists of all points on lines joining points on the sides of the "triangle" ABC. Intuitively, Q_2 is the plane we are looking for. As soon as we have the next basic assumption at our disposal, we can prove that this is indeed the case. But to state that assumption we must first know that there exists a plane. Therefore we define

$$Q_{n+1} = \{X : X \in YZ \text{ with } Y \in Q_n \text{ and } Z \in Q_n\}$$

for $n = 2, 3, \ldots$. The union Q_∞ of all Q_n, $n = 1, 2, \ldots$ is the plane through A, B, and C. We show as follows that Q_∞ satisfies the conditions of Definition 5.4. It is clear that Q_0 lies in Q_∞. If the points D and E lie in Q_∞, there are k and l such that $D \in Q_k$ and $E \in Q_l$. We may, and do, assume that $l \geq k$. By the definition of Q_{l+1} we then have $DE \subset Q_{l+1}$, hence also $DE \subset Q_\infty$. It is clear that Q_{n+1} is the set of all points we need to add to Q_n in order to satisfy (2). Consequently, (3) is also satisfied.

From this theorem we can easily deduce that through any line and point outside the line there is a plane. Moreover, it follows that two lines with a common point are coplanar.

Basic Assumption 5.6. *For every plane α in W there is an isometric surjection φ from α to the coordinate plane $\mathbb{R}^2$.*

We note that there already exists an isometric bijection from $\mathbb{R}^2$ to the plane V; see Theorem 1.27. This basic assumption makes it possible to apply the usual geometry of the coordinate plane to every plane; see for example the proof of the next theorem. Moreover, we can use the coordinate plane $\mathbb{R}^2$ to compare properties of figures in different planes. Let us now give the definition of congruence.

Definition 5.7. *We call two figures F_1 and F_2 in 3-space congruent, denoted by $F_1 \cong F_2$, if there is an isometric surjection $\mathcal{H}$ from F_1 to F_2.*

Later, in Sect. 5.4, we will see that for any two congruent figures there is always an isometry of the *whole* 3-space W that maps one of the figures into the other.

Theorem 5.8. *Through any three noncollinear points there is exactly one plane.*

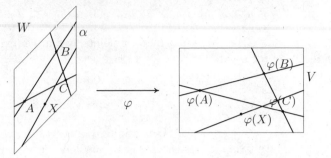

Fig. 5.1. There is exactly one plane through A, B, and C

Proof. Let A, B, and C be noncollinear points. Let α and β be two planes through these points; see Fig. 5.1. There is exactly one line AB through A and B. By Definition 5.4 this line lies in the planes α and β. The same holds for the lines BC and CA. Let X be an arbitrary point of α. By Basic Assumption 5.6 there is an isometric surjection from α to $\mathbb{R}^2$. An arbitrary line in $\mathbb{R}^2$ passing through $\varphi(X)$ meets at least two of the three lines $\varphi(A)\varphi(B)$, $\varphi(B)\varphi(C)$, and $\varphi(C)\varphi(A)$. It follows that an arbitrary line l through X in the plane α meets at least two of the three lines AB, BC, and CA. Without loss of generality, we may, and do, assume that l meets AC at D and BC at E. Since l is determined by D and E, and both D and E lie in β, l also lies in β. Hence $X \in \beta$ and consequently $\alpha \subseteq \beta$. By interchanging α and β we see that also $\beta \subseteq \alpha$ and therefore $\beta = \alpha$.

Let us now consider the position of lines in 3-space with respect to each other. Given two lines l and m in 3-space, we distinguish two cases: either the lines are coplanar, or they are not.

Definition 5.9. *Two non-coplanar lines are called* skew. *Two coplanar lines l and m are called parallel if either $l = m$ or $l \cap m = \emptyset$; otherwise they are called* intersecting *lines.*

Since skew lines do not lie in the same plane, they cannot have a common point. The classification of pairs of lines as skew, parallel, and intersecting is therefore disjoint. We indicate that two lines l and m are parallel by writing $l \mathbin{//} m$. Since every plane has an isometric surjection to $\mathbb{R}^2$, two intersecting lines have a unique intersection point (Theorem 1.19). The following theorem gives an important consequence of Basic Assumption 5.6.

Theorem 5.10. *For every point P and every line m there is exactly one line through P and parallel to m.*

Proof. If P lies on m, then by definition the line can only be m. Let us now assume that P does not lie on m. There is then a unique plane α through P and m; see Theorem 5.8.

5.1 The Basic Assumptions of Solid Geometry

By definition, the parallel line we are looking for must lie in α. On the one hand, there is an isometric surjection from the plane α to $\mathbb{R}^2$. Since there is also an isometric surjection from $\mathbb{R}^2$ to the coordinate plane V, there is an isometric surjection from α to V. On the other hand, Basic Assumption 1.20 states that in V, for every point and every line there is a unique line through the point and parallel to the given line. Consequently, the same holds in α.

Definition 5.11. *We say that a line m and a plane α are* parallel *if either m lies in α or m and α have no point in common. If the line m and the plane α are not parallel, we say that they* intersect *each other.*

Since there is exactly one line through any two distinct points, a line m and a plane α that intersect have exactly one point in common. We call this point the *intersection point* of m and α.

Definition 5.12. *We say that two planes α and β are* parallel *if either $\alpha = \beta$ or α and β have no point in common. If the planes α and β are not parallel, we say that they* intersect *each other.*

We indicate that two planes α and β are parallel by writing $\alpha \mathbin{/\mkern-6mu/} \beta$.

The space W that we want to describe is three-dimensional. We must fix this in basic assumptions. The following basic assumption guarantees that the dimension of W is at most three. We mention the result from linear algebra that in n-dimensional space with $n \geq 4$ two planes may have only a single point in common.

Basic Assumption 5.13. *Two intersecting planes α and β meet in a straight line. This line is called the* intersection line *of the planes.*

This basic assumption has the following very useful consequence.

Theorem 5.14 (Three-planes theorem). *We are given three planes α, β, and γ such that any two intersect each other. Let l be the intersection line of α and β, m that of β and γ, and n that of γ and α. The intersection lines l, m, and n either are parallel to one another or meet at a single point.*

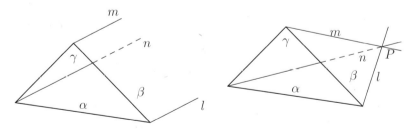

Fig. 5.2. The three-planes theorem

Proof. Consider the intersection lines l and m. These lie in the plane β; see Fig. 5.2. There are two possibilities. l and m intersect or $l \,//\, m$. If l and m intersect, they have one point in common, say P. The point P lies in α and γ, and therefore on n. Hence P is the common point of the intersection lines. If $l \,//\, m$, we must also have $l \,//\, n$ and $m \,//\, n$. Indeed, if $l = m$, l is the common intersection line of α, β, and γ. If $l \,//\, m$, then if one of the other pairs is not parallel, the same arguments as above show that l and m have a common point, which is impossible.

Our first application of the three-planes theorem is in the proof of the following property.

Property 5.15. *The parallel relation between lines in 3-space is reflexive, symmetric, and transitive.*

The reflexivity and the symmetry follow almost directly from the definition. The deduction of the transitivity is more delicate. Let k, l, and m be lines with $k \,//\, l$ and $l \,//\, m$. Let P be a point on m. Consider the planes α through k and l, β through k and P, and γ through l and P. By the three-planes theorem, the intersection line s of β and γ is parallel to k and l. Since s passes through P, Theorem 5.10 implies that $s = m$. Hence $m \,//\, k$, since $s \,//\, k$. This proves the transitivity.

Theorem 5.16. *For every point P and every plane α there is exactly one plane through P and parallel to α.*

Proof. If P lies in α, the proof is trivial. Let us assume that P does not lie in α. Consider Fig. 5.3. Choose two intersecting lines l and m in the plane α.

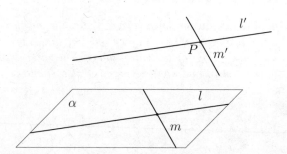

Fig. 5.3. There is a unique plane through P and parallel to α

By Theorem 5.10 there is a unique line l' through P with $l' \,//\, l$, and a unique line m' through P with $m' \,//\, m$. Moreover, there is a unique plane β through l' and m'. We claim that $\alpha \,//\, \beta$ and prove this by contradiction. If α and β are not parallel, then by Basic Assumption 5.13 they intersect in a line, which we call s. In the plane α, the line s meets the line l or the line m, since the geometry of the plane α is the same as that of the coordinate plane. Assume that s and l intersect each other. Let γ be the plane through l and l'. By the three-planes theorem, the intersection lines s of α and β, l of α and γ, and l' of β and γ have a common point, because l and s intersect each other. This is impossible, because l and l' are parallel. Consequently, our assumption that α and β are not parallel must be wrong. To complete the proof, we must show that the plane through P and parallel to α is unique. Let β' be a plane with the same property. Then by Theorem 5.10, the intersection line of β' and the plane through P and l must coincide with l'. In particular, l' lies in β'. For the same reason m' lies in β'. It now follows from Theorem 5.8 that $\beta = \beta'$.

Using Theorem 5.16, we can prove that the parallel relation between planes in 3-space is (reflexive, symmetric, and) transitive (compare with Theorem 1.20).

Perpendicular Position

In plane geometry two positions of straight lines with respect to each other play an important role: parallel and perpendicular. This is also the case in solid geometry. We have already talked about parallel lines. We now turn our attention to perpendicular lines.

Definition 5.17. *Let l and m be lines, and let C' be an arbitrary point in 3-space. By Theorem 5.10 there are unique lines l' and m' through C' that are parallel to l and m, respectively. We say that l and m are perpendicular to each other if $l' \perp m'$. We indicate this by writing $l \perp m$. If $l \perp m$ and l and m do not intersect, we also say that the lines are perpendicular skew lines.*

To justify this definition we must check a number of things. First we need to verify that the lines l' and m' through C' are coplanar. Once we know this we can use the isometric surjection from the corresponding plane to the coordinate plane to determine whether the lines l' and m' are perpendicular to each other. Finally, we must check that the notion we define does not depend on the choice of the point C'. The following proposition provides this result.

Proposition 5.18. *Let l and m be two lines. Let C' be a point, and let l' and m' be lines through C' that are parallel to l and m, respectively. Let C'' be another point, and let l'' and m'' be lines through C'' that are parallel to l and m, respectively. Then $l' \perp m'$ if and only if $l'' \perp m''$.*

Proof. Consider Fig. 5.4. The plane α' through l' and m' is parallel to the plane α'' through l'' and m''; see Exercise 5.3. We choose points A' and A'' on l' and l'', respectively, such that $A'C'C''A''$ is a parallelogram. Likewise, we choose points B' and B'' on m' and m'', respectively, such that $B'C'C''B''$ is a parallelogram. Then $A'A''B''B'$ is also a parallelogram, because $A'A'' // B'B''$ by the transitivity of the parallel relation, and $A'B' // A''B''$ because $\alpha' // \alpha''$.

5 SOLID GEOMETRY

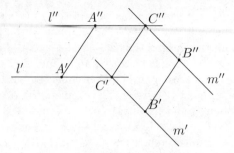

Fig. 5.4. $l' \perp m'$ if and only if $l'' \perp m''$

In particular, we have the following equalities of distances: $A'B' = A''B''$, $B'C' = B''C''$, and $C'A' = C''A''$. Since there are isometric surjections from the planes α' and α'' to the coordinate plane $\mathbb{R}^2$, we can use the congruence criterion **SSS** to deduce that $\triangle A'B'C' \cong \triangle A''B''C''$. The conclusion of the proposition immediately follows.

Let us now define when a line and a plane are perpendicular to each other.

Definition 5.19. *We say that the line l is perpendicular to the plane α if it is perpendicular to every line in α. We indicate this by writing $l \perp \alpha$.*

If a line l and a plane α are perpendicular to each other, they intersect. We can prove this by contradiction. If l and α were parallel, α would contain lines parallel to l; let m be such a line. In the plane β through l and m we have both $l \,/\!/\, m$ and $l \perp m$, which is impossible.

Let us now fix the existence of planes and lines that are perpendicular to each other in a basic assumption.

Basic Assumption 5.20. *For every point P and every plane α there is a line l through P that is perpendicular to α.*

We will show shortly that the line l is unique; consequently, we call l the *perpendicular* through P on α. Moreover, we note that l cannot lie in the plane α because in that case every line in α is perpendicular to l. This in turn implies that l is perpendicular to itself, which is impossible. In particular, this basic assumption says that 3-space is more than only the plane; for example, it contains four noncoplanar points.

Theorem 5.21. *If the lines k and l are both perpendicular to the plane α, then $k \,/\!/\, l$. Through every point there is exactly one line perpendicular to α.*

The *foot* of a perpendicular on a plane is its intersection point with the plane.

5.1 The Basic Assumptions of Solid Geometry

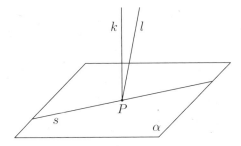

Fig. 5.5. Perpendiculars onto α are parallel

Proof. Consider Fig. 5.5. Since the lines k and l are perpendicular to α, they intersect α. We assume that k and l pass through the same point P of α and prove that $k = l$. The general case follows by moving the lines toward P in a parallel motion. The proof is by contradiction. Suppose that k and l are distinct lines. Since they pass through P, they lie in one plane, which we call β. Since P lies in the planes α and β, by Basic Assumption 5.13, α and β have an intersection line s. Since s lies in α, $k \perp s$ and $l \perp s$. Since the plane β contains only one line through P that is perpendicular to s, it follows that $k = l$.

Let n be a line and let α be a plane. Verifying that $n \perp \alpha$ using the definition is too much work: we would need to check that n is perpendicular to every line in α. The following theorem gives us a way out.

Theorem 5.22. *Let n be a line perpendicular to two intersecting lines k and l in the plane α. Then n is perpendicular to every line in α, hence $n \perp \alpha$.*

Proof. Let C be the intersection point of k and l. We may, and do, assume that n passes through C. Let m be a line in α. To prove that $n \perp m$ we may, and do, assume that m also passes through C. See Fig. 5.6. We choose

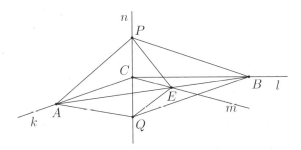

Fig. 5.6. If $n \perp k$ and $n \perp l$ then $n \perp \alpha$

points P and Q on the line n equidistant from C, but on different sides of it. We then choose points A and B on the lines k and l, respectively, such that the lines AB and m intersect. Since $n \perp l$ and $PC = CQ$, we have $PB = BQ$. Likewise, $PA = AQ$. We can compare the images of $\triangle PAB$ and $\triangle QAB$ in the coordinate plane $\mathbb{R}^2$. By the congruence criterion **SSS** we have $\triangle PAB \cong \triangle QAB$, hence $PE = EQ$. In the isosceles triangle PEQ, the median EC from E is also the altitude (Sect. 2.2, Exercise 2.13). Hence $n \perp m$.

Here is a simple application of the theorem.

Theorem 5.23. *For every line l and every point P there is a unique plane α through P such that $l \perp \alpha$.*

Proof. Let β be a plane through l and P. In β we take the line k through P and perpendicular to l. Let Q be the intersection point of k and l, and let n be the perpendicular through Q on β. Let α be the plane through k and n. By the theorem we just proved we have $l \perp \alpha$. The uniqueness of α is a consequence of the following, more general, statement.

Theorem 5.24. *Two planes α and β that are perpendicular to the same line n are parallel.*

Proof. A plane γ through n meets the planes α and β in parallel lines. Using this we can find intersecting lines l and m in α and l' and m' in β such that $l \,/\!/\, l'$ and $m \,/\!/\, m'$. As in the proof of Theorem 5.16, we can then show that $\alpha \,/\!/\, \beta$.

We use this theorem to define the dihedral angle between two intersecting planes; see Fig. 5.7. Let s be the intersection line of the planes α and β. Let γ

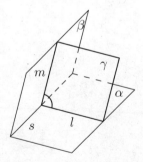

Fig. 5.7. A dihedral angle between the planes α and β

be a plane perpendicular to s. We call γ a *normal plane* for α and β. Let l be the intersection line of γ and α, and m that of γ and β. We call the angles between l and m the *dihedral angles*. We can easily see that the size of the dihedral angles does not depend on the choice of the normal plane. If the dihedral angles are equal to $\pi/2$, we say that α and β are *perpendicular to each other*.

We now turn our attention to two special solid figures; see Fig. 5.8. On the

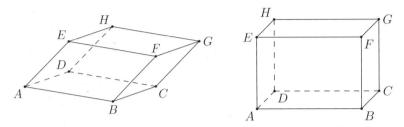

Fig. 5.8. (a) A parallelepiped; (b) a cuboid

left we have a *parallelepiped* $ABCD \cdot EFGH$; these are eight points divided into two ordered quadruples $\{A, B, C, D\}$ and $\{E, F, G, H\}$ whose points lie in distinct parallel planes in such a way that the lines AE, BF, CG, and DH joining corresponding points in the quadruples are parallel to each other. The points A, B, and so on are called the *vertices* of the parallelepiped. The twelve line segments $[AB]$, $[BC]$, $[CD]$, $[DA]$, $[AE]$, $[BF]$, $[CG]$, $[DH]$, $[EF]$, $[FG]$, $[GH]$, $[HE]$ are called the *edges*. The vertices A, B, F, and E lie in one plane and are, in this order, the vertices of a parallelogram. The interior of this parallelogram is called a *face* of the parallelepiped. The six faces are $ABCD$, $EFGH$, $ABFE$, $BCGF$, $CDHG$, $DAEH$. We call BG, CF, and so on, *face diagonals*, and $[AG]$, $[BH]$, $[CE]$, $[DF]$ *space diagonals*.

A parallelepiped is called a *cuboid* if there is a vertex where the extensions of the edges are perpendicular to each other. It is not difficult to see that if this property holds at one vertex, it holds at all vertices. For the cuboid $ABCD \cdot EFGH$, the Pythagorean theorem gives

$$AG^2 = AB^2 + BC^2 + CG^2 = AB^2 + AD^2 + AE^2 .$$

Exercises

5.1. Let α and β be parallel planes. If the plane γ intersects α, it also intersects β. Moreover, the intersection lines are parallel.

5.2. The parallel relation between planes is reflexive, symmetric, and transitive.

5.3. Given planes α and α', let l and m be intersecting lines in the plane α and let l' and m' be lines in the plane α' such that $l \mathbin{/\mkern-5mu/} l'$ and $m \mathbin{/\mkern-5mu/} m'$. Then $\alpha \mathbin{/\mkern-5mu/} \alpha'$.
Hint: see the proof of Theorem 5.16.

5.4. Let l and m be lines and let α be a plane. If $l \,/\!/\, m$ and $l \perp \alpha$, then $m \perp \alpha$.

5.5. If E is the midpoint of the line segment $[AB]$, the plane α through E and perpendicular to AB is called the *perpendicular bisecting plane* of $[AB]$. Show that $\alpha = \{\, X : XA = XB \,\}$.

5.2 Local Coordinates, the Inner and Outer Products

In the previous chapters we have seen that coordinates are an important tool for studying plane geometry. In solid geometry too they are essential. The *vector space* $\mathbb{R}^3$ is the set of all ordered triples (x_1, x_2, x_3) of real numbers. We call such an ordered triple (x_1, x_2, x_3) a *vector*, and often denote it by the associated bold letter: $\mathbf{x} = (x_1, x_2, x_3)$. We call the numbers x_1, x_2, and x_3 the *components* of the vector $\mathbf{x}$. We say that vectors $\mathbf{x}$ and $\mathbf{y}$ are *equal* if and only if the corresponding components are equal: $x_i = y_i$ for $i = 1, 2, 3$. We define the *norm* of a vector $\mathbf{x}$ in $\mathbb{R}^3$ to be

$$\|\mathbf{x}\| = \sqrt{x_1^2 + x_2^2 + x_3^2}\,.$$

Figure 5.9 shows how we can picture $\mathbb{R}^3$. As is commonly done, the vectors are drawn as arrows starting at the *origin*, the intersection point of the axes. We often call the set $\{\,(x_1, 0, 0) : x_1 \in \mathbb{R}\,\}$ the x_1-axis. The x_2-axis and

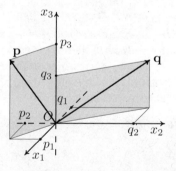

Fig. 5.9. Vectors in $\mathbb{R}^3$

x_3-axis have analogous definitions. The algebraic operations of *addition* and *scalar multiplication* in $\mathbb{R}^3$ are defined on the components. For all vectors $\mathbf{a}$, $\mathbf{b}$ and every real number λ we have

$$\mathbf{a} + \mathbf{b} = (a_1, a_2, a_3) + (b_1, b_2, b_3) = (a_1 + b_1, a_2 + b_2, a_3 + b_3),$$
$$\lambda \mathbf{a} = \lambda(a_1, a_2, a_3) = (\lambda a_1, \lambda a_2, \lambda a_3)\,.$$

We call the vector $\mathbf{a} + \mathbf{b}$ the *sum* of the vectors $\mathbf{a}$ and $\mathbf{b}$. We call the vector $\lambda \mathbf{a}$ the *scalar product* of the real number λ and the vector $\mathbf{a}$; we also say that

5.2 Local Coordinates, the Inner and Outer Products

$\lambda\mathbf{a}$ is a scalar multiple of the vector $\mathbf{a}$. The set $\mathbb{R}^3$ is a group with respect to addition. The unit element of the group is the vector $\mathbf{o} = (0,0,0)$; we call it the *zero vector*. We can check the validity of these statements on the components. For every λ in $\mathbb{R}$ and every $\mathbf{a}$ in $\mathbb{R}^2$ we have

$$\|\lambda\mathbf{a}\| = \sqrt{\lambda^2 a_1^2 + \lambda^2 a_2^2 + \lambda^2 a_3^2} = |\lambda|\,\|\mathbf{a}\|\ .$$

Later we will discuss the geometric interpretation of the algebraic operations. We define

$$\rho(\mathbf{p},\mathbf{q}) = \|\mathbf{q} - \mathbf{p}\|\ .$$

We will show that ρ is a metric, and that $\mathbb{R}^3$ with this metric is an isometric copy of the 3-space W. Therefore we will call $\mathbb{R}^3$ the *coordinate 3-space*. As for the coordinate plane, we will usually denote the points in the coordinate 3-space with bold lowercase letters. There is much liberty in the choice of a coordinate system, as we see in the following theorem.

Theorem 5.25. *The function ρ is a metric. We call the vector space $\mathbb{R}^3$ with metric ρ the coordinate 3-space.*
For every point C of the 3-space W and for every triple (l_1, l_2, l_3) of lines through C that are pairwise perpendicular to each other, there is an isometric surjection Φ from W to the coordinate 3-space $\mathbb{R}^3$ such that $\Phi(l_i)$ coincides with the x_i-axis, for $i = 1, 2, 3$, and $\Phi(C) = \mathbf{o}$. All points P and Q of W satisfy $PQ = \rho(\Phi(P), \Phi(Q))$.

We call $(\mathbb{R}^3, \rho)$ a *coordinate system* at the point C. We sometimes call the components of the vectors the *local coordinates* at C.

Proof. For a given triple of lines (l_1, l_2, l_3) we define, for $i = 1, 2, 3$, projections π_i from W onto l_i. See Fig. 5.10 (**a**). By Theorem 5.16 there is exactly

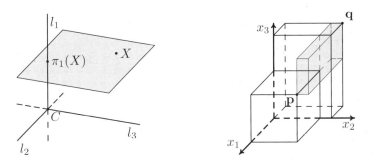

Fig. 5.10. Fixing a coordinate system at C: (**a**) projection onto l_1; (**b**) images under Φ

one plane through X and parallel to the plane through l_2 and l_3; $\pi_1(X)$ is the intersection point of this plane with the line l_1. We define the projections π_2

and π_3 in a similar way. The three-planes theorem implies that if $X \neq Y$, there is an i in $\{1, 2, 3\}$ such that $\pi_i(X) \neq \pi_i(Y)$. By Basic Assumption 5.2 there is an isometric surjection φ_1 from the line l_1 to the x_1-axis. By, if necessary, composing the isometric surjection with a translation, we have $\varphi_1(C) = 0$. There are, in fact, two isometric surjections with this property. Likewise, for $i = 2, 3$, there are isometric surjections φ_i from the line l_i to the x_i-axis such that $\varphi_i(C) = 0$. For an arbitrary point X of W we now define

$$\Phi(X) = (\varphi_1(\pi_1(X)), \varphi_2(\pi_2(X)), \varphi_3(\pi_3(X))) = \mathbf{x}.$$

Figure 5.10 (b) shows the images $\mathbf{p}$ and $\mathbf{q}$ of the points P and Q, respectively. These are diametrically opposite vertices of the shaded cuboid. The points P and Q in W are diametrically opposite vertices of a cuboid whose edges are each parallel to one of the lines l_1, l_2, l_3. The square of the length of the space diagonal PQ is equal to the sum of the squares of the lengths of the edges that meet at P. Hence

$$PQ^2 = \pi_1(P)\pi_1(Q)^2 + \pi_2(P)\pi_2(Q)^2 + \pi_3(P)\pi_3(Q)^2.$$

The edges of the shaded cuboid in Fig. 5.10 (b) have lengths

$$|q_i - p_i| = |\varphi_i(\pi_i(Q)) - \varphi_i(\pi_i(P))| = \pi_i(Q)\pi_i(P), \text{ for } i = 1, 2, 3.$$

Consequently,

$$\rho(\mathbf{p}, \mathbf{q})^2 = (p_1 - q_1)^2 + (p_2 - q_2)^2 + (p_3 - q_3)^2.$$

It follows that $\rho(\mathbf{p}, \mathbf{q}) = PQ$ for all P and Q in W. Since Φ is a bijection, ρ is a metric and Φ is an isometric surjection.

Parametric Equations

The form of the parametric equation of the straight line is the same as in plane geometry. If $\mathbf{p}$ and $\mathbf{q}$ are distinct points of the coordinate 3-space, the line $\mathbf{pq}$ has *parametric equation*

$$\boxed{\mathbf{x} = (1-\lambda)\mathbf{p} + \lambda\mathbf{q}, \quad \lambda \in \mathbb{R}.} \tag{5.1}$$

This means that the map that sends $\lambda \in \mathbb{R}$ onto the point $\mathbf{x}$ using (5.1) is a bijection. The geometric meaning of λ is the same as in plane geometry: the lengths of the line segments $[\mathbf{px}]$ and $[\mathbf{xq}]$ are in the ratio $\lambda : (1 - \lambda)$. The deduction of these properties is almost the same as in plane geometry. Only the interpretation is different; we are now dealing with vectors in $\mathbb{R}^3$.

We can use this to deduce a number of properties of the tetrahedron. The role of the tetrahedron in solid geometry is similar to that of the triangle in plane geometry, and many notions and results related to the triangle have analogues for the tetrahedron. The *tetrahedron*, Fig. 5.11 (a), consists of four

5.2 Local Coordinates, the Inner and Outer Products

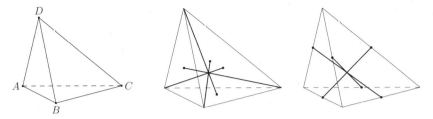

Fig. 5.11. (a) A tetrahedron; (b) its medians; (c) the line segments joining the midpoints of opposite edges

noncoplanar points A, B, C, and D, called the *vertices* of the tetrahedron, together with the line segments $[AB]$, $[BC]$, $[AC]$, $[AD]$, $[BD]$, $[CD]$, called the *edges* of the tetrahedron, and the triangles ABC, ABD, BCD, ACD with their interiors, called the *faces* of the tetrahedron. We say that $[AB]$ and $[CD]$ are *opposite* edges of the tetrahedron, and that the face ABC is *opposite* the vertex D.

Example 5.26 (Centroid). The line joining a vertex of the tetrahedron to the centroid of the opposite face is called a *median* of the tetrahedron; see Fig. 5.11 (b). We will show the following: *the medians meet at one point*, which we call the *centroid* of the tetrahedron. The centroid bisects the line segments joining the midpoints of opposite edges. To prove the properties of the centroid we fix a coordinate system. Let $\mathbf{a}$, $\mathbf{b}$, $\mathbf{c}$, and $\mathbf{d}$ be the vertices of the tetrahedron. The centroid of the face with vertices $\mathbf{b}$, $\mathbf{c}$, and $\mathbf{d}$ is $(\mathbf{b}+\mathbf{c}+\mathbf{d})/3$. The parametric equation of the median from $\mathbf{a}$ is therefore

$$\mathbf{x} = (1-\lambda)\mathbf{a} + \lambda\tfrac{1}{3}(\mathbf{b}+\mathbf{c}+\mathbf{d}) \ .$$

Substituting $\lambda = 3/4$ gives the point $\mathbf{z} = (\mathbf{a}+\mathbf{b}+\mathbf{c}+\mathbf{d})/4$ on the median through $\mathbf{a}$. Since the formula for $\mathbf{z}$ is symmetric in the vertices, it follows that $\mathbf{z}$ lies on all four medians. The midpoint of $[\mathbf{ab}]$ is $(\mathbf{a}+\mathbf{b})/2$, that of $[\mathbf{cd}]$ is $(\mathbf{c}+\mathbf{d})/2$; this implies that $\mathbf{z}$ is the midpoint of the line that joins $(\mathbf{a}+\mathbf{b})/2$ and $(\mathbf{c}+\mathbf{d})/2$.

As in Sect. 1.5, we can use (5.1) to show that for every vector $\mathbf{a} \neq \mathbf{o}$, the vectors $\mathbf{a}$ and $\lambda\mathbf{a}$ lie on the same line through $\mathbf{o}$; we call this line the *span* of $\mathbf{a}$. By substituting $\mathbf{q} = \mathbf{p} + \mathbf{a}$ in (5.1), we obtain the parametric equation

$$\boxed{\mathbf{x} = \mathbf{p} + \lambda\mathbf{a}, \quad \lambda \in \mathbb{R} \ .}$$

Since we assumed that $\mathbf{p}$ and $\mathbf{q}$ are distinct, $\mathbf{a} \neq \mathbf{o}$.

Let us now deduce the parametric equation of a plane; see Fig. 5.12. Let P, Q, and R be three noncollinear points. We choose a coordinate system in which the three chosen points become $\mathbf{p}$, $\mathbf{q}$, and $\mathbf{r}$. The plane α through $\mathbf{p}$, $\mathbf{q}$, and $\mathbf{r}$ has parametric equation

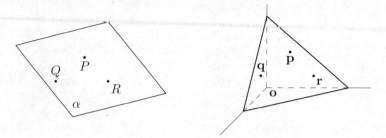

Fig. 5.12. Determining an equation of the plane through P, Q, and R

$$\mathbf{x} = (1 - \lambda - \mu)\mathbf{p} + \lambda\mathbf{q} + \mu\mathbf{r}, \quad \lambda \in \mathbb{R}, \ \mu \in \mathbb{R}. \tag{5.2}$$

We will show that the map Φ that sends $(\lambda, \mu) \in \mathbb{R}^2$ to the point $\mathbf{x}$ defined in (5.2) is a bijection from $\mathbb{R}^2$ to α. We first note that the lines $l = \mathbf{pq}$ and $m = \mathbf{pr}$ meet at $\mathbf{p}$ and are distinct. The line l is the image of $\{(\lambda, 0) : \lambda \in \mathbb{R}\}$ under Φ, and the line m is the image of $\{(0, \mu) : \mu \in \mathbb{R}\}$. We rewrite the equation as

$$\mathbf{x} = \mathbf{p} + \lambda(\mathbf{q} - \mathbf{p}) + \mu(\mathbf{r} - \mathbf{p}), \quad \lambda \in \mathbb{R}, \ \mu \in \mathbb{R}.$$

With $\mathbf{q} - \mathbf{p} = \mathbf{a}$ and $\mathbf{r} - \mathbf{p} = \mathbf{b}$ this gives

$$\mathbf{x} = \mathbf{p} + \lambda\mathbf{a} + \mu\mathbf{b}, \quad \lambda \in \mathbb{R}, \ \mu \in \mathbb{R}.$$

For a fixed λ and μ this point $\mathbf{x}$ is the midpoint of the line segment between $\mathbf{p} + 2\lambda\mathbf{a}$ and $\mathbf{p} + 2\mu\mathbf{b}$. Since these two points lie on l and m, respectively, and therefore in α, $\mathbf{x}$ also lies in α. Conversely, given a point $\mathbf{y}$ in α, we draw a line through $\mathbf{y}$ that intersects the lines l and m; the intersection points are $\mathbf{p} + \lambda_0\mathbf{a}$ and $\mathbf{p} + \mu_0\mathbf{b}$, respectively. For a suitable choice of ν,

$$\mathbf{y} = (1 - \nu)(\mathbf{p} + \lambda_0\mathbf{a}) + \nu(\mathbf{p} + \mu_0\mathbf{b}) = \mathbf{p} + (1 - \nu)\lambda_0\mathbf{a} + \nu\mu_0\mathbf{b}.$$

The second representation of $\mathbf{y}$ implies that $\mathbf{y}$ is in the image of Φ; Φ therefore maps $\mathbb{R}^2$ *onto* α, that is, Φ is surjective. Finally, if a point $\mathbf{y}$ satisfies

$$\mathbf{y} = \mathbf{p} + \lambda_1\mathbf{a} + \mu_1\mathbf{b} = \mathbf{p} + \lambda_2\mathbf{a} + \mu_2\mathbf{b},$$

then $(\lambda_1 - \lambda_2)\mathbf{a} = (\mu_2 - \mu_1)\mathbf{b}$. Since $\mathbf{a}$ and $\mathbf{b}$ have different spans, it follows that $\lambda_1 = \lambda_2$ and $\mu_1 = \mu_2$. This shows the injectivity of Φ.

The above immediately implies that in 3-space, as in the plane, the sum of two vectors $\mathbf{a}$ and $\mathbf{b}$ is given by the parallelogram rule. Indeed, the points $\mathbf{o}$, $\mathbf{b}$, $\mathbf{a} + \mathbf{b}$, $\mathbf{a}$ lie in a single plane. Since there is an isometric surjection from this plane to $\mathbb{R}^2$, we can conclude from our knowledge of the plane that in the given order the points are the vertices of a parallelogram. As in the plane, vector addition is associative; see Fig. 5.13 for the geometric meaning of this associativity.

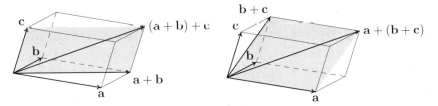

Fig. 5.13. The associativity of vector addition

The Inner Product

We now define the inner product of vectors in $\mathbb{R}^3$.

Definition 5.27. *The inner product of the vectors* $\mathbf{x}$ *and* $\mathbf{y}$ *in* $\mathbb{R}^3$ *is given by*

$$\langle \mathbf{x}, \mathbf{y} \rangle = x_1 y_1 + x_2 y_2 + x_3 y_3 \ .$$

The inner product is not a vector, but a real number. The rules for computing the inner product are the same as those for the inner product in $\mathbb{R}^2$; in particular, $\langle \mathbf{a}, \mathbf{a} \rangle = \|\mathbf{a}\|^2$ for every $\mathbf{a}$ in $\mathbb{R}^3$. As in Sect. 1.6, we can prove the following *triangle inequality* and *Cauchy–Bunyakovskiĭ–Schwarz inequality*: for all $\mathbf{x}$ and $\mathbf{y}$ in $\mathbb{R}^3$,

$$\|\mathbf{x} + \mathbf{y}\| \leq \|\mathbf{x}\| + \|\mathbf{y}\| \quad \text{and} \quad |\langle \mathbf{x}, \mathbf{y} \rangle| \leq \|\mathbf{x}\| \, \|\mathbf{y}\| \ .$$

Theorem 5.28. *The lines* $l : \mathbf{x} = \mathbf{p} + \lambda \mathbf{a}$ *and* $m : \mathbf{x} = \mathbf{q} + \mu \mathbf{b}$ *are perpendicular to each other if and only if* $\langle \mathbf{a}, \mathbf{b} \rangle = 0$.

Proof. The lines $l' : \mathbf{x} = \lambda \mathbf{a}$ and $m' : \mathbf{x} = \mu \mathbf{b}$ pass through $\mathbf{o}$ and are parallel to l and m, respectively. To determine whether l and m are perpendicular to each other, it suffices to do this for the lines l' and m'. Since l' and m' are coplanar, $l' \perp m'$ if and only if all λ_0 and μ_0 satisfy

$$\|\lambda_0 \mathbf{a}\|^2 + \|\mu_0 \mathbf{b}\|^2 = \|\lambda_0 \mathbf{a} - \mu_0 \mathbf{b}\|^2 \qquad \text{(Pythagoras)} \ .$$

By expanding the right-hand side of this formula, we see that it holds if and only if $\langle \mathbf{a}, \mathbf{b} \rangle = 0$.

The inner product comes in handy for deducing a number of important properties of the sphere.

Example 5.29 (The sphere). In analogy to the circle, the sphere is defined as follows. Let M be a point in 3-space, and let r be a positive number. The sphere $\odot(M, r)$ is the set of all points at a distance r from M: $\odot(M, r) = \{ X : MX = r \}$. The point M is called the *center* of the sphere and r is its *radius*.

If we fix a coordinate system at the center M of the sphere, the equation becomes
$$\|\mathbf{x}\| = r \text{ , that is, } x_1^2 + x_2^2 + x_3^2 = r^2 \text{ .}$$
Given $\mathbb{S} = \odot(M, r)$, we define the *interior* in $\mathbb{S}$ and the *exterior* ex $\mathbb{S}$ by
$$\text{in}\,\mathbb{S} = \{\,\mathbf{x} : \|\mathbf{x}\| < r\,\} \quad \text{and} \quad \text{ex}\,\mathbb{S} = \{\,\mathbf{x} : \|\mathbf{x}\| > r\,\} \text{ .}$$

The following properties of the sphere are analogous to those deduced in Sect. 4.1 for the circle. The proof can be copied almost literally. The word *convex* has the same meaning as in plane geometry; a set is *convex* if the line segment joining any two points in the set is also contained in the set.

Theorem 5.30. *The sphere $\mathbb{S} = \odot(M, r)$ has the following properties:*

1. *Every line has zero, one, or two intersection points with $\mathbb{S}$.*
2. *The interior of $\mathbb{S}$ is convex.*
3. *If the points P and Q lie on $\mathbb{S}$ and are distinct; every point of $[PQ]$ other than P and Q lies in the interior of $\mathbb{S}$.*
4. *If the point P lies in the interior of $\mathbb{S}$ and the point Q is different from P, the line PQ and the sphere $\mathbb{S}$ have two intersection points.*
5. *If the point P lies in the interior of $\mathbb{S}$ and the point Q in its exterior, one of the intersection points of PQ and $\mathbb{S}$ lies on the line segment $[PQ]$.*

Example 5.31 (Tangent plane). Let $\mathbb{S} = \odot(M, r)$ be a sphere. A line with exactly one intersection point with $\mathbb{S}$ is called a *tangent line*. A plane with exactly one intersection point with $\mathbb{S}$ is called a *tangent plane*. The unique intersection point in question is called the *point of contact*. If P is a point on the sphere $\mathbb{S} = \odot(M, r)$, every line through P that is perpendicular to MP is a tangent line. Indeed, using the Pythagorean theorem we easily see that every point Q on a line through P that is perpendicular to MP satisfies $MQ > r$ if $P \neq Q$. Moreover, every line tangent to the sphere is perpendicular to the line joining M to the point of contact. We easily see this by considering the plane through the tangent line and the center M; this plane meets the sphere in a circle and the tangent line in question is also tangent to this circle. The tangent plane with point of contact P is the plane through P and perpendicular to MP, Theorem 5.23. In a coordinate system the equation of the tangent plane is $\langle \mathbf{x} - \mathbf{p}, \mathbf{p} - \mathbf{m} \rangle = 0$. We can rewrite this equation as
$$\langle \mathbf{x}, \mathbf{p} - \mathbf{m} \rangle = \langle \mathbf{p}, \mathbf{p} - \mathbf{m} \rangle \text{ .}$$

More generally (see Fig. 5.14), if the span of the vector $\mathbf{b}$ is perpendicular to a plane α, the equation of this plane has the form
$$\langle \mathbf{b}, \mathbf{x} \rangle - c = 0 \text{ , that is, } b_1 x_1 + b_2 x_2 + b_3 x_3 = c \text{ .}$$

The distance from a point $\mathbf{q}$ to this plane is $|\langle \mathbf{b}, \mathbf{q} \rangle - c|/\|\mathbf{b}\|$. This formula can be deduced in the same way as the formula in Sect. 1.6 giving the distance from a point to a line.

5.2 Local Coordinates, the Inner and Outer Products

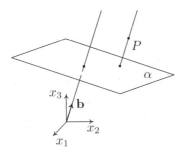

Fig. 5.14. The distance from a point to a plane

If two vectors **a** and **b** have different spans, these vectors and **o** are coplanar. As was done for the plane, Sect. 4.2, Exercise 4.14, we can deduce that $\langle \mathbf{a}, \mathbf{b} \rangle$ is equal to $\|\mathbf{a}\| \|\mathbf{b}\| \cos\varphi$, where φ is the angle between the vectors **a** and **b**. We can also deduce that the area of the parallelogram spanned by **a** and **b** is equal to $\|\mathbf{a}\| \|\mathbf{b}\| |\sin\varphi|$.

The Outer Product

The vectors $\mathbf{a} = (a_1, a_2, a_3)$ and $\mathbf{b} = (b_1, b_2, b_3)$ span a plane γ with parametric equation

$$\mathbf{x} = \mathbf{p} + \lambda \mathbf{a} + \mu \mathbf{b} \ .$$

We are looking for the normal equation $n_1 x_1 + n_2 x_2 + n_3 x_3 = d$ of this plane. That is, we are looking for a vector $\mathbf{n} = (n_1, n_2, n_3)$ perpendicular to the plane γ or, which is equivalent, satisfying $\langle \mathbf{n}, \mathbf{a} \rangle = 0$ and $\langle \mathbf{n}, \mathbf{b} \rangle = 0$; see Theorem 5.22. With what we know about determinants (Sect. 4.5, in particular Exercise 4.42), we can determine **n**:

$$\mathbf{n} = \left(\begin{vmatrix} a_2 & a_3 \\ b_2 & b_3 \end{vmatrix}, -\begin{vmatrix} a_1 & a_3 \\ b_1 & b_3 \end{vmatrix}, \begin{vmatrix} a_1 & a_2 \\ b_1 & b_2 \end{vmatrix} \right) \ .$$

Consequently, $\langle \mathbf{a}, \mathbf{n} \rangle$ is equal to the determinant of a 3×3 matrix with two equal rows, and is therefore equal to 0. Likewise, $\langle \mathbf{b}, \mathbf{n} \rangle = 0$. This leads to the following definition.

Definition 5.32. *The* vector product *or* outer product *of vectors* **a** *and* **b** *is*

$$\mathbf{a} \times \mathbf{b} = \left(\begin{vmatrix} a_2 & a_3 \\ b_2 & b_3 \end{vmatrix}, -\begin{vmatrix} a_1 & a_3 \\ b_1 & b_3 \end{vmatrix}, \begin{vmatrix} a_1 & a_2 \\ b_1 & b_2 \end{vmatrix} \right) \ .$$

The rules for computing determinants imply a number of important properties of the outer product.

1. If **a** and **b** have the same span, then $\mathbf{a} \times \mathbf{b} = \mathbf{o}$.

2. The distributive law holds:
$$\mathbf{a} \times (\mathbf{b} + \mathbf{c}) = \mathbf{a} \times \mathbf{b} + \mathbf{a} \times \mathbf{c}.$$

3. The outer product is anticommutative:
$$\mathbf{a} \times \mathbf{b} = -\mathbf{b} \times \mathbf{a}, \text{ for all } \mathbf{a}, \mathbf{b} \in \mathbb{R}^3.$$

4. For the *standard basis vectors* $\mathbf{e}_1 = (1,0,0)$, $\mathbf{e}_2 = (0,1,0)$, and $\mathbf{e}_3 = (0,0,1)$ we obtain
$$\mathbf{e}_1 \times \mathbf{e}_2 = \mathbf{e}_3, \quad \mathbf{e}_2 \times \mathbf{e}_3 = \mathbf{e}_1, \quad \mathbf{e}_3 \times \mathbf{e}_1 = \mathbf{e}_2.$$

We will now prove the following theorem.

Theorem 5.33. *The norm $\|\mathbf{a} \times \mathbf{b}\|$ of the outer product of the vectors $\mathbf{a}$ and $\mathbf{b}$ is equal to the area of the parallelogram spanned by $\mathbf{a}$ and $\mathbf{b}$.*

Proof. The proof is a short computation. It follows from the definition that
$$\mathbf{a} \times \mathbf{b} = (a_2 b_3 - a_3 b_2, -a_1 b_3 + a_3 b_1, a_1 b_2 - a_2 b_1).$$

This gives
$$\begin{aligned}
\|\mathbf{a} \times \mathbf{b}\|^2 &= a_2^2 b_3^2 + a_3^2 b_2^2 - 2 a_2 b_3 a_3 b_2 \\
&\quad + a_1^2 b_3^2 + a_3^2 b_1^2 - 2 a_1 b_3 a_3 b_1 \\
&\quad + a_1^2 b_2^2 + a_2^2 b_1^2 - 2 a_1 b_2 a_2 b_1 \\
&= (a_1^2 + a_2^2 + a_3^2)(b_1^2 + b_2^2 + b_3^2) - (a_1 b_1 + a_2 b_2 + a_3 b_3)^2 \\
&= \|\mathbf{a}\|^2 \|\mathbf{b}\|^2 - (\langle \mathbf{a}, \mathbf{b} \rangle)^2 \\
&= \|\mathbf{a}\|^2 \|\mathbf{b}\|^2 - \|\mathbf{a}\|^2 \|\mathbf{b}\|^2 \cos^2 \varphi \\
&= \|\mathbf{a}\|^2 \|\mathbf{b}\|^2 \sin^2 \varphi.
\end{aligned}$$

The result now follows from the formula for the area of a parallelogram.

The combinations of inner and outer products prove to be interesting. We know from remarks made prior to the definition of the outer product that $\langle \mathbf{a}, (\mathbf{a} \times \mathbf{b}) \rangle = 0$. Let us determine $\langle (\mathbf{a} \times \mathbf{b}), \mathbf{c} \rangle$. We have

$$\begin{aligned}
\langle (\mathbf{a} \times \mathbf{b}), \mathbf{c} \rangle &= \begin{vmatrix} a_2 & a_3 \\ b_2 & b_3 \end{vmatrix} c_1 - \begin{vmatrix} a_1 & a_3 \\ b_1 & b_3 \end{vmatrix} c_2 + \begin{vmatrix} a_1 & a_2 \\ b_1 & b_2 \end{vmatrix} c_3 \\
&= \begin{vmatrix} c_1 & c_2 & c_3 \\ a_1 & a_2 & a_3 \\ b_1 & b_2 & b_3 \end{vmatrix} = \begin{vmatrix} a_1 & a_2 & a_3 \\ b_1 & b_2 & b_3 \\ c_1 & c_2 & c_3 \end{vmatrix} \\
&= \det(\mathbf{abc}).
\end{aligned}$$

The last line of this computation introduces a shortened notation for the determinant; this is often easier to work with than a square filled with numbers. Using the rules for computing determinants, we see that

$$\langle (\mathbf{a} \times \mathbf{b}), \mathbf{c} \rangle = \langle (\mathbf{b} \times \mathbf{c}), \mathbf{a} \rangle = \langle (\mathbf{c} \times \mathbf{a}), \mathbf{b} \rangle$$
$$= -\langle (\mathbf{b} \times \mathbf{a}), \mathbf{c} \rangle = -\langle (\mathbf{c} \times \mathbf{b}), \mathbf{a} \rangle = -\langle (\mathbf{a} \times \mathbf{c}), \mathbf{b} \rangle \ .$$

In particular,

$$\langle (\mathbf{e}_1 \times \mathbf{e}_2), \mathbf{e}_3 \rangle = \langle (\mathbf{e}_2 \times \mathbf{e}_3), \mathbf{e}_1 \rangle = \langle (\mathbf{e}_3 \times \mathbf{e}_1), \mathbf{e}_2 \rangle = 1,$$
$$\langle (\mathbf{e}_2 \times \mathbf{e}_1), \mathbf{e}_3 \rangle = \langle (\mathbf{e}_3 \times \mathbf{e}_2), \mathbf{e}_1 \rangle = \langle (\mathbf{e}_1 \times \mathbf{e}_3), \mathbf{e}_2 \rangle = -1 \ .$$

Using these computations we can prove the following statement concerning determinants.

Theorem 5.34. *The volume of the parallelepiped spanned by the vectors* $\mathbf{a}$, $\mathbf{b}$, *and* $\mathbf{c}$ *is equal to* $|\det(\mathbf{abc})|$.

Proof. Consider the parallelepiped $OAEB \cdot CFGH$; see Fig. 5.15. Choose the

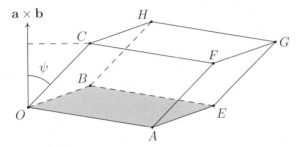

Fig. 5.15. The volume of the parallelepiped is $|\det(\mathbf{abc})|$

origin of the coordinate system at the vertex O. As usual, we denote the vectors associated to the vertices by the corresponding bold lowercase letters. It should be clear why we say that this parallelepiped is spanned by $\mathbf{a}$, $\mathbf{b}$, and $\mathbf{c}$. Since $\det(\mathbf{abc}) = \langle (\mathbf{a} \times \mathbf{b}), \mathbf{c} \rangle$, we obtain

$$\det(\mathbf{abc}) = \|\mathbf{a} \times \mathbf{b}\| \|\mathbf{c}\| \cos \psi \ ,$$

where ψ is the angle between the vectors $\mathbf{a} \times \mathbf{b}$ and $\mathbf{c}$. Note that $\|\mathbf{c}\| |\cos \psi|$ is the height of the parallelepiped. Theorem 5.33 says that $\|\mathbf{a} \times \mathbf{b}\|$ is the area of the parallelogram spanned by $\mathbf{a}$ and $\mathbf{b}$, that is, the parallelogram $OAEB$. Since the product of these two numbers is exactly the volume of the parallelepiped, this completes the proof.

Exercises

5.6. Let $ABCD$ be a tetrahedron. The *altitude* from A is the perpendicular from A onto the opposite face BCD.

The altitudes of a tetrahedron pass through one point if and only if opposite edges are perpendicular to each other.

If the altitudes pass through one point, we call this point the *orthocenter* of the tetrahedron; the tetrahedron itself is then called *orthocentric*.

5.7. Let A, B, C, D be points in 3-space. Let E, F, G, H be the respective centers of the line segments $[AB]$, $[BC]$, $[CD]$, and $[DA]$. Then E, F, G, H either are collinear or are, in this order, the vertices of a (planar) parallelogram.

5.8. Consider the parallelepiped $ABCD \cdot EFGH$.

(a) The sum of the squares of the lengths of the four space diagonals is equal to the sum of the squares of the lengths of the edges, that is,

$$AG^2 + BH^2 + CE^2 + DF^2 = 4\left(AB^2 + AD^2 + AE^2\right).$$

(b) In a cuboid the square of the length of a space diagonal is equal to the sum of the squares of the lengths of the three edges that meet at one vertex. In other words, in the cuboid $ABCD \cdot EFGH$ we have

$$AG^2 = AB^2 + AD^2 + AE^2.$$

(c) Let k, l, m be three lines that are pairwise perpendicular to each other and meet at O. Let n be a line through O, and let α, β, and γ be the angles between the line n and the lines k, l, m, respectively. Then

$$\cos^2 \alpha + \cos^2 \beta + \cos^2 \gamma = 1.$$

5.9. Use the inner product to show that if two pairs of opposite edges of a tetrahedron are perpendicular, this also holds for the third pair, that is, the tetrahedron is orthocentric.

5.10. For all vectors $\mathbf{a}$ and $\mathbf{b}$ in $\mathbb{R}^3$ and every real number λ we have $(\lambda \mathbf{a}) \times \mathbf{b} = \lambda(\mathbf{a} \times \mathbf{b})$.

5.11. The outer product satisfies the following surprising relations:

$$(\mathbf{a} + \mathbf{b}) \times (\mathbf{a} - \mathbf{b}) = -2(\mathbf{a} \times \mathbf{b}),$$
$$(\mathbf{a} + \mathbf{b}) \times (\mathbf{a} + \mathbf{b}) = \mathbf{o}.$$

5.12. The outer product is not associative; in general,

$$\mathbf{a} \times (\mathbf{b} \times \mathbf{c}) \neq (\mathbf{a} \times \mathbf{b}) \times \mathbf{c}.$$

This is a consequence of the following observations. Let $\mathbf{b} \times \mathbf{c} = \mathbf{u}$ and $\mathbf{a} \times \mathbf{u} = \mathbf{v}$.

(a) We have $\langle \mathbf{v}, \mathbf{u} \rangle = 0$; hence there are real numbers λ and μ such that $\mathbf{v} = \lambda \mathbf{b} + \mu \mathbf{c}$.
Hint: the vector $\mathbf{u}$ is perpendicular to the plane through the points $\mathbf{o}$, $\mathbf{b}$, and $\mathbf{c}$.

(b) We have $\langle \mathbf{a}, \mathbf{v} \rangle = 0$; hence there is a real number κ such that
$$\mathbf{a} \times (\mathbf{b} \times \mathbf{c}) = \mathbf{v} = \kappa(\langle \mathbf{a}, \mathbf{c} \rangle \mathbf{b} - \langle \mathbf{a}, \mathbf{b} \rangle \mathbf{c}) \ .$$

(c) By computing one coordinate, for example the first, of both the left- and right-hand sides, we obtain $\kappa = 1$.

(d) Note that $(\mathbf{a} \times \mathbf{b}) \times \mathbf{c} = -\mathbf{c} \times (\mathbf{a} \times \mathbf{b})$. With the above this implies that the outer product is not associative.

(e) The *Jacobi identity* states that
$$\mathbf{a} \times (\mathbf{b} \times \mathbf{c}) + \mathbf{b} \times (\mathbf{c} \times \mathbf{a}) + \mathbf{c} \times (\mathbf{a} \times \mathbf{b}) = \mathbf{o} \ .$$

5.3 What Exactly Do I See?

In this section we briefly discuss two methods for representing solid figures in the plane. These are parallel projection and central projection, or perspective. Almost all maps we see daily, including TV images, are in perspective, but technical drawings and figures in geometry books are usually made using parallel projections. The systematic study of the representation of solid figures is known as *descriptive geometry*; reference [5] is a standard work on this part of geometry. Before the rise of computer graphics, descriptive geometry was a mandatory part of architecture studies [57]; see also [58]. Since drawing figures is now mostly done using computer programs, the interest in descriptive geometry as part of technical studies has almost disappeared.

Definition 5.35. *Let τ be a plane in the 3-space W and let r be a line that is not parallel to τ. The parallel projection $\pi \colon W \to \tau$ from W onto the image plane τ in the direction r is defined as follows. For every point P of W, $\pi(P)$ is the intersection point of τ with the line through P and parallel to r; see Fig. 5.16 (a).*

How do we find the point $\pi(P)$? By Theorem 5.10, every point P admits a unique line l through P and parallel to r. This line is not parallel to τ; the intersection point of l and τ is $\pi(P)$. A line parallel to r is called a *projection line*. If $r \perp \tau$, we speak of an *orthogonal projection*; otherwise, we speak of an *oblique projection*.

Let us begin by discussing a number of important properties of parallel projections.

Theorem 5.36. *The image of a projection line under a parallel projection is a point. The image of any other straight line is a straight line.*

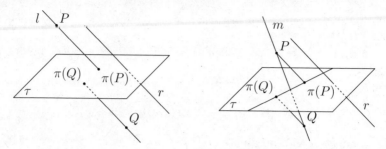

Fig. 5.16. Oblique projections: **(a)** of points; **(b)** of lines

Proof. Let τ be the image plane and r the direction of the parallel projection π. The first result of the theorem follows from the fact that r is not parallel to τ. Next, let m be a line that is not a projection line. Choose a point P on m and a line n through P and parallel to r. The lines m and n are distinct and there is a unique plane through both lines. This plane intersects τ in $\pi(m)$.

The theorem immediately implies that if P and Q are points on the line m and $\pi(P) \neq \pi(Q)$, then $\pi(m)$ is the line through $\pi(P)$ and $\pi(Q)$; see Fig. 5.16 **(b)**.

Theorem 5.37. *Parallel projection maps parallel lines l and m that are not projection lines onto parallel lines.*

Proof. Choose a point P on l and a point Q on m. We draw a line r_1 through P and parallel to r, and a line r_2 through Q and parallel to r. The plane α through l and r_1 and the plane β through m and r_2 are parallel; Sect. 5.1, Exercise 5.3. By Sect. 5.1, Exercise 5.1, the intersection lines of α and β with τ are parallel.

Theorem 5.38. *Parallel projection contracts or expands line segments on parallel lines by a fixed ratio. In other words, the ratio between line segments on parallel lines is invariant under parallel projection.*

In particular, a parallel projection maps the midpoint of a line segment onto the midpoint of the image of the line segment.

Proof. Let τ be the image plane and r the projection direction. The images under the parallel projection onto τ with direction r are indicated by primes in Fig. 5.17. Let l and m be parallel lines. By the previous theorem, l' and m' are also parallel. Moreover, all projection lines are parallel. We distinguish two cases. If $l \parallel \tau$, then $m \parallel \tau$ and the quadrilaterals $ABB'A'$, $BCC'B'$, and $PQQ'P'$ are parallelograms. In that case $AB = A'B'$, $BC = B'C'$, and $PQ = P'Q'$. If neither l nor m is parallel to τ, these lines intersect τ at S and T, respectively. Using triangle similarities, we can show that

$$AB : A'B' = BC : B'C' = PQ : P'Q'.$$

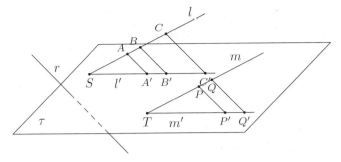

Fig. 5.17. Invariance of ratios on parallel lines

The proof also implies that if the plane α is parallel to τ, π maps α isometrically onto τ.

Example 5.39 (Military projection). Our first example of an oblique projection is the *military projection*. This is characterized by the angle between the projection direction and the image plane τ, which is $\pi/4$. This is the angle between a projection line and its perpendicular projection on τ; see Fig. 5.18, on the left. This angle is the same for all projection lines. The name for this

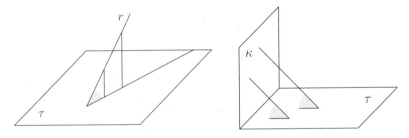

Fig. 5.18. The angle between the direction and the image plane

type of projection is related to its use in the military. Such a projection reproduces the layout of the projected object accurately; the image plane gives a scaled-down image of the area we want to chart. Figure 5.19 shows how this applies to the area around a fantasy church.

Earlier theorems imply that the distances and the heights of the buildings in the image plane are realistic, up to scaling. If we consider Fig. 5.18, on the right, we see that it makes no difference whether we project onto τ or onto the plane κ perpendicular to τ; in other words, the dihedral angle of τ and κ is $\pi/2$. This is because the projection lines and the image plane make an angle of $\pi/4$ with each other; the projection lines give an isometry between τ and κ. If Fig. 5.19 is seen as the result of a projection onto κ, we speak of a *cavalier projection*. A *cavalier* is an artificial hill behind a fortification whose great height offers a good view of the surroundings, namely a cavalier projection.

280   5 SOLID GEOMETRY

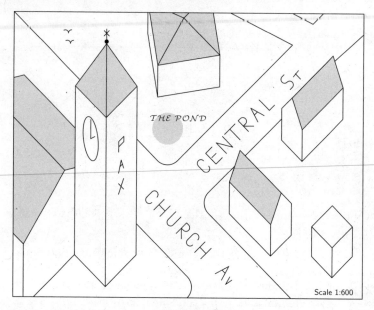

Fig. 5.19. Military projection of the "Peace Church" (*pax* is Latin for peace)

On the lower right of Fig. 5.19 we see a cube, that is, a rhomboid with equal edges, that sits on the image plane. The form of the projection of a cube tells us much about the advantages and disadvantages of the projection method in question. With a bit of knowledge of solid geometry you can deduce the projections of other solid images from that of the cube. Figure 5.20 shows different parallel projections of a cube.

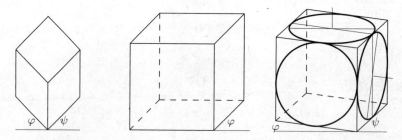

Fig. 5.20. Parallel projections of a cube: (**a**) military; (**b**) cabinet; (**c**) 42/7

Figure 5.20 (**a**) shows a military projection of a cube, as in Fig. 5.19. It is essential that in the image, too, all edges have the same length. The angles φ and ψ are both chosen equal to $\pi/4$.

Figure 5.20 (**b**) is a *cabinet projection* of the cube. This drawing method is very useful and simple, and is almost always used in geometry books and in technical drawings. The ratio of the edges is now 1 : 1 : 0.5 and the angle φ is usually equal to $\pi/6$ or $\pi/4$. If, for example, we want to draw the space diagonals, it is not a good idea to choose the angle φ equal to $\pi/4$, as in that case one of the diagonals disappears partially behind one of the edges. Which parallel projection gives this image for the cube? We can represent it as follows: the cube is placed in front of the image plane and the direction of the projection is from the upper right front corner to the lower left back corner.

Finally, Fig. 5.20 (c) shows the projection that is the norm for technical drawings (ISO 5456-3). This is slightly different from the cabinet projection. The ratio of the edges is still 1 : 1 : 0.5, but the angles ψ and φ with the horizontal line are now approximately 42° (≈ 0.73 rad) respectively 7° (≈ 0.12 rad). This projection method is more work than the usual one, but is preferable if the objects that are drawn contain circles. As we see in Fig. 5.20 (c), a circle on the front face of the cube is mapped onto a curve that is very close to a circle; consequently, you can just as well draw a real circle in the front face. Circles that lie on the top face or on a side are shown as ellipses. If we look more closely at the ellipse on the top face, we see that the major axis is horizontal; this is due to the size of φ, as explained in Sect. 4.4, Exercise 4.37. In the ellipse on the side, the angle between the major axis and the normal vector of the face is also a right angle in the figure. This gives the circles on the top and side faces of the cube a natural look.

To construct one of the drawings in Fig. 5.20 we need to know only how to draw the edges that meet at one of the vertices. Indeed, by Theorems 5.37 and 5.38, parallel projections map parallel lines onto parallel lines, and the ratios of the lengths of line segments on parallel lines are invariant. Knowing this, we can easily extend the drawing of three edges coming together at one vertex to a drawing of a cube. Above we mentioned how the cabinet projection can be obtained as a parallel projection. There is a theorem to corroborate this, by Pohlke (1810–1876): *Suppose we are given three line segments in the image plane with one common endpoint. Suppose moreover that the three line segments are not part of the same line. Then there is a cube and a, possibly oblique, parallel projection onto the image plane such that three suitably chosen edges of the cube that meet at one vertex are mapped onto these line segments.* We would have a hard time proving this theorem without knowledge of descriptive and projective geometry. Therefore we refer to [5]. The parallel projections are classified according to the number of edges of equal length that meet at one vertex in the image of a cube: *isometric* projections have three edges of equal length, as in the military projection; *dimetric* projections have two edges of equal length, as in technical drawings; and in *trimetric* projections the edges at a vertex all have different lengths. See Fig. 5.21.

Oblique projections can also be found in paintings. In [59], Reutersvärd calls oblique projection *Japanese perspective*, because many Japanese prints show the characteristic properties of parallel projection, that is, parallel lines are represented by parallel lines, and ratios between line segments on parallel

Fig. 5.21. Parallel projections: (**a**) isometric; (**b**) dimetric; (**c**) trimetric

lines are drawn to scale. Another, especially interesting, example of the use of parallel projection in painting is the horizontal scroll painting *The Kangxi Emperor's Southern Inspection Tour*. Scroll 12 is shown in [49, pp. 124–129]. This last scroll shows the return of the emperor after an inspection trip, in 1689, and was drawn under the direction of Wang Hui, one of the most famous painters of that time. Because of its unusual dimensions, 2 × 76 ft, the total painting consists of many parts, which were put together in a sophisticated way. Each part has the characteristic properties of a parallel projection.

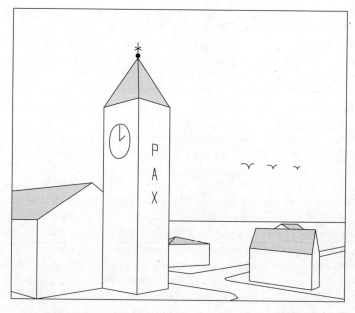

Fig. 5.22. Central projection of the Peace Church

Central Projections

In the Renaissance a completely different method was developed to make realistic representations of three-dimensional scenes [20]. This new method is visible in the works of Alberti (1404–1472) and Dürer (1471–1528). The underlying mathematical theory is the study of perspective; see, for example, [74], [57]. The first book is for beginners, the second for more advanced readers.

Definition 5.40. *Let τ be a plane in the 3-space W and let O be a point outside τ. Let β be the plane through O and parallel to τ (Theorem 5.16). The central projection with center O is the map $\pi\colon W \setminus \beta \to \tau$ that maps every point P in $W \setminus \beta$ to the intersection point of OP with τ; see Fig. 5.23 (a).*

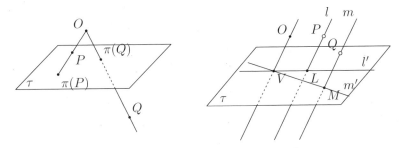

Fig. 5.23. Central projection: (a) of points; (b) of lines

The central projection is very different from the parallel projection. Figure 5.23 (b) shows the projections $\pi(l) = l'$ and $\pi(m) = m'$ of parallel lines l and m, respectively. We have assumed that l and m are not parallel to τ and that neither l nor m passes through the center O. The plane through O and l meets the image plane in l'. The plane through O and parallel to τ meets l at P. The line through the center and parallel to l meets the projection plane at V.

The central projection maps the set $l \setminus \{P\}$ onto $l' \setminus \{V\}$. If we move along the line l from the "lower left" to P and then from P to the "upper right," then in the image we move along l' from V to the right to the "point at infinity" of l', and after passing P we come back from the "point at infinity" on the left and move toward V. The intersection point L of l and V is mapped onto itself. The point V is called the *vanishing point* of the line l. We can determine the image of m in the same manner. The images $\pi(l)$ and $\pi(m)$ come together at the vanishing point V. This implies that in general, the images of parallel lines under a central projection meet at a common vanishing point. This does not hold for a line parallel to τ or passing through the center.

If on l we set a row of equidistant points, their images on l' converge at the vanishing point V of the line. Thus the characteristic properties of the parallel projection are missing in the central projection. For applications, we can imagine the center O as the eye of a painter or onlooker, or the lens of a digital camera, and the image plane as the easel, the painting on the wall, respectively the screen of the camera. We call the resulting image an image *in perspective*; see, for example, Fig. 5.22. The horizon is a special line of vanishing points, those of the lines parallel to the "horizontal" plane through the center. Dimensions are distorted; an object in the foreground is shown larger than if it were in the background (consider the birds in the figure).

There also exist more exotic projection methods. They are related to questions such as, "how would the world look through a curved mirror?" or rather the converse question, "how do you make a drawing such that when looking at it through a curved mirror you obtain a normal image?" This helps train your spatial insight, but it is also simply much fun to take a closer look at these projection methods [45].

Exercise

5.13. The drawing at the beginning of Chap. 5 shows one of the standard examples of the so-called *impossible figures*. At each vertex the drawing brings to mind a separate solid object. However, these objects do not fit together!

The first impossible figures were designed by the Swedish artist Oscar Reutersvärd; see, for example, [59], [60] for an impression of his work. Reference [17] is a very instructive book on impossible figures.

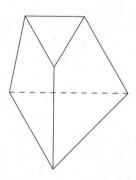

And now the exercise: the object shown here, bounded by five flat faces, is drawn using a parallel projection. Why is this an impossible figure?
Hint: [17, pp. 90, 94], the three-planes theorem, or Sect. 5.1, Exercises 5.1 and 5.3.

5.4 Transformations of Three-Space

In this section we analyze the isometries of the 3-space W. Our approach is similar to that of Chap. 2. An important result for us is Theorem 2.25, Sect. 2.4, which states that the set of all isometries of a metric space is a group under the composition operation. This holds in particular for the set of isometries of W. Our aim is to analyze the structure of this group. We have already seen that in the plane, the reflections are the building blocks for all other isometries. Something similar holds for the isometries of W.

For simplicity of notation, unless stated otherwise, all isometries in the remainder of this chapter are isometries of the 3-space W.

5.4 Transformations of Three-Space

Reflection in a Plane

We begin with the definition of reflection in a plane. This shows many similarities to the definition of a reflection of the plane in a line; see Definition 2.1.

Definition 5.41. *Let α be a plane in W. The* reflection in α *is the map $\mathcal{S}_\alpha$ from W to itself with the following properties:*

1. *For every point X of α, $\mathcal{S}_\alpha(X) = X$.*
2. *For every point X outside α, $\mathcal{S}_\alpha(X)$ is the unique point for which α is the perpendicular bisector of the line segment $[X\,\mathcal{S}_\alpha(X)]$.*

The plane α called is the mirror *of the reflection $\mathcal{S}_\alpha$.*

If X does not lie on the mirror α, we find $\mathcal{S}_\alpha(X)$ by dropping the perpendicular from X on α; the point $\mathcal{S}_\alpha(X)$ lies on this perpendicular, has the same distance from the foot as X, but lies on the other side of the foot from X. The fixed points of the reflection $\mathcal{S}_\alpha$ are the points of α. This implies that two reflections $\mathcal{S}_\alpha$ and $\mathcal{S}_\beta$ are equal if and only if $\alpha = \beta$. Furthermore, it immediately follows from the definition that $\mathcal{S}_\alpha \circ \mathcal{S}_\alpha = \mathrm{id}_W$, which shows the bijectivity of $\mathcal{S}_\alpha$.

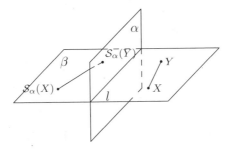

Fig. 5.24. The reflection $\mathcal{S}_\alpha$ is an isometry

We see as follows that a reflection is an isometry. For given points X and Y we consider the plane β through X and Y and perpendicular to α. Let l be the intersection line of α and β. The action of $\mathcal{S}_\alpha$ on the plane β is none other than the reflection $\mathcal{S}_l^\beta$ of the plane β in the line l; see Fig. 5.24. We have

$$\mathcal{S}_\alpha(X)\mathcal{S}_\alpha(Y) = \mathcal{S}_l^\beta(X)\mathcal{S}_l^\beta(Y) = XY \ .$$

We will now prove that an isometry is determined by its action on four noncoplanar points. The following preparatory theorems are the analogues for 3-space of Theorems 2.2 and 2.3.

Theorem 5.42. *An isometry $\mathcal{F}$ with at least three noncollinear fixed points is either a reflection or the identity map id_W.*

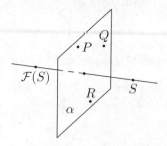

Fig. 5.25. An isometry of W with three fixed points

Proof. Consider Fig. 5.25. Let P, Q, and R be three noncollinear fixed points of $\mathcal{F}$. Assume that $\mathcal{F} \neq \mathrm{id}_W$; then there is a point S with $\mathcal{F}(S) \neq S$. Every fixed point X of $\mathcal{F}$ satisfies

$$XS = \mathcal{F}(X)\mathcal{F}(S) = X\mathcal{F}(S).$$

This implies that X lies in the perpendicular bisector plane α of the line segment $[S\,\mathcal{F}(S)]$. This holds in particular for the points P, Q, and R, which implies that α is the plane through P, Q, and R. A point Y outside α is not a fixed point. Consequently, as above, α is the perpendicular bisector plane of the line segment $[Y\,\mathcal{F}(Y)]$. By the definition of $\mathcal{S}_\alpha$, it follows that $\mathcal{F} = \mathcal{S}_\alpha$, which proves that $\mathcal{F}$ is a reflection.

As a corollary we obtain the following important statement.

Theorem 5.43. *An isometry with four noncoplanar fixed points is the identity.*

Proof. Let $\mathcal{F}$ be an isometry with four noncoplanar fixed points A, B, C, and D. Since the points A, B, and C are noncollinear, the last theorem implies that $\mathcal{F}$ is either a reflection or the identity. If $\mathcal{F}$ were a reflection, all fixed points of $\mathcal{F}$ would lie on the mirror. Since the points A, B, C, and D are noncoplanar, it follows that $\mathcal{F}$ cannot be a reflection.

We can now prove that an isometry is determined by its action on four noncoplanar points, as stated above.

Theorem 5.44. *Every isometry is determined by its action on four noncoplanar points. More precisely, given noncoplanar points A, B, C, and D, if isometries $\mathcal{F}$ and $\mathcal{G}$ coincide on the set $\{A, B, C\, D\}$, then $\mathcal{F} = \mathcal{G}$.*

Proof. Let $\mathcal{F}$ and $\mathcal{G}$ be as in the theorem. Define $\mathcal{H}$ as the composition $\mathcal{F}$ followed by $\mathcal{G}^{-1}$. By the last theorem, $\mathcal{H}$ is the identity. It follows that $\mathcal{F} = \mathcal{G}$.

The following theorem tells us that as in plane geometry, the reflections are the building blocks of the isometries.

5.4 Transformations of Three-Space

Theorem 5.45. *Every isometry is the composition of at most four reflections. If an isometry $\mathcal{F}$ has a fixed point, it can be written as the composition of at most three reflections.*

Proof. Let $\mathcal{F}$ be an isometry. Choose four noncoplanar points A, B, C, and D, where A is one of the fixed points of $\mathcal{F}$ if it has any. Since an isometry maps lines to lines and planes to planes, the images $\mathcal{F}(A)$, $\mathcal{F}(B)$, $\mathcal{F}(C)$, and $\mathcal{F}(D)$ are also noncoplanar. We will now define reflections whose composition is $\mathcal{F}$. Let $\mathcal{S}_1$ be the reflection in the perpendicular bisector plane of $[A\mathcal{F}(A)]$. This maps A to $\mathcal{F}(A)$; if A is a fixed point, we take $\mathcal{S}_1 = \mathrm{id}_W$. By the definition of $\mathcal{S}_1$, the following distances are equal:

$$\mathcal{S}_1(A)\mathcal{S}_1(B) = \mathcal{F}(A)\mathcal{F}(B) = AB \ .$$

We see that the point $\mathcal{F}(A) = \mathcal{S}_1(A)$ is equidistant from $\mathcal{F}(B)$ and $\mathcal{S}_1(B)$. Consequently, $\mathcal{F}(A)$ lies on the perpendicular bisector plane of $[\mathcal{S}_1(B)\,\mathcal{F}_1(B)]$. For the same reason, $\mathcal{F}(A)$ lies in the perpendicular bisector planes of $[\mathcal{S}_1(C)\,\mathcal{F}_1(C)]$ and $[\mathcal{S}_1(D)\,\mathcal{F}_1(D)]$. Let $\mathcal{S}_2$ be the reflection in the perpendicular bisector plane of $[\mathcal{S}_1(B)\mathcal{F}(B)]$. If $\mathcal{S}_1(B)$ is equal to $\mathcal{F}(B)$, we take $\mathcal{S}_2 = \mathrm{id}_W$. The map $\mathcal{S}_2$ leaves $\mathcal{F}(A) = \mathcal{S}_1(A)$ in place and sends $\mathcal{S}_1(B)$ to $\mathcal{F}(B)$. Continuing in the same fashion gives us at most four reflections whose composition $\mathcal{G}$ satisfies $\mathcal{G}(A) = \mathcal{F}(A)$, $\mathcal{G}(B) = \mathcal{F}(B)$, $\mathcal{G}(C) = \mathcal{F}(C)$, and $\mathcal{G}(D) = \mathcal{F}(D)$. Theorem 5.44 now implies that $\mathcal{G} = \mathcal{F}$.

Using this result we can prove the following in the same way as Theorem 2.6.

Theorem 5.46. *If F_1 and F_2 are two congruent figures in W, and $\mathcal{H} \colon F_1 \to F_2$ is an isometric surjection, there is an isometry $\mathcal{G} \colon W \to W$ with $\mathcal{G}(X) = \mathcal{H}(X)$ for every X in F_1.*

The Orientation of a Tripod

Isometries may lead to changes in orientation. We consider the effect of a reflection $\mathcal{S}$ with mirror α on a solid tripod; see Fig. 5.26. The reflection maps the tripod $PABC$ onto the tripod $\mathcal{S}(P)\mathcal{S}(A)\mathcal{S}(B)\mathcal{S}(C)$. The image resembles the original tripod in many ways; the images are congruent. However, it is impossible to transform the tripod $PABC$ into the tripod $\mathcal{S}(P)\mathcal{S}(A)\mathcal{S}(B)\mathcal{S}(C)$ in a continuous movement. We therefore say that the tripods have different orientations. The difference is sometimes also explained as follows: if we place a right-handed corkscrew along PC, with its point at P, and turn it in the direction from A to B, it moves toward C. We say that the tripod $PABC$ is oriented according to the right-hand corkscrew rule. The converse happens to tripod $\mathcal{S}(P)\mathcal{S}(A)\mathcal{S}(B)\mathcal{S}(C)$: if we place a corkscrew along $\mathcal{S}(P)\mathcal{S}(C)$, with its point at $\mathcal{S}(P)$, and turn from $\mathcal{S}(A)$ to $\mathcal{S}(B)$, the corkscrew moves away from $\mathcal{S}(C)$. There are more examples of this phenomenon: the difference between left-handed and right-handed sugar molecules lies in their orientation.

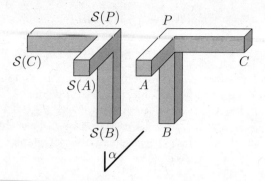

Fig. 5.26. The effect of a reflection on a tripod

To make this more precise mathematically, we first define a tripod. It replaces the angle that had an essential role in the definition of orientation in plane geometry. If P, A, B, and C are four noncoplanar points, the *tripod* $PABC$ is the union of the line segments $[PA]$, $[PB]$, and $[PC]$. We define the orientation using the determinant. We choose a coordinate system; see Fig. 5.27. The unit vectors along the axes are $\mathbf{e}_1 = (1,0,0)$, $\mathbf{e}_2 = (0,1,0)$,

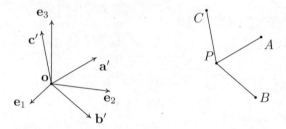

Fig. 5.27. Determining the orientation of the tripod $PABC$

and $\mathbf{e}_3 = (0,0,1)$. The vectors corresponding to the points P, A, B, and C are $\mathbf{p}$, $\mathbf{a}$, $\mathbf{b}$, and $\mathbf{c}$. Let $\mathbf{a}' = \mathbf{a} - \mathbf{p}$, $\mathbf{b}' = \mathbf{b} - \mathbf{p}$, and $\mathbf{c}' = \mathbf{c} - \mathbf{p}$. We say that the tripod $PABC$ has a *positive orientation* if $\det(\mathbf{a}'\mathbf{b}'\mathbf{c}') > 0$, and that it has a *negative orientation* if $\det(\mathbf{a}'\mathbf{b}'\mathbf{c}') < 0$. Since P, A, B, and C are noncoplanar, $\det(\mathbf{a}'\mathbf{b}'\mathbf{c}') \neq 0$ (Theorem 5.34) and the tripod $PABC$ has either a positive of a negative orientation. We also note that the tripod $\mathbf{oe}_1\mathbf{e}_2\mathbf{e}_3$ has a positive orientation. The tripods $PABC$, $PCAB$, and $PBCA$ have the same orientation; $PBAC$, $PCBA$, and $PACB$ have the opposite orientation. The orientation of a tripod depends on the choice of the coordinate system. We will now prove that a reflection reverses the orientation of a tripod; this statement holds for every coordinate system.

Theorem and Definition 5.47. *Every reflection $\mathcal{S}$ reverses orientation: the orientation of a tripod $PABC$ is the opposite of that of $\mathcal{S}(P)\mathcal{S}(A)\mathcal{S}(B)\mathcal{S}(C)$.*

5.4 Transformations of Three-Space

Proof. Consider Fig. 5.28. This is very similar to Fig. 2.15, but must be interpreted differently. What we see in the figure is a projection of 3-space onto a plane perpendicular to the mirror α; we see the mirror α as a line! We could call this figure a top view.

Let us fix a coordinate system. In this coordinate system α has equation

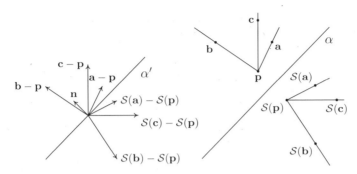

Fig. 5.28. A reflection is orientation-reversing (top view)

$\langle \mathbf{n}, \mathbf{x} \rangle - d = 0$; we may, and do, assume that $\|\mathbf{n}\| = 1$. We want to compare the orientations of the tripods $\mathbf{pabc}$ and $\mathcal{S}(\mathbf{p})\mathcal{S}(\mathbf{a})\mathcal{S}(\mathbf{b})\mathcal{S}(\mathbf{c})$. Let $\mathbf{a}' = \mathbf{a} - \mathbf{p}$, $\mathbf{b}' = \mathbf{b} - \mathbf{p}$, and $\mathbf{c}' = \mathbf{c} - \mathbf{p}$. Note that $\mathcal{S}(\mathbf{a}) - \mathcal{S}(\mathbf{p})$, $\mathcal{S}(\mathbf{b}) - \mathcal{S}(\mathbf{p})$, and $\mathcal{S}(\mathbf{c}) - \mathcal{S}(\mathbf{p})$ arise from $\mathbf{a}'$, $\mathbf{b}'$, and $\mathbf{c}'$, respectively, through the reflection $\mathcal{S}'$ in the plane α' with equation $\langle \mathbf{n}, \mathbf{x} \rangle = 0$. Hence $\mathcal{S}'(\mathbf{a}') = \mathcal{S}(\mathbf{a}) - \mathcal{S}(\mathbf{p})$, $\mathcal{S}'(\mathbf{b}') = \mathcal{S}(\mathbf{b}) - \mathcal{S}(\mathbf{p})$, and $\mathcal{S}'(\mathbf{c}') = \mathcal{S}(\mathbf{c}) - \mathcal{S}(\mathbf{p})$. We want to show that

$$\det(\mathbf{a}'\mathbf{b}'\mathbf{c}') = -\det(\mathcal{S}'(\mathbf{a}')\mathcal{S}'(\mathbf{b}')\mathcal{S}'(\mathbf{c}')) . \tag{5.3}$$

We have

$$\mathcal{S}'(\mathbf{a}') = \mathbf{a}' - 2\langle \mathbf{a}', \mathbf{n} \rangle \mathbf{n}, \quad \mathcal{S}'(\mathbf{b}') = \mathbf{b}' - 2\langle \mathbf{b}', \mathbf{n} \rangle \mathbf{n}, \text{ and } \mathcal{S}'(\mathbf{c}') = \mathbf{c}' - 2\langle \mathbf{c}', \mathbf{n} \rangle \mathbf{n} .$$

To prove this, we note that $\langle \mathbf{a}', \mathbf{n} \rangle$ is the distance from $\mathbf{a}'$ to the plane α'. The formula we want to prove now follows by computing the determinants. The following considerations allow us to avoid much computation. The vectors $\mathbf{a}'$, $\mathbf{b}'$, and $\mathbf{c}'$ cannot all three lie in the plane α'; say that $\mathbf{c}'$ does not lie in α'. Then for a suitable choice of λ the vector $\mathbf{a}'' = \mathbf{a}' - \lambda \mathbf{c}'$ lies in the plane α'. If we replace $\mathbf{a}'$ by $\mathbf{a}''$, the determinant on the left-hand side of (5.3) does not change. By carrying out the mirror operation, we see that we can also replace $\mathcal{S}'(\mathbf{a}')$ by $\mathbf{a}''$ without changing the determinant on the right-hand side of (5.3). Likewise, we can replace $\mathbf{b}'$ by a vector $\mathbf{b}''$ in α' and then replace $\mathbf{c}$ by a vector perpendicular to α'. Using these vectors, the verification of (5.3) is simple.

We will use Table 5.1 to discuss different types of isometries, and will show that the list it gives is complete. We call an isometry $\mathcal{F}$ *direct* if the orientation

Table 5.1. The different types of isometries

isometry	number of fixed points	orientation-preserving	min. number of reflections
identity	∞	yes	0
reflection	∞	no	1
rotation*	∞	yes	2
translation*	0	yes	2
glide reflection	0	no	3
point reflection	1	no	3
improper rotation	1	no	3
screw	0	yes	4

* other than the identity

of any tripod $PABC$ is the same as that of $\mathcal{F}(P)\mathcal{F}(A)\mathcal{F}(B)\mathcal{F}(C)$. If this is not the case we call it *indirect*. A direct isometry is also called a *motion*. It follows from Theorems 5.47 and 5.45 that an isometry is either direct or indirect.

Translations and Rotations

Let us first determine which isometries can be written as a product of two reflections. These are direct isometries. Let $\mathcal{F}$ be an isometry that is the composition $\mathcal{S}_\alpha \circ \mathcal{S}_\beta$ of the reflections $\mathcal{S}_\beta$ followed by $\mathcal{S}_\alpha$. If $\alpha = \beta$, $\mathcal{F}$ is the identity map. If $\alpha /\!/ \beta$, $\mathcal{F}$ is a translation. If the planes α and β intersect, $\mathcal{F}$ is a rotation. We will use the descriptions given here as definitions of translation and rotation. We characterize translations and rotations using the results of Sect. 2.4. Let us begin with the translation.

Theorem and Definition 5.48. *Let $\mathcal{F}$ be an isometry. The following statements are equivalent:*

1. *$\mathcal{F}$ is a translation.*
2. *$\mathcal{F}$ is the composition of two reflections with parallel mirrors.*
3. *In every coordinate system, $\mathcal{F}(\mathbf{x}) = \mathbf{x} + \mathcal{F}(\mathbf{o})$ for all $\mathbf{x}$ in the coordinate space.*
4. *For all points X and Y such that X, Y, and $\mathcal{F}(X)$ are noncollinear, the quadrilateral $XY\mathcal{F}(Y)\mathcal{F}(X)$ is a parallelogram.*

Proof. We choose (2) as definition of a translation; consequently, the first two statements are equivalent. If the translation $\mathcal{F}$ is the composition of reflections $\mathcal{S}_\alpha \circ \mathcal{S}_\beta$, then in every coordinate system the *translation vector* $\mathcal{F}(\mathbf{o})$

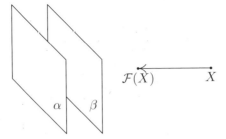

Fig. 5.29. The translation $\mathcal{F}$ as composition of two reflections

is perpendicular to the planes α and β, and its length is twice the distance between the planes. The same holds for every vector $\mathcal{F}(\mathbf{x}) - \mathbf{x}$. This shows that (3) follows from (2). If (3) holds, then in a chosen coordinate system, $\mathcal{F}(\mathbf{x}) = \mathbf{x} + \mathcal{F}(\mathbf{o})$ for all $\mathbf{x}$. It follows that $\mathcal{F}(\mathbf{x}) - \mathbf{x} = \mathcal{F}(\mathbf{y}) - \mathbf{y}$, for all $\mathbf{x}$ and $\mathbf{y}$, and also that $\mathbf{y}\mathbf{x}\mathcal{F}(\mathbf{x})\mathcal{F}(\mathbf{y})$ is a parallelogram whenever $\mathbf{x}$, $\mathbf{y}$, and $\mathcal{F}(\mathbf{x})$ are noncollinear. Thus (4) follows from (3). It is not difficult to show that the converses also hold: (3) follows from (4), and (2) from (3). Section 2.4 gives more information on translations.

The following theorem follows directly using (3).

Theorem 5.49. *The translations form a subgroup of the group $\mathcal{I}(W)$ of all isometries of the 3-space W. This subgroup is isomorphic to $\mathbb{R}^3$.*

We mentioned above that a rotation is the composition of two reflections with intersecting mirrors. There is no direct generalization of the definition

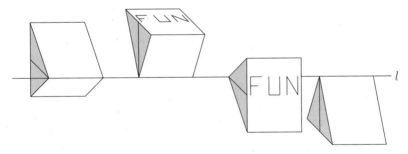

Fig. 5.30. A rotation with rotation axis l

of a rotation of the plane V to 3-space. The difficulty lies in the fact that the notion of oriented rotation angle introduced after Theorem 2.30 cannot be extended to 3-space. In 3-space, whether you see the orientation of a rotation of a plane as clockwise or counterclockwise depends on the side of the plane you are on.

Theorem and Definition 5.50. *Let $\mathcal{F}$ be an isometry. The following statements are equivalent:*

1. *$\mathcal{F}$ is a rotation.*
2. *$\mathcal{F}$ is the composition of two reflections with intersecting mirrors.*
3. *$\mathcal{F}$ preserves orientation and has a line of fixed points. This line is called its rotation axis.*

Proof. We choose (2) as the definition of a rotation. The equivalence of (1) and (2) follows from the definition. Statement (3) is a trivial consequence of (2). If (3) holds, we first choose a plane γ perpendicular to the rotation axis l. Since $\mathcal{F}$ preserves the distances between points, we see that $\mathcal{F}$ maps the plane γ onto itself. The action of $\mathcal{F}$ on γ is orientation-preserving; the intersection point of γ and l is a fixed point. Using the classification in Table 5.1, we see that the action of $\mathcal{F}$ on γ is that of a rotation. Choose two lines m and n in γ such that the composition of the reflection in n followed by that in m is equal to this rotation. Let α be the plane through l and m, and β that through l and n. On both the line l and in the plane γ, the action of the composition of reflections $\mathcal{S}_\alpha \circ \mathcal{S}_\beta$ corresponds to $\mathcal{F}$. By Theorem 5.44 this composition is therefore equal to $\mathcal{F}$.

Reflections in a Point, Improper Rotations, and Glide Reflections

The isometries that are the product of three reflections are indirect. We can divide them into three types, which we will now discuss. Let us begin with the reflection in a point.

Definition 5.51. *Let P be a point in 3-space. The* reflection in P *is the map $\dot{\mathcal{S}}_P$ from 3-space onto itself given by*

1. *$\dot{\mathcal{S}}_P(P) = P$.*
2. *For every point X other than P, $\dot{\mathcal{S}}_P(X)$ is the unique point such that P is the midpoint of the line segment $[X\dot{\mathcal{S}}_P]$.*

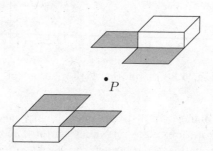

Fig. 5.31. The reflection $\dot{\mathcal{S}}_P$ in P

It immediately follows from the definition that $\dot{\mathcal{S}}_P \circ \dot{\mathcal{S}}_P = \mathrm{id}_W$. By fixing a coordinate system at P, we directly see that a reflection in a point is composed of three reflections. The reflections $\mathcal{S}_1$, $\mathcal{S}_2$, and $\mathcal{S}_3$ in respectively the x_2x_3-, x_1x_3-, and x_1x_2-planes map the point $\mathbf{x} = (x_1, x_2, x_3)$ as follows:

$$(x_1, x_2, x_3) \xrightarrow{\mathcal{S}_1} (-x_1, x_2, x_3) \xrightarrow{\mathcal{S}_2} (-x_1, -x_2, x_3) \xrightarrow{\mathcal{S}_3} (-x_1, -x_2, -x_3);$$

the reflection in the origin P has the same result. We have no restrictions in choosing the axes of the coordinate system, and therefore also have none in choosing the mirrors.

Theorem 5.52. *A reflection in a point is the composition of three reflections whose mirrors are pairwise perpendicular to each other.*

The definition of the next type of isometry implies that it is the composition of three reflections.

Definition 5.53. *An* improper rotation *is the composition of a rotation and a reflection with mirror perpendicular to the axis of the rotation.*

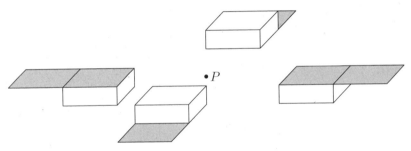

Fig. 5.32. An improper rotation with fixed point P

Figure 5.32 shows us that the order in which the rotation and reflection are carried out does not matter. It follows from the characterization of the rotation that an isometry $\mathcal{F}$ is an improper rotation if and only if $\mathcal{F}$ is the composition of three reflections with two mirrors perpendicular to the third.

The last type of isometry that can be written as a composition of three reflections is the glide reflection.

Definition 5.54. *A* glide reflection *of the 3-space W is the composition of a reflection and a translation with a nonzero vector whose span is parallel to the mirror.*

It follows directly from the definition that the order in which the reflection and translation are carried out does not matter. The glide reflection can also be described as an isometry that is the composition of three reflections of

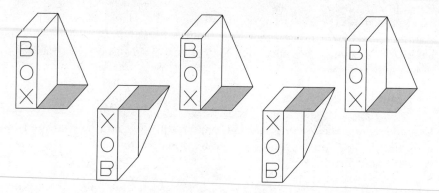

Fig. 5.33. A glide reflection

which two have noncoinciding parallel mirrors that are perpendicular to the mirror of the third reflection.

The following theorem states that there are no other types of isometries that can be written as a product of three reflections.

Theorem 5.55. *Every indirect isometry of the 3-space W is either a reflection, a reflection in a point, an improper rotation, or a glide reflection.*

Proof. By Theorem 5.45, an indirect isometry $\mathcal{F}$ is the product of either one or three reflections. In the first case, $\mathcal{F}$ is a reflection. Let us now consider the case that $\mathcal{F}$ is the product of three reflections: $\mathcal{F} = \mathcal{S}_\alpha \circ \mathcal{S}_\beta \circ \mathcal{S}_\gamma$. We may, and do, assume that none of the mirrors α, β, and γ coincide. By the three-planes theorem there are two possibilities: either the three mirrors have a common point or their intersection lines are parallel to each other. We begin with the latter. Consider a plane δ perpendicular to the intersection lines we just mentioned. The isometry $\mathcal{F}$ maps the plane δ onto itself, and the action of $\mathcal{F}$ on δ is the product of three reflections in the plane δ, that is, it is either a reflection or a glide reflection of the plane δ. Since $\mathcal{F}$ is an isometry, it must be either a reflection or a glide reflection of W. Indeed, if κ is a plane parallel to δ, the action of $\mathcal{F}$ on κ must follow the action of $\mathcal{F}$ on δ.

We now consider the case that the three mirrors have a common point P. We write the reflection $\dot{\mathcal{S}}_P$ in P as the product of three reflections $\mathcal{S}_\mu$ followed by $\mathcal{S}_\lambda$, followed by $\mathcal{S}_\kappa$, with mirrors pairwise perpendicular to each other and meeting at P. Then

$$\mathcal{F} \circ \dot{\mathcal{S}}_P = \mathcal{S}_\alpha \circ \mathcal{S}_\beta \circ \mathcal{S}_\gamma \circ \mathcal{S}_\kappa \circ \mathcal{S}_\lambda \circ \mathcal{S}_\mu .$$

The map $\mathcal{G}$ on the right-hand side is the product of six reflections whose mirrors all contain the point P; it is a direct isometry. Since there is a fixed point, $\mathcal{G}$ can be written as a product of at most three reflections (Theorem 5.45), and therefore as a product of at most two reflections. Consequently, $\mathcal{G}$ is either the identity map or a rotation $\mathcal{R}$. In the first case, $\mathcal{F} = \dot{\mathcal{S}}_P$, and $\mathcal{F}$ is a reflection in a point. In the second case, we may, and do, assume that $\mathcal{R}$ is not the identity. We then have $\mathcal{F} \circ \dot{\mathcal{S}}_P = \mathcal{R}$. By multiplying both sides of this equation by $\dot{\mathcal{S}}_P$, we obtain $\mathcal{F} = \mathcal{R} \circ \dot{\mathcal{S}}_P$. We then write $\dot{\mathcal{S}}_P$ as the product $\mathcal{S}_\tau$ followed by $\mathcal{S}_\sigma$ followed by $\mathcal{S}_\rho$ of reflections with pairwise perpendicular mirrors. We can choose the reflections such that the intersection line of ρ and σ coincides with the axis of $\mathcal{R}$. The map $\mathcal{R} \circ \mathcal{S}_\rho \circ \mathcal{S}_\sigma$ is then the product of four reflections with a common axis; by considering a plane perpendicular to this axis, for example the plane τ, we immediately see that the map is a rotation, say $\mathcal{R}_1$. Thus we find that

$$\mathcal{F} = \mathcal{R} \circ \dot{\mathcal{S}}_P = \mathcal{R} \circ \mathcal{S}_\rho \circ \mathcal{S}_\sigma \circ \mathcal{S}_\tau = \mathcal{R}_1 \circ \mathcal{S}_\tau \, ;$$

in other words, $\mathcal{F}$ is an improper rotation.

The Screw

We complete our study of the isometries of 3-space by studying the isometries that are the product of four reflections. These isometries are all of one type, the screw.

Definition 5.56. *A* screw *is the composition of a rotation and a translation with a nonzero vector whose span is parallel to the axis of the rotation.*

It follows from the definition that the order in which the rotation and translation are carried out does not matter. Unlike a rotation, a screw can

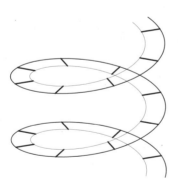

Fig. 5.34. A screw

be oriented; the explanation of the notion of orientation using the corkscrew rule was based on this. When you turn a right-handed corkscrew *into* a cork, the corkscrew turns clockwise. In the real world, most screws turn clockwise when being tightened. We can also describe a screw as an isometry that is the composition of four reflections of which two have nonidentical parallel mirror planes that are perpendicular to the two other mirrors, which also do not coincide. There are no other types of compositions of four reflections. This is a direct consequence of the following theorem.

Theorem 5.57. *A direct isometry, or motion, $\mathcal{F}$ of W is either the identity, a translation, a rotation, or a screw.*

Proof. A direct isometry is the product of an even number of reflections. We may, and do, assume that we are dealing with two or four reflections. Taking into account what we have already discussed, it suffices to analyze the composition of four reflections. Let $\mathcal{T}$ be the translation with translation vector $\mathcal{F}(\mathbf{o})$. Then $\mathcal{T}^{-1} \circ \mathcal{F}$ is a motion with a fixed point, namely $\mathbf{o}$, and therefore, as we saw in the proof of the last theorem, a rotation. We call this rotation $\mathcal{R}$. It follows that $\mathcal{F} = \mathcal{T} \circ \mathcal{R}$. If $\mathcal{R}$ is the identity, $\mathcal{F}$ is a translation. Let us assume that this is not the case. We can decompose the vector $\mathcal{T}(\mathbf{o})$ into two vectors, one along the rotation axis and one perpendicular to it. We write this as $\mathcal{T} = \mathcal{T}_{//} \circ \mathcal{T}_{\perp}$, where the translation vector of $\mathcal{T}_{//}$, or rather its span, is parallel to the axis, while that of $\mathcal{T}_{\perp}$ is perpendicular to the axis. Then

$$\mathcal{F} = \mathcal{T}_{//} \circ \mathcal{T}_{\perp} \circ \mathcal{R}\,.$$

The composition $\mathcal{T}_{\perp} \circ \mathcal{R}$ is again a rotation whose axis is parallel to that of $\mathcal{R}$. This easily follows from the observation that planes perpendicular to the axis of $\mathcal{R}$ are mapped onto themselves by $\mathcal{T}_{\perp} \circ \mathcal{R}$ and the analysis of the action of this map on those planes. It now follows from the definition that the composition of $\mathcal{T}_{//}$ and $\mathcal{T}_{\perp} \circ \mathcal{R}$ is a screw.

Similarities

As in plane geometry, it is often useful to consider not only isometries but also similarities. What we stated in Sect. 2.5 about dilations of the plane can be repeated almost literally for dilations of 3-space. The definition of a *central dilation*, Definition 2.43, can also be seen as the definition of a spatial map. The analogue holds for Theorem 2.44.

Theorem and Definition 5.58. *The following statements concerning a bijective map $\mathcal{F}$ of the 3-space W onto itself are equivalent:*

1. *$\mathcal{F}$ is a dilation.*
2. *All X and Y in W with $X \neq Y$ satisfy $XY // \mathcal{F}(X)\mathcal{F}(Y)$.*
3. *$\mathcal{F}$ is a central dilation or a translation.*

Proof. The equivalence of (1) and (2) is the definition of dilation, Definition 2.40. We have already noted that both a central dilation and a translation have property (2). The proof that (3) follows from (2) is almost the same as the proof of the equivalent statement of Theorem 2.45. We only need to remark that for every pair of points P and Q in 3-space, the points P, Q, $\mathcal{F}(P)$, and $\mathcal{F}(Q)$ lie in one plane, because $\mathcal{F}(P)\mathcal{F}(Q) // PQ$. Consequently, the division into two cases in the proof of Theorem 2.45 can also be applied here.

We call two figures F_1 and F_2 in 3-space *similar*, denoted by $F_1 \sim F_2$, if there exist an isometry $\mathcal{H}$ and a central dilation $\mathcal{V}$ such that $\mathcal{V} \circ \mathcal{H}(F_1) = F_2$, or $\mathcal{V} \circ \mathcal{H}(F_2) = F_1$. The composition $\mathcal{V} \circ \mathcal{H}$ of the central dilation $\mathcal{V}$ and the isometry $\mathcal{H}$ is called a *similarity*.

Exercises

5.14. The rotations around a given axis form a subgroup of $\mathcal{I}(W)$.
Hint: Fix a coordinate system such that the x_3-axis coincides with the common rotation axis. Then note that the x_3-coordinate is invariant under rotations. In this way, this exercise on transformations of 3-space reduces to a problem concerning rotations in the $x_1 x_2$-plane.

5.15. Reflections $\mathcal{S}_\alpha$ and $\mathcal{S}_\beta$ with $\alpha \neq \beta$ *commute*, that is, $\mathcal{S}_\alpha \circ \mathcal{S}_\beta = \mathcal{S}_\beta \circ \mathcal{S}_\alpha$, if and only if α and β are perpendicular to each other.
Hint: consider the actions restricted to the plane perpendicular to the intersection line of α and β.

5.16. Let P be a point and let α be a plane through P. The restriction of the action of $\dot{\mathcal{S}}_P$ to α is also a reflection in a point.

5.17. The motions form a subgroup of $\mathcal{I}(W)$.

5.18. We are given a rotation $\mathcal{R}$ and a translation $\mathcal{T}$. As in the proof of Theorem 5.57, we write $\mathcal{T} = \mathcal{T}_{//} \circ \mathcal{T}_\perp$; the span of the vector $\mathcal{T}_{//}$ is parallel to the axis of $\mathcal{R}$, while that of $\mathcal{T}_\perp$ is perpendicular to this axis.

(a) We have $\mathcal{T}_{//} \circ \mathcal{R} = \mathcal{R} \circ \mathcal{T}_{//}$.
(b) The map $\mathcal{R} \circ \mathcal{T}_\perp \circ \mathcal{R}^{-1}$ is a translation.
   Hint: consider a plane perpendicular to the axis of $\mathcal{R}$.
(c) If $\mathcal{G}$ is a direct isometry, and $\mathcal{T}$ is an arbitrary translation, $\mathcal{G} \circ \mathcal{T} \circ \mathcal{G}^{-1}$ is a translation. This property expresses the fact that the subgroup of the translations is a *normal subgroup* of $\mathcal{I}(W)^+$.

5.19. Let P be the fixed point of both the reflection $\mathcal{G}$ in a point and the improper rotation $\mathcal{F}$. Then $\mathcal{F} \circ \mathcal{G} = \mathcal{G} \circ \mathcal{F}$. The same result holds if $\mathcal{F}$ is a reflection and P lies in the mirror, and also if $\mathcal{F}$ is a rotation and P lies on the rotation axis.

5.20. If $\mathcal{R}$ is a rotation around the axis l, and $\mathcal{F}$ is an isometry, then $\mathcal{F} \circ \mathcal{R} \circ \mathcal{F}^{-1}$ is a rotation around the axis $\mathcal{F}(l)$.

5.21. Let P, Q, and R be noncollinear fixed points of an isometry $\mathcal{F}$. Then every point of the plane through P, Q, and R is a fixed point of $\mathcal{F}$.

5.22. Let $\mathcal{R}_i$ be a rotation with axis l_i, $i = 1, 2$. If l_1 and l_2 intersect, the composition $\mathcal{R}_2 \circ \mathcal{R}_1$ is a rotation.
Hint: use Theorem 5.45 or write $\mathcal{R}_i$ as the composition of two suitable reflections.

5.5 Symmetry and Regular Polyhedra

In this section we study the symmetry of solid figures. The *symmetry group* of a figure F is the group $\mathcal{I}(F)$ of all isometries that map F onto itself. By studying the symmetry of solid figures we almost automatically obtain an overview of all finite subgroups of $\mathcal{I}(W)$. Such an overview is very important for crystallography. We will come back to this later, following up on what we wrote in Sect. 3.5. The symmetry groups contain both direct and indirect isometries. Recall that the direct isometries are also called motions. We discussed the general structure of the symmetry groups in Theorem 3.6. That theorem holds almost unchanged for spatial symmetry groups. Because of its importance we repeat the theorem here.

Theorem 5.59. *Let G be a subgroup of $\mathcal{I}(W)$; let G^+ be the set of direct isometries of G, and let G^- be the set of indirect isometries of G. Then G^+ is a subgroup of G. Moreover, if $G^- \neq \emptyset$, say $\mathcal{H} \in G^-$, the map ι_r given by $\mathcal{F} \stackrel{\iota_r}{\leadsto} \mathcal{F} \circ \mathcal{H}$ is a bijection from G^+ to G^-. The map ι_l given by $\mathcal{F} \stackrel{\iota_l}{\leadsto} \mathcal{H} \circ \mathcal{F}$ is also a bijection from G^+ to G^-.*

The figures we are going to study are the *convex polyhedra*; these are bounded parts of 3-space that arise by intersecting half-spaces. We note that if K is a convex polyhedron, the subgroup G^+ of all motions in $\mathcal{I}(K)$ can contain only rotations, including the identity. This is because a figure that is invariant under translations or screws must be unbounded. When G^+ contains only rotations, we call it the *rotation group* of K.

We consider only intersections of finitely many half-spaces. We call the part of the half-plane that is bounded by these half-spaces and belongs to the polyhedron a *face* of the polyhedron. If two faces have a common point, the intersection is either a line segment or a point; such a line segment is called an *edge*. If two edges have a common point, that point is called a *vertex*. A polyhedron with all vertices but one lying in one plane is called a *pyramid*; the plane in question is the *base plane*, and the point lying outside it is the *apex*. We call the pyramid regular if the face in the plane is a regular polygon and the foot of the perpendicular from the apex to the base plane lies in the center of that polygon; see Fig. 5.40 (**a**).

Regular Polyhedra

A *regular polyhedron* of type $\{p,q\}$ is a convex polyhedron such that every face is a regular p-gon and q edges meet at every vertex.

Let us assume that there exists a regular polyhedron of type $\{p,q\}$. Every angle of the regular p-gons that make up its faces is equal to $(1 - 2/p)\pi$. Together, the q planes that meet at one vertex make an angle of size $q(1-2/p)\pi$. Since we have a convex polyhedron, a necessary condition for the existence of a regular polyhedron of type $\{p,q\}$ is

5.5 Symmetry and Regular Polyhedra

$$q\left(1-\frac{2}{p}\right)\pi < 2\pi, \text{ that is, } (p-2)(q-2) < 4.$$

Consequently, only the following types are possible: $\{3,3\}$, $\{4,3\}$, $\{3,4\}$, $\{5,3\}$, $\{3,5\}$. We will show that there is a polyhedron of each type; this polyhedron is unique up to similarity transformations. In the same order as the types listed above, the polyhedra are the *tetrahedron*, the *cube*, the *octahedron*, the *dodecahedron*, and the *icosahedron*; see Fig. 5.35. These regular polyhedra are called the *Platonic solids*. According to Plato, the four elements water, fire, air, and earth have the form of respectively an icosahedron, a tetrahedron, an octahedron, and a cube, and the surrounding cosmos has the form of a dodecahedron. Kepler associated the twelve faces of the dodecahedron with the twelve signs of the zodiac. Much more significantly, Kepler associated the distances from the planets to the sun with relationships of nested Platonic solids. Regular polyhedra also occur in the work of Dalí [15]. In the *Last Supper* (1955, National Gallery of Art, Washington), the main theme is contained in a regular dodecahedron. The dodecahedron supports the space, while its edges give the space its structure.

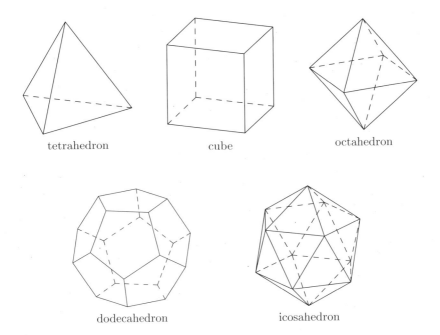

Fig. 5.35. Regular polyhedra

How do we know that there is only one regular polyhedron of each type, up to similarity transformations? This is connected with the *Euler formula*: if a convex polyhedron has F faces, E edges, and V vertices, then

$$F - E + V = 2.$$

This formula comes from topology and holds not only for convex polyhedra, but more generally for figures that can be unfolded in the plane. To clarify what we mean by this, we have unfolded the dodecahedron in the plane in Fig. 5.36 (a). In (b), we see a plane figure made up of points and curves joining the points. The Euler formula also holds for such a figure: F is the number of sections into which the plane is divided, E the number of curve segments, and V the number of given points. We can use the Euler formula

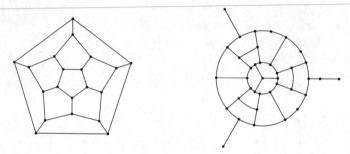

Fig. 5.36. The Euler formula: (a) unfolded dodecahdron; (b) points and curves

to prove that up to scaling, there is only one polyhedron of type $\{5,3\}$. If F is the number of faces of such a polyhedron, we can find a formula for F as follows. First we count the number of edges: each face has 5 edges; this gives $5F$ edges. We have counted every edge twice, since each edge is part of two faces. We obtain $R = (5/2)F$. Then we count the vertices: each face has 5 vertices; this gives $5F$ vertices. We have counted every vertex 3 times, since three edges meet at each vertex. We obtain $H = (5/3)F$. If we now substitute this in the Euler formula, we obtain $F = 12$. Consequently, every polyhedron of type $\{5,3\}$ is a dodecahedron. The only freedom left in the construction of a *regular* dodecahedron is the length of the edges.

The deduction of the Euler formula is not simple. The following sketch of the proof relies heavily on intuition. First we consider a line segment together with its two endpoints: the figure contains one region, one curved segment, and two points. The Euler formula holds for this simple figure. The idea of the proof of the Euler formula is to extend the figure step by step, each time adding a line segment with endpoint, if possible, and checking that the formula still holds. This is shown in Fig. 5.37 for Fig. 5.36 (b), where we start in the lower left corner.

Semiregular Polyhedra

In addition to the regular polyhedra, there are also *semiregular* polyhedra, also known as *Archimedean solids*. The faces of a semiregular polyhedron are

Fig. 5.37. The Euler formula remains valid at every step

regular polygons that are not necessarily all of the same form, though the same number of faces comes together at every vertex. We consider two types of semiregular polyhedra. There are many more types; see, for example,[30] or [73]. Reference [47] is a colorful book on regular polyhedra, semiregular polyhedra, and much more. For a scientific work on polyhedra, see [13], which is clearly written and therefore very accessible. The standard text on this subject is [10].

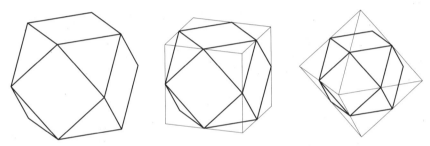

Fig. 5.38. The semiregular cuboctahedron: (**a**) by itself; (**b**) as a truncated cube; (**c**) as a truncated octahedron

The first semiregular polyhedron we consider is the semiregular *cuboctahedron*. It has type $\{3, 4, 3, 4\}$: at every vertex four faces meet, successively a triangle, a square, a triangle, and a square; see Fig. 5.38 (**a**). The semiregular cuboctahedron is bounded by six squares and eight equilateral triangles. As (**b**) shows, we can construct the semiregular cuboctahedron out of a cube by cutting off pieces (truncating the cube). Figure (**c**) shows that the semiregular cuboctahedron is also a truncated octahedron.

The second semiregular polyhedron we consider is the truncated icosahedron, bounded by twelve pentagons and twenty hexagons; see Fig. 5.39. Every pentagon is surrounded by hexagons, and every hexagon is surrounded by alternating pentagons and hexagons. The truncated icosahedron has type $\{5, 6, 6\}$: at every vertex a pentagon and two hexagons meet. As the name suggests, the figure can be obtained by truncating an icosahedron. At the end of the 1980s, a new carbon molecule was discovered, C_{60}, which consists of sixty atoms arranged in a truncated icosahedron. Different representations of carbon C were already known: among others diamond, the crystalline form, and graphite. In graphite three of the four joins between the carbon atoms are

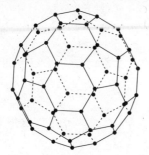 

Fig. 5.39. Two versions of the buckyball

used to make layers, which in turn are kept together by the remaining joins. It now turned out that sixty carbon atoms can also organize themselves into a truncated icosahedron. The sixty atoms take the positions of the vertices; three of the four joins bind each atom to three of its neighbors, while the fourth join lies on the edge between the hexagons. This representation of carbon is used to study isolated atoms of other elements; it is as if those atoms are imprisoned in a cage, formed by the C_{60} molecule. Other carbon molecules with a similar structure and many more atoms are also known. The structure of the carbon molecule out of pentagons and hexagons is very similar to that of the so-called geodesic domes of the American engineer and philosopher R. Buckminster Fuller. To honor Fuller, the carbon molecules mentioned here are called fullerenes. The molecule C_{60} is also called the buckyball. Another representation of the truncated icosahedron is the soccer ball.

Symmetry Groups

We will now determine the rotation groups and the symmetry groups of a number of figures. These are, successively, the regular n-sided pyramid, the n-sided bipyramid, the tetrahedron, the cube, the octahedron, the semiregular cuboctahedron, the dodecahedron, and the icosahedron.

The dihedral group D_n. Let us first consider a regular pyramid K with n sides, $n \geq 3$. We assume that the length of the slanted edges is not equal to that of the edges in the base plane; see Fig. 5.40 (**a**). If they were equal and $n = 3$, we would have a tetrahedron. The figure is then more regular than the pyramid we are considering and also has a different symmetry group. Since K is a bounded solid figure, the symmetry group of K cannot contain translations, glide reflections, or screws. The direct isometries of K are therefore rotations. The rotation group of K consists of all rotations of 3-space over multiples of $2\pi/n$ with the altitude of the pyramid as axis. The rotation group is isomorphic to the rotation group of the regular n-gon in the base plane, that is, isomorphic to C_n. The full symmetry group of K is D_n, which is isomorphic to the symmetry group of the regular n-gon. The relation between D_n and C_n

 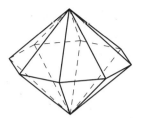 

Fig. 5.40. (a) A pyramid; (b) a bipyramid; (c) a bipyramid on a lens

is described in Theorem 5.59. Let us introduce another figure that is close to the pyramid. It is the *bipyramid*, obtained as the union of the pyramid and its mirror image in the base plane; see Fig. 5.40 (b). We call this a regular n-sided bipyramid, even though the figure has $2n$ faces. The *rotation group* of this figure is isomorphic to D_n, which helps explain the name we give to this group. We call D_n the *dihedral group*, from the Greek *dihedron* (solid with two faces): in 3-space, we can see the regular n-gon, whose symmetry group it is, as a degenerate bipyramid. If $\mathcal{R}$ is a rotation of the bipyramid that interchanges its apices, every element of the rotation group D_n of the bipyramid either is equal to an element of the rotation group C_n of the n-sided pyramid or is obtained by multiplying such an element by $\mathcal{R}$. We first came across the group D_2 in Example 3.5, as the symmetry group of the letter H. We can also see D_2 as the rotation group of the 2-sided bipyramid constructed from the regular 2-gon, the lens; see Fig. 5.40 (c).

Intermezzo: Permutation Groups

In what follows we need a number of properties of permutations. For the sake of completeness we present the most important properties.

Definition 5.60. *Let X be a set with n elements, where $n \geq 1$. A bijection $X \to X$ is called a* permutation *of n elements.*

The set of all permutations of n elements is denoted by S_n. If we put the discrete metric on X, the permutations are isometries; see Theorem 1.7. Therefore we can also see S_n as a group (Theorem 2.25); the group operation is composition. This group is called the *symmetric group*.

Property 5.61. *The number of elements of S_n is $n!$.*

Indeed, if we want to make a bijection f from the set X with n elements to itself, we must first place the elements in some order. There are n possibilities for the image of the first element under the map f. This leaves $n-1$ possibilities for the image of the second element, and so on. Thus we find that S_n has
$$n! = n \times (n-1) \times (n-2) \times \cdots \times 3 \times 2 \times 1$$

elements. We distinguish *even* and *odd* permutations. In order to avoid complicated notation, we illustrate these notions using an example, the set

$$X = \{ A, B, C, D, E, F \}$$

consisting of six elements, and the permutation g. To give the action of g on X, we write the elements and their images under each other, as follows:

$$\begin{array}{cccccc} A & B & C & D & E & F \\ g(A) & g(B) & g(C) & g(D) & g(E) & g(F) \end{array}$$

Let us take an explicit example:

$$\begin{array}{cccccc} A & B & C & D & E & F \\ D & A & B & C & F & E \end{array}.$$

The elements of the set X are written in the top row, alphabetically. To determine whether g is even or odd, we look for the pairs of elements that have been *transposed*, that is, whose order has been interchanged by g. First we have the pair (A, B): A is to the left of B, but $D = g(A)$ is to the right of $A = g(B)$. We also find the transposed pairs (A, C), (A, D), (E, F). Because we have an even number of *transpositions*, we call g an *even permutation*; if we have an odd number of transpositions, we speak of an *odd permutation*. Whether a permutation is even or odd is intrinsic and does not depend on the order in which we place the elements of X. If we were to write them in the order D, E, F, A, B, C, the example above would become

$$\begin{array}{cccccc} D & E & F & A & B & C \\ C & F & E & D & A & B \end{array},$$

which, up to ordering, is the same as before. If we now determine which pairs are interchanged, we find the pairs (D, E) (because in this ordering F is to the left of C), (D, F), (D, A), (D, B), (D, C), (E, F), (E, A), (F, A). With this ordering of the elements of X, we also find an even number of transpositions. More generally, we can show that the order in which the elements of X are written does not affect the parity of the permutation. To show this, interchange two adjacent columns, and note that every ordering of the columns can be obtained by repeatedly interchanging two adjacent columns.

It immediately follows from the definition that the product of two even or of two odd permutations is an even permutation. The product of an even and an odd permutation is an odd permutation.

The tetrahedral group A_4. We will show that the rotation group of the tetrahedron T is isomorphic to A_4, the subgroup of S_4 consisting of the even permutations. We use Fig. 5.41 to study the group:

1. There are rotations about the altitudes, over multiples of $2\pi/3$. This give nine rotations, where the identity map is counted only once.

5.5 Symmetry and Regular Polyhedra

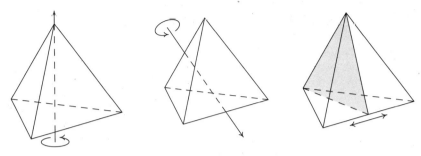

Fig. 5.41. The symmetry group of the tetrahedron: (**a**) rotations about altitudes; (**b**) rotations about lines joining the midpoints of opposite edges; (**c**) reflections

2. There are rotations over π about the lines joining the midpoints of opposite edges; see Fig. 5.41 (**b**). This give three rotations.

In all, we have 12 rotations, $\mathcal{R}_1, \ldots, \mathcal{R}_{12}$. Next consider a reflection $\mathcal{S}$ in a perpendicular bisector plane of an edge. The symmetries

$$\begin{array}{cccc} \mathcal{R}_1 & \mathcal{R}_2 & \ldots & \mathcal{R}_{12} \\ \mathcal{R}_1 \circ \mathcal{S} & \mathcal{R}_2 \circ \mathcal{S} & \ldots & \mathcal{R}_{12} \circ \mathcal{S} \end{array}$$

are all distinct, as we can easily check. We therefore already have 24 elements of the symmetry group. But every element of the symmetry group induces a permutation of the four vertices of the tetrahedron, with different transformations inducing different permutations; this follows from Theorem 5.44. Consequently, the symmetry group does not have more elements than S_4, which has 24. This implies that we have found the whole symmetry group: the symmetry group $\mathcal{I}(T)$ of the tetrahedron is isomorphic to S_4. Every rotation induces an even permutation of the vertices, and there are as many rotations as there are even permutations. The rotation group is therefore isomorphic to the subgroup A_4 of the even permutations in S_4. Moreover, this group contains the Klein four-group V_4. Indeed, together with the identity, the rotations mentioned in (2) form a subgroup that is isomorphic to V_4. See also Exercise 5.27.

The cube and octahedron are *dual polyhedra*. Consider the centers of the faces of the cube. If two faces of the cube have a common edge, join the centers of those faces using a line segment. This gives rise to the octahedron. Likewise, we can build the cube out of the octahedron. The dodecahedron and icosahedron are also dual polyhedra. Dual polyhedra have the same symmetry groups.

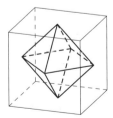

Fig. 5.42. Dual polyhedra

The octahedral group S_4. The cube and the octahedron are dual polyhedra. This implies that they have the same symmetry group. This symmetry group is also equal to that of the semiregular cuboctahedron, by what we said about that polyhedron when discussing Fig. 5.38. We will show that the rotation group of the cube is isomorphic to S_4, the group of permutations of four elements. We have drawn two tetrahedra in the cube $ABCDEFGH$; see Fig. 5.43, on the left. The tetrahedron $ACFH$ is drawn with thick lines,

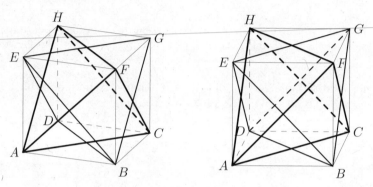

Fig. 5.43. Two tetrahedra in a cube

the other one, $BDEG$, is drawn with slightly thinner lines; of the latter, the edge DG lies behind the edge AF, making it difficult to see. To obtain a better view of the situation, we have also drawn the cube on the right, using a different parallel projection. Every isometry that maps the tetrahedron $ACFH$ onto itself also maps the cube onto itself. The symmetry group of the tetrahedron $ACFH$ is isomorphic to S_4 and has 24 elements. Moreover, every isometry that maps the cube onto itself also either maps each of the tetrahedra onto itself or interchanges the tetrahedra. It follows that the symmetry group of the cube has 48 elements. Since the symmetry group of the cube also contains indirect isometries, the rotation group of the cube, that is, the octahedral group, must have 24 elements. Let us describe this rotation group. As we did for the tetrahedral group, we will write the elements of the symmetry group of the tetrahedron $ACFH$ as follows:

$$\begin{array}{cccc} \mathcal{R}_1 & \mathcal{R}_2 & \ldots & \mathcal{R}_{12} \\ \mathcal{R}_1 \circ \mathcal{S} & \mathcal{R}_2 \circ \mathcal{S} & \ldots & \mathcal{R}_{12} \circ \mathcal{S} \ . \end{array}$$

The $\mathcal{R}_i$, $i = 1, \ldots, n$, are the rotations, and $\mathcal{S}$ is a fixed reflection in a perpendicular bisector plane of an edge of the tetrahedron $ACFH$. Let $\mathcal{J}$ be the reflection in the center of the cube. All isometries in the following list are direct:

$$\begin{array}{cccc} \mathcal{R}_1 & \mathcal{R}_2 & \ldots & \mathcal{R}_{12} \\ \mathcal{R}_1 \circ \mathcal{S} \circ \mathcal{J} & \mathcal{R}_2 \circ \mathcal{S} \circ \mathcal{J} & \ldots & \mathcal{R}_{12} \circ \mathcal{S} \circ \mathcal{J} \ . \end{array}$$

These transformations are all distinct. To see this, note that the transformations $\mathcal{R}_i$ map each tetrahedron onto itself, while the transformations $\mathcal{R}_i \circ \mathcal{S} \circ \mathcal{J}$ interchange the tetrahedra. These are all the rotations in the symmetry group of the cube. We define a map φ between the isometries in the two lists above, that is, between the isometry group of $ACFH$ and the rotation group of the cube:

$$\mathcal{R}_i \xrightarrow{\varphi} \mathcal{R}_i, \qquad i = 1, \ldots, 12,$$
$$\mathcal{R}_i \circ \mathcal{S} \xrightarrow{\varphi} \mathcal{R}_i \circ \mathcal{S} \circ \mathcal{J}, \quad i = 1, \ldots, 12.$$

It follows from the above that this map is bijective. Moreover, this map turns out to be an isomorphism. Obviously, when checking this, we need to distinguish a number of cases. We will discuss only the most difficult one. Let $\mathcal{F} = \mathcal{R}_i \circ \mathcal{S}$ and $\mathcal{G} = \mathcal{R}_j \circ \mathcal{S}$. We will prove that $\varphi(\mathcal{F} \circ \mathcal{G}) = \varphi(\mathcal{F}) \circ \varphi(\mathcal{G})$. As a product of two indirect isometries, the map $\mathcal{F} \circ \mathcal{G}$ is direct; hence $\mathcal{F} \circ \mathcal{G} = \mathcal{R}_k$ for some k. Since $\varphi(\mathcal{R}_k) = \mathcal{R}_k$, we have

$$\varphi(\mathcal{F} \circ \mathcal{G}) = \mathcal{F} \circ \mathcal{G} = \mathcal{R}_i \circ \mathcal{S} \circ \mathcal{R}_j \circ \mathcal{S},$$

while

$$\varphi(\mathcal{F}) \circ \varphi(\mathcal{G}) = \underbrace{\mathcal{R}_i \circ \mathcal{S} \circ \mathcal{J}} \circ \underbrace{\mathcal{R}_j \circ \mathcal{S} \circ \mathcal{J}}.$$

But the reflection $\mathcal{J}$ in a point commutes with all isometries in the symmetry group of the tetrahedron $ACFH$ (see, for example, Exercise 5.19). Therefore the right-hand sides are equal; consequently, so are the left-hand sides.

The dodecahedral group A_5. The dodecahedron and icosahedron are dual polyhedra; see Fig. 5.42. In particular, the dodecahedron and icosahedron have the same symmetry group. The buckyball also has the same symmetry group (Exercise 5.31). We will show that the rotation group of the dodecahedron, and therefore that of the icosahedron and of the buckyball, is isomorphic to A_5, the group of even permutations of five elements. Let us begin by listing the rotations. We proceed as we did for the tetrahedral group. Recall that the dodecahedron has 12 faces, 30 edges, and 20 vertices. We find the following rotations:

1. The identity map is the unit element of the rotation group.
2. There are 24 rotations over multiples of $2\pi/5$ about axes that join the centers of (diametrically) opposite faces.
3. There are 20 rotations over multiples of $2\pi/3$ about axes that join (diametrically) opposite vertices.
4. There are 15 rotations over π about axes that join the midpoints of (diametrically) opposite edges.

In all, we find 60 rotations. Figure 5.44.(b) shows a special cube inside the dodecahedron. There are five such cubes. Each of the rotations of the dodecahedron mentioned above induces an even permutation of the cubes, hence an even permutation of five elements. Different rotations induce different permutations. The number of even permutations of five elements is 60. Apparently,

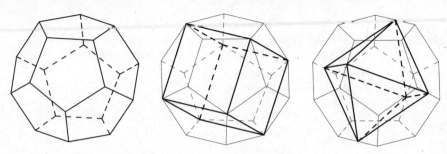

Fig. 5.44. The dodecahedron: (**a**) by itself; (**b**) with an inscribed cube; (**c**) with an inscribed octahedron

we have found all rotations, and the dodecahedral group, the rotation group of the dodecahedron, is isomorphic to A_5, the group of even permutations of five elements. There are also five octahedra in the dodecahedron; one of these is drawn in Fig. 5.44 (**c**). Every rotation of the dodecahedron induces an even permutation of the octahedra.

Recall that a *motion* is another name for a direct isometry of the 3-space W, that is, a translation, rotation, or screw (Table 5.1). The motions form a subgroup of $\mathcal{I}(W)$. The following theorem tells us that we have seen all finite subgroups of the group of motions.

Theorem 5.62. *Every finite subgroup of the group of motions is isomorphic to one of the following groups:*

1. *The cyclic group C_n, $n = 1, 2, \ldots$*
2. *The dihedral group D_n, $n = 1, 2, \ldots$*
3. *The tetrahedral group A_4*
4. *The octahedral group S_4*
5. *The dodecahedral group A_5*

We see that in addition to the two infinite series of C_n's and D_n's there are only three finite motion groups. This important theorem on the structure of the finite motion groups, and therefore also on the structure of 3-space, is the basis of crystallography. The proof of the theorem needs some preparation. We first consider the analogue of the first part of Theorem 3.7.

Theorem 5.63. *If G is a finite subgroup of $\mathcal{I}(W)$, the elements of G have a common fixed point.*

Proof. Let us choose a coordinate system. We write $G = \{\mathcal{F}_1, \ldots, \mathcal{F}_n\}$. For every $\mathbf{x}$ we let

$$G\mathbf{x} = \{\mathcal{F}_1(\mathbf{x}), \ldots, \mathcal{F}_n(\mathbf{x})\}$$

denote the *orbit* of the point $\mathbf{x}$ under the action of G. The centroid $\mathbf{z}$ of the orbit is given by

$$\mathbf{z} = \frac{\mathcal{F}_1(\mathbf{x}) + \cdots + \mathcal{F}_n(\mathbf{x})}{n} = \sum_{i=1}^{n} \frac{\mathcal{F}_i(\mathbf{x})}{n}$$

We claim that $\mathbf{z}$ is a common fixed point of the elements of G. Indeed, every isometry is the composition of reflections, and a reflection maps the centroid of an orbit to the centroid of the reflected orbit. Therefore we have

$$\mathcal{F}_j(\mathbf{z}) = \mathcal{F}_j \circ \left(\sum_{i=1}^{n} \frac{\mathcal{F}_i(\mathbf{x})}{n} \right) = \sum_{i=1}^{n} \frac{\mathcal{F}_j \circ \mathcal{F}_i(\mathbf{x})}{n}, \quad i = 1, \ldots, n.$$

Multiplication on the left by $\mathcal{F}_j$ defines a permutation of the group G. The sets $\{\mathcal{F}_j \circ \mathcal{F}_1, \ldots, \mathcal{F}_j \circ \mathcal{F}_n\}$ and $\{\mathcal{F}_1, \ldots, \mathcal{F}_n\}$ are therefore equal, up to permutation. Consequently,

$$\mathcal{F}_j(\mathbf{z}) = \sum_{i=1}^{n} \frac{\mathcal{F}_j \circ \mathcal{F}_i(\mathbf{x})}{n} = \sum_{i=1}^{n} \frac{\mathcal{F}_i(\mathbf{x})}{n} = \mathbf{z}, \quad j = 1, \ldots, n.$$

An important notion for the proof of Theorem 5.62 is that of a pole, which we will now define. Let $\mathbb{S}^2$ be the *unit sphere* in the coordinate space $\mathbb{R}^3$:

$$\mathbb{S}^2 = \{\mathbf{x} : \mathbf{x} \in \mathbb{R}^3 \text{ and } \|\mathbf{x}\| = 1\}.$$

Let G be a finite subgroup of the group of motions of $\mathbb{R}^3$ whose elements have the origin $\mathbf{o}$ as a common fixed point. The elements of G are then rotations; see Table 5.1. A point $\mathbf{p}$ of $\mathbb{S}^2$ is called a *pole* of G if there is an $\mathcal{R}$ in G such that $\mathcal{R} \neq \mathrm{id}$ and $\mathcal{R}(\mathbf{p}) = \mathbf{p}$. Every $\mathcal{R}$ in G admits two poles, namely the intersection points of the axis of $\mathcal{R}$ with $\mathbb{S}^2$. If $\mathbf{p}$ is a pole of G, then

$$G^{\mathbf{p}} = \{\mathcal{R} : \mathcal{R} \in G, \ \mathcal{R}(\mathbf{p}) = \mathbf{p}\}$$

is a subgroup of G. This is not difficult to prove. We call $G^{\mathbf{p}}$ the *isotropy group* of $\mathbf{p}$.

Proposition 5.64. *Let $\mathbf{p}$ be a pole of a finite group G of rotations of $\mathbb{R}^3$ with fixed point $\mathbf{o}$. The isotropy group of $\mathbf{p}$ is a cyclic group.*

Proof. Let k be the number of elements of $G^{\mathbf{p}}$. We may, and do, assume that $k \geq 2$. Let $\mathcal{R}_i$, $i = 1, \ldots, k$, be the elements of $G^{\mathbf{p}}$, where $\mathcal{R}_1 = \mathrm{id}$, and let $\vartheta_i \in [0, 2\pi)$ be the rotation angle of $\mathcal{R}_i$, $i = 2, \ldots, k$. We renumber the $\mathcal{R}_i$ so that $\vartheta_2 < \vartheta_3 < \cdots < \vartheta_k$. For $3 \leq j \leq k$ we write $\vartheta_j = m_j \vartheta_2 + \varphi_j$ with m_j a positive integer and $0 \leq \varphi_j < \vartheta_2$. The rotation over φ_j is an element of $G^{\mathbf{p}}$, since $\mathcal{R}_j \mathcal{R}_2^{-m_j} \in G^{\mathbf{p}}$. Since ϑ_2 is the smallest rotation angle, we must have $\varphi_j = 0$. It follows that $\vartheta_j = (j-1) \times (2\pi/k)$ for $2 \leq j \leq k$.

This proposition leads us to introduce the notion of *k-tuple pole*: $\mathbf{p}$ is a *k-tuple pole* of G if $G^{\mathbf{p}}$ is isomorphic to C_k, the cyclic group with k elements.

Proposition 5.65. *Let $\mathbf{p}$ be a k-tuple pole of a finite group G of rotations of $\mathbb{R}^3$ with fixed point $\mathbf{o}$. Let n be the number of elements of G. The orbit of $\mathbf{p}$ consists of exactly n/k poles, and each of these poles is k-tuple.*

Proof. We write $G = \{\mathcal{R}_1, \ldots, \mathcal{R}_n\}$. If $\mathbf{q}$ lies in the orbit of $\mathbf{p}$, we have $\mathbf{q} = \mathcal{R}_j(\mathbf{p})$ for some j. For every $\mathcal{R}_i$ in $G^{\mathbf{p}}$, $\mathcal{R}_j \circ \mathcal{R}_i \circ \mathcal{R}_j^{-1}$ is an element van $G^{\mathbf{q}}$ (Sect. 5.4, Exercise 5.20), and every element of $G^{\mathbf{q}}$ can be obtained in this way. Moreover, $\mathcal{R}_j \circ \mathcal{R}_{i_1} \circ \mathcal{R}_j^{-1} \neq \mathcal{R}_j \circ \mathcal{R}_{i_2} \circ \mathcal{R}_j^{-1}$ for $i_1 \neq i_2$. It follows that $\mathbf{q}$ is a k-tuple pole, and that $G^{\mathbf{q}} = \{\mathcal{R}_j \circ \mathcal{R}_i \circ \mathcal{R}_j^{-1} : \mathcal{R}_i \in G^{\mathbf{p}}\}$. Since $\mathbf{q} = \mathcal{R}_j(\mathbf{p})$, we also have $\mathbf{q} = \mathcal{R}_j \circ \mathcal{R}_i(\mathbf{p})$ for every $\mathcal{R}_i$ in $G^{\mathbf{p}}$. Note that $\mathcal{R}_j \circ \mathcal{R}_{i_1} \neq \mathcal{R}_j \circ \mathcal{R}_{i_2}$ for $i_1 \neq i_2$. Moreover, if $\mathbf{q} = \mathcal{R}_m(\mathbf{p})$, then $\mathcal{R}_j^{-1} \circ \mathcal{R}_m \in G^{\mathbf{p}}$ and $\mathcal{R}_m = \mathcal{R}_j \circ (\mathcal{R}_j^{-1} \circ \mathcal{R}_m)$; consequently, there are exactly k elements of G that map $\mathbf{p}$ onto $\mathbf{q}$. This implies that the orbit of $\mathbf{p}$ consists of exactly n/k poles that are each k-tuple.

We can now prove Theorem 5.62.

Proof of Theorem 5.62. Let n be the number of elements of G. We may, and do, assume that $n \geq 2$. By Theorem 5.63, the elements of G have a common fixed point. We choose this as origin of a coordinate system. Let us consider the action of the group G on the sphere $\mathbb{S}^2$. In this proof we will call every element of G that is not the identity map a *true rotation*. The group G has $(n-1)$ true rotations. Let us count the true rotations in a different way. If $\mathbf{p}$ is a k-tuple pole of G, the isotropy group $G^{\mathbf{p}}$ has $(k-1)$ true rotations. Every rotation has two poles; if we distribute the true rotations equally over pole and antipole, one k-tuple pole represents $(k-1)/2$ true rotations. By Proposition 5.65, the orbit of a k-tuple pole therefore has $(n/k) \times (k-1)/2$ true rotations. The poles of G may lie on different orbits. Summing over all orbits gives

$$n - 1 = \sum_{\text{orbits}} \tfrac{1}{2} n \left(1 - \tfrac{1}{k}\right),$$

where the value of k depends on the orbit. We can rewrite the formula as

$$2 - \tfrac{2}{n} = \sum_{\text{orbits}} \left(1 - \tfrac{1}{k}\right). \tag{5.4}$$

Since $n \geq 2$ and $k \geq 2$, we have $1 - (1/k) \geq 1/2$. The left-hand side of (5.4) is less than 2, which implies that the number of orbits over which we take the sum is either 2 or 3.

- The number of orbits is 2. Equation (5.4) becomes

$$2 - \tfrac{2}{n} = 1 - \tfrac{1}{k_1} + 1 - \tfrac{1}{k_2},$$

that is,

$$2 = \tfrac{n}{k_1} + \tfrac{n}{k_2}.$$

Since n/k_i is an integer, this is possible only if $k_1 = k_2 = n$. Hence there is an n-tuple pole and an n-tuple antipole. The group G is isomorphic to C_n (case (1) of the theorem).

- The number of orbits is 3. Equation (5.4) becomes

$$2 - \tfrac{2}{n} = 1 - \tfrac{1}{k_1} + 1 - \tfrac{1}{k_2} + 1 - \tfrac{1}{k_3},$$

that is, $\quad n + 2 = \tfrac{n}{k_1} + \tfrac{n}{k_2} + \tfrac{n}{k_3}.$

We immediately see that not all k_i can be greater than 3; let us assume that $k_3 = 2$. Some computation gives

$$(k_1 - 2)(k_2 - 2) = 4\left(1 - \tfrac{k_1 k_2}{n}\right).$$

Since the right-hand side is less than 4, assuming that $k_1 \leq k_2$, we obtain the following cases:

1. $k_1 = 2$, $k_2 = p$, $k_3 = 2$, $n = 2p$;
2. $k_1 = 3$, $k_2 = 3$, $k_3 = 2$, $n = 12$;
3. $k_1 = 3$, $k_2 = 4$, $k_3 = 2$, $n = 24$;
4. $k_1 = 3$, $k_2 = 5$, $k_3 = 2$, $n = 60$.

Successively, these cases give the dihedral groups D_n, the tetrahedral group A_4, the octahedral group S_4, and the dodecahedral group A_5. We will check this explicitly for one case; see [46] for the rest.

We consider case (4). The 5-tuple poles lie in one orbit. Since $n = 60$ and $k_2 = 5$, there are 12 of them, in 6 pairs of pole and antipole. Let $\mathbf{p}_n$ and $\mathbf{p}_z$ be such a pair of poles, which we will call the north pole and the south pole. Let $\mathcal{R}$ be a true rotation from G around the north–south axis. The rotation $\mathcal{R}$ has no fixed points on $\mathbb{S}^2$ other than the north pole and south pole, and must therefore interchange the remaining 5-tuple poles; see Sect. 5.4, Exercise 5.20. These poles cannot all lie on the equator; if that were the case, what would they map the north and south poles to? Thus we find five of the poles in the northern hemisphere, and five in the southern hemisphere. The rotation $\mathcal{R}$ around the north–south axis permutes the 5-tuple poles in the northern hemisphere, and also those in the southern hemisphere. Note that any of the 5-tuple poles can play the role of north pole. The action of G is determined by its action on the icosahedron; hence G must be a subgroup of the icosahedral group, which in turn is equal to the dodecahedral group. Since $n = 60$, G must equal the icosahedral group.

Crystallography

Using the classification of the finite rotation groups given above, we can list all finite subgroups of the symmetry group $\mathcal{I}(W)$ of 3-space. This is very important for crystallography. A crystal is in fact a periodic tiling of 3-space, also called a honeycomb. Mathematically, a *periodic tiling* of 3-space is given by a subset D such that the set of translations in the symmetry group $\mathcal{I}(D)$ is equal to

$$\{\,T_1^k \circ T_2^l \circ T_3^m : k \in \mathbb{Z},\, l \in \mathbb{Z},\, m \in \mathbb{Z}\,\}$$

for noncoplanar translation vectors T_1, T_2, T_3. The aim of crystallography is to obtain an overview of the different periodic tilings of 3-space and to develop methods to determine which periodic tiling corresponds to each crystal. The classification of the periodic tilings of 3-space uses the symmetry groups. In all, 230 symmetry groups can be distinguished. Of these, 65 consist strictly of direct isometries. Of those 65, 22 belong to 11 *enantiomorphic pairs*. In such an enantiomorphic pair, the lattices (see below) are mirror images of each other. We are not going to classify the tilings of 3-space. Considering the work done in Chap. 3 in order to classify the periodic tilings of the plane, we expect this to be too much work. We can, however, show that if a tiling of 3-space is invariant under a rotation over $2\pi/n$, n must be equal to 2, 3, 4, or 6; in particular, there can be no 5-fold rotation symmetry.

The definition of a spatial lattice is analogous to that of a lattice in the plane; see Definition 3.24. Let P be a point in 3-space. We take three translations T_1, T_2, and T_3 with noncoplanar translation vectors. The set

$$\{\,T_1^k \circ T_2^l \circ T_3^m(P) : k \in \mathbb{Z},\, l \in \mathbb{Z},\, m \in \mathbb{Z}\,\}$$

is called a *spatial lattice* of P. In the case of a periodic tiling, we of course consider the lattice $\{\,\mathcal{T}(P) : \mathcal{T} \in \mathcal{I}(D)\,\}$; see Fig. 5.45.

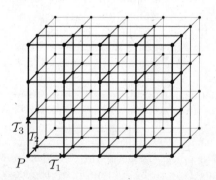

Fig. 5.45. A spatial lattice

Theorem 5.66. *Let D be a periodic tiling of 3-space. Let $\mathcal{R}$ be a rotation other than the identity in the symmetry group $\mathcal{I}(D)$ of D. Then*

1. *$\mathcal{R}$ maps the lattice of a point on its rotation axis onto itself.*
2. *$\mathcal{R}$ is a rotation over $2\pi/n$ with n equal to 2, 3, 4, or 6.*

Proof. Consider the lattice B of a point C on the rotation axis of $\mathcal{R}$. We fix a coordinate system at C; that is, $C = \mathbf{o}$. As in Chap. 3, we can show that B is *discrete*: there is a number $d > 0$ such that the distance between any two lattice points is greater than or equal to d. We will now show that the lattice B is invariant under $\mathcal{R}$. Let $\mathbf{q}$ be a point of B. By the definition of B, there is a translation $\mathcal{T}$ with $\mathcal{T}(\mathbf{o}) = \mathbf{q}$. The isometry $\mathcal{R} \circ \mathcal{T} \circ \mathcal{R}^{-1}$ belongs to $\mathcal{I}(D)$ and is a translation; see Sect. 5.4, Exercise 5.18. This implies that $\mathcal{R} \circ \mathcal{T} \circ \mathcal{R}^{-1}(\mathbf{o})$ belongs to B. Since

$$\mathcal{R} \circ \mathcal{T} \circ \mathcal{R}^{-1}(\mathbf{o}) = \mathcal{R} \circ \mathcal{T}(\mathbf{o}) = \mathcal{R}(\mathbf{q}),$$

we see that $\mathcal{R}(\mathbf{q}) \in B$. The lattice is therefore invariant under $\mathcal{R}$, that is, $\mathcal{R}(B) \subset B$. Since this holds also for $\mathcal{R}^{-1}$, we see that $\mathcal{R}$ maps the lattice B bijectively onto itself.

For the second part of the theorem, we note that every point of B lies on the axis of a rotation in $\mathcal{I}(D)$; this axis is parallel to the axis l of $\mathcal{R}$. Indeed, if $\mathbf{q} = \mathcal{T}(\mathbf{o})$ with $\mathcal{T} \in \mathcal{I}(D)$, then $\mathcal{T} \circ \mathcal{R} \circ \mathcal{T}^{-1}$ is a rotation with axis $\mathcal{T}(l)$. Let us now project the 3-space onto a plane U perpendicular to l, by a projection parallel to l. The lattice B is mapped onto a set B^*; every point of B^* lies on a rotation axis parallel to l. It follows that B^* is invariant under the rotations mentioned above. Since B is discrete, B^* must be a lattice. The second result now follows from Theorem 3.28.

Exercises

5.23. The set of all even permutations of a set with n elements is a subgroup of S_n. We denote it by A_n.

5.24. A well-known problem in recreational mathematics that can be solved using the Euler formula is that of the *utility graph*. Three utility companies, W(ater), G(as), and E(lectricity), must be connected to three houses A, B, and C. This has to be done in such a way that the cables lie in one plane and do not intersect. Is this possible? Let us assume that it is possible. In general, the cables are not straight lines, but smooth curves. The Euler formula must hold for this plane figure.

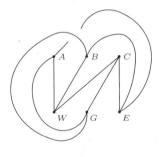

(a) The number of connections E by curves and the number of vertices V are easy to find. Using this we can show that the number of regions F must equal 5.
(b) Every region is bounded by at least four edges. It follows that $F \leq 4.5$. Conclusion?

5.25. In the plane, we are given five points that we want to join pairwise using nonintersecting curves. Is this possible? In 3-space it is possible: the vertices of the tetrahedron and its centroid can be connected using nonintersecting line segments. This figure is called the *complete 5-graph* K_5.

5.26. There is a link between orientations of tripods and permutations. Let $PABC$ be a tripod. If g is a permutation of $\{A, B, C\}$, the tripods $PABC$ and $Pg(A)g(B)g(C)$ have the same orientation if and only if g is an even permutation.

5.27. Consider the following permutations of $\{A, B, C, D\}$:

$$ABCD, \quad ABCD, \quad ABCD, \quad ABCD,$$
$$ABCD, \quad BADC, \quad CDAB, \quad DCBA.$$

These permutations form a subgroup of S_4 that is isomorphic to V_4.
Hint: consider a tetrahedron with vertices A, B, C, D and view the permutations as rotations of the tetrahedron.

5.28. A *tiling* of the plane of type $\{p, q\}$ is a covering of the plane by regular p-gons with only edge-points in common, such that any point that belongs to more than two p-gons belongs to exactly q p-gons.
(a) We have $(p-2)(q-2) = 4$.
(b) The possible types are $\{3,6\}$, $\{4,4\}$, and $\{6,3\}$.

5.29. Let D be a regular tiling of 3-space. Consider the lattice B of a point C. Let $\mathcal{J}$ be the reflection in the point C. Suppose there is a rotation $\mathcal{R}$ over $2\pi/n$ such that the improper rotation $\mathcal{J} \circ \mathcal{R}$ maps the lattice B onto itself. Then n is equal to 2, 3, 4, or 6.
Hint: $\mathcal{J}$ maps B onto itself.

5.30. There is a close link between the golden ratio $1 + \tau$ (Sect. 2.5, Exercise 2.48) and the icosahedron. The twelve vertices of three golden rectangles, namely
$(0, -1, \tau)$, $(0, 1, \tau)$, $(0, 1, -\tau)$, $(0, -1, -\tau)$,
$(-\tau, 0, 1)$, $(\tau, 0, 1)$, $(\tau, 0, -1)$, $(-\tau, 0, -1)$,
¡$(1, \tau, 0)$, $(1, -\tau, 0)$, $(-1, -\tau, 0)$, $(-1, \tau, 0)$,
are the vertices of an icosahedron.

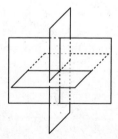

5.31. Why does the buckyball have the same symmetry group as the dodecahedron?
Hint: consider the pentagons.

5.6 Quadrics and Ruled Surfaces

In this section we study *quadrics*. These are surfaces that, in a suitable coordinate system, can be described by a quadratic equation. For the moment we will work in a fixed coordinate system. We will in general denote the coordinates of $\mathbf{x}$ by x, y, and z instead of x_1, x_2, and x_3, respectively. This makes complicated formulas easier to read.

5.6 Quadrics and Ruled Surfaces

We begin by describing the surface of revolution that arises by rotating a hyperbola around an axis.

We consider the following hyperbola in the yz-plane:

$$\frac{y^2}{a^2} - \frac{z^2}{c^2} = 1; \quad x = 0. \tag{5.5}$$

See Fig. 5.46 (a). The hyperbola is symmetric with respect to the y- and z-axes. We rotate this hyperbola around the z-axis. The figure that arises is called a *one-sheeted circular hyperboloid*. Let us deduce the equation of this surface of revolution. The distance from the point $P(x, y, z)$ to the z-axis is $\sqrt{x^2 + y^2}$; see Fig. 5.46 (a). The point $P(x, y, z)$ lies on the one-sheeted circular hyperboloid if and only if the point $(0, \sqrt{x^2 + y^2}, z)$ lies on the hyperbola (5.5), that is, if

$$\frac{x^2}{a^2} + \frac{y^2}{a^2} - \frac{z^2}{c^2} = 1. \tag{5.6}$$

This is the equation of the one-sheeted circular hyperboloid.

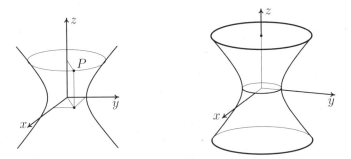

Fig. 5.46. (a) A hyperbola; (b) the one-sheeted hyperboloid it generates

By changing the scale on the y-axis, that is, replacing y by $(a/b)y$, we obtain the equation of the more general one-sheeted elliptic hyperboloid:

$$\frac{x^2}{a^2} + \frac{y^2}{b^2} - \frac{z^2}{c^2} = 1. \tag{5.7}$$

We can place the hyperbola in two essentially different positions in the yz-plane, both symmetric with respect to the z-axis. We have just seen one way; the other is sketched in Fig. 5.47 (a). In this case the equation of the hyperbola is

$$\frac{z^2}{c^2} - \frac{y^2}{a^2} = 1; \quad x = 0. \tag{5.8}$$

In rotating around the z-axis, the hyperbola describes a *two-sheeted circular hyperboloid*; see Fig. 5.47 (b). By scaling we obtain the equation of the more general *two-sheeted elliptic hyperboloid*:

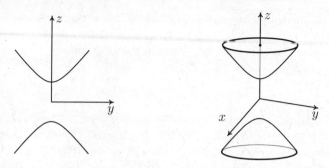

Fig. 5.47. (a) A hyperbola; (b) the two-sheeted hyperboloid it generates

$$-\frac{x^2}{a^2} - \frac{y^2}{b^2} + \frac{z^2}{c^2} = 1. \tag{5.9}$$

The one-sheeted hyperboloid and the two-sheeted hyperboloid are examples of *quadrics* in $\mathbb{R}^3$; the equation of such a surface has degree two. We can also rotate quadratic curves other than the hyperbola. Examples of resulting quadrics are the *ellipsoid*, with equation

$$\frac{x^2}{a^2} + \frac{y^2}{b^2} + \frac{z^2}{c^2} = 1,$$

and the *elliptic paraboloid*, with equation

$$\frac{x^2}{a^2} + \frac{y^2}{b^2} = 2z.$$

These are sketched in Fig. 5.48. The construction of the ellipsoid and the

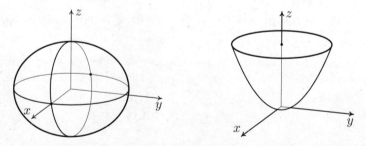

Fig. 5.48. (a) The ellipsoid; (b) the elliptic paraboloid

elliptic paraboloid is similar to that of the hyperboloids.

Straight Lines on Quadrics

We will now study the existence of straight lines on quadrics. Since the ellipsoid is bounded, we immediately see that no straight lines lie completely in the ellipsoid. Though it is more difficult to prove, we can also see that there are no straight lines on the two-sheeted hyperboloid. No arguments are needed

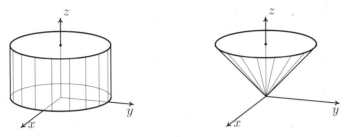

Fig. 5.49. (a) The elliptic cylinder; (b) a (half) cone

to convince us that there are straight lines lying entirely in the quadrics the *cylinder* and the *cone*. Figure 5.49 (a) shows an *elliptic cylinder*. Its equation is

$$\frac{x^2}{a^2} + \frac{y^2}{b^2} = 1 \ .$$

Figure 5.49 (b) shows an *elliptic cone*. Its equation is

$$\frac{x^2}{a^2} + \frac{y^2}{b^2} - \frac{z^2}{c^2} = 0 \ .$$

The elliptic cylinder in (a) intersects the xy-plane in an ellipse. Let us make two remarks. First, there is no reason to give one of the coordinates a special role; we can also define a cylinder over an ellipse in another coordinate plane. Second, instead of an ellipse we can also take a hyperbola or parabola and form a cylinder with that. This gives rise to the *hyperbolic cylinder* and the

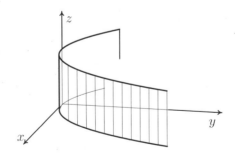

Fig. 5.50. The parabolic cylinder

parabolic cylinder. See Fig. 5.50 for a sketch of the latter. The equation of the hyperbolic cylinder is

$$\frac{x^2}{a^2} - \frac{y^2}{b^2} = 1 \ ,$$

and that of the parabolic cylinder is

$$y^2 = 2px .$$

An obvious question is now whether there are also more types of cones. The cone in Fig. 5.49 is a cone over an ellipse. Is there also a cone over a hyperbola or over a parabola? We will come back to this question when we consider the intersection of a quadric and a plane.

Straight Lines on the One-Sheeted Hyperboloid

Let us now consider the one-sheeted hyperboloid. It is not directly clear from the illustrations that there are straight lines on the one-sheeted hyperboloid. We can rewrite equation (5.6) of this surface of revolution as follows:

$$\frac{x^2}{a^2} - \frac{z^2}{c^2} = 1 - \frac{y^2}{b^2} ,$$

that is,

$$\left(\frac{x}{a} - \frac{z}{c}\right)\left(\frac{x}{a} + \frac{z}{c}\right) = \left(1 - \frac{y}{b}\right)\left(1 + \frac{y}{b}\right) . \qquad (5.10)$$

Consider the following systems of straight lines:

$$\begin{cases} \lambda\left(\frac{x}{a} + \frac{z}{c}\right) = \mu\left(1 + \frac{y}{b}\right) , \\ \mu\left(\frac{x}{a} - \frac{z}{c}\right) = \lambda\left(1 - \frac{y}{b}\right) ; \end{cases} \lambda \in \mathbb{R},\ \mu \in \mathbb{R},\ |\lambda| + |\mu| > 0 , \qquad (5.11)$$

and

$$\begin{cases} \rho\left(\frac{x}{a} + \frac{z}{c}\right) = \sigma\left(1 - \frac{y}{b}\right) , \\ \sigma\left(\frac{x}{a} - \frac{z}{c}\right) = \rho\left(1 + \frac{y}{b}\right) ; \end{cases} \rho \in \mathbb{R},\ \sigma \in \mathbb{R},\ |\rho| + |\sigma| > 0 . \qquad (5.12)$$

For every choice of λ and μ, (5.11) defines a straight line as the intersection of two planes. Likewise, for every choice of ρ and σ, (5.12) represents a straight line. Let us first state the most important property of these systems.

Property 5.67. *Every line in either system lies entirely in the one-sheeted hyperboloid (5.10).*

Proof. Consider the line in (5.11) corresponding to the parameter values λ_0 and μ_0:

$$\begin{cases} \lambda_0\left(\frac{x}{a} + \frac{z}{c}\right) = \mu_0\left(1 + \frac{y}{b}\right) , \\ \mu_0\left(\frac{x}{a} - \frac{z}{c}\right) = \lambda_0\left(1 - \frac{y}{b}\right) . \end{cases} \qquad (5.13)$$

If $\mathbf{x} = (x, y, z)$ is a point on this line, x, y, and z satisfy (5.13) and therefore also

$$\lambda_0\mu_0\left(\frac{x}{a} + \frac{z}{c}\right)\left(\frac{x}{a} - \frac{z}{c}\right) = \lambda_0\mu_0\left(1 + \frac{y}{b}\right)\left(1 - \frac{y}{b}\right) . \qquad (5.14)$$

If $\lambda_0\mu_0 \neq 0$, we can divide both members of (5.14) by $\lambda_0\mu_0$, and we see that $\mathbf{x}$ lies on the one-sheeted hyperboloid (5.10). If, for example, $\lambda_0 = 0$ but $\mu_0 \neq 0$, (5.13) implies that

$$1 + \tfrac{y}{b} = 0 \quad \text{and} \quad \tfrac{x}{a} - \tfrac{z}{c} = 0,$$

and therefore that x, y, and z satisfy (5.10).

Figure 5.51 shows the two systems of lines on the one-sheeted hyperboloid. We will now deduce a number of properties of these systems of lines.

Fig. 5.51. Two systems of lines on the one-sheeted hyperboloid

Property 5.68. *Through every point of the one-sheeted hyperboloid (5.10) there is exactly one line of the system (5.11) and one line of the system (5.12).*

Proof. If $\mathbf{p} = (p, q, r)$ lies on the one-sheeted hyperboloid (5.10), we have

$$\left(\tfrac{p}{a} - \tfrac{r}{c}\right)\left(\tfrac{p}{a} + \tfrac{r}{c}\right) = \left(1 - \tfrac{q}{b}\right)\left(1 + \tfrac{q}{b}\right).$$

Let us first assume that the left-hand side and right-hand side of the equation are both nonzero. For

$$\mu = 1 \quad \text{and} \quad \lambda = \left(\tfrac{p}{a} - \tfrac{r}{c}\right)\left(1 - \tfrac{q}{b}\right)^{-1}$$

we have a line from system (5.11) that passes through $\mathbf{p}$. For

$$\sigma = 1 \quad \text{and} \quad \rho = \left(\tfrac{p}{a} - \tfrac{r}{c}\right)\left(1 + \tfrac{q}{b}\right)^{-1}$$

we have a line from system (5.12) that passes through $\mathbf{p}$. The property we will deduce shortly shows that there is not more than one line from each system passing through $\mathbf{p}$. If, for example, $1 - q/b = 0$, then $1 + q/b \neq 0$. For

$$\lambda = 1 \quad \text{and} \quad \mu = \left(\tfrac{p}{a} + \tfrac{r}{c}\right)\left(1 + \tfrac{q}{b}\right)^{-1}$$

we have a line from system (5.11) that passes through $\mathbf{p}$. For

$$\sigma = 1 \quad \text{and} \quad \rho = \left(\tfrac{p}{a} - \tfrac{r}{c}\right)\left(1 + \tfrac{q}{b}\right)^{-1}$$

we have a line from system (5.12).

Property 5.69. *Two distinct lines from one system are skew.*

Proof. In this proof we use a number of results from linear algebra without further explanation. We take two distinct lines from the system (5.11), say the line l_1 corresponding to λ_1, μ_1 and the line l_2 corresponding to λ_2, μ_2. Since the lines are distinct, we have $\lambda_1 \mu_2 \neq \lambda_2 \mu_1$. Let us first show that the lines do not intersect. Suppose that they do intersect, at $X = (x_1, x_2, x_3)$. This point is a solution of the following system of equations (the order is important):

$$\begin{cases} \lambda_1 \frac{x}{a} - \mu_1 \frac{y}{b} + \lambda_1 \frac{z}{c} = \mu_1, \\ \lambda_2 \frac{x}{a} - \mu_2 \frac{y}{b} + \lambda_2 \frac{z}{c} = \mu_2, \\ \mu_1 \frac{x}{a} + \lambda_1 \frac{y}{b} - \mu_1 \frac{z}{c} = \lambda_1, \\ \mu_2 \frac{x}{a} + \lambda_2 \frac{y}{b} - \mu_2 \frac{z}{c} = \lambda_2. \end{cases} \tag{5.15}$$

The augmented matrix associated to this system is

$$A = \begin{pmatrix} \lambda_1 & -\mu_1 & \lambda_1 & \mu_1 \\ \lambda_2 & -\mu_2 & \lambda_2 & \mu_2 \\ \mu_1 & \lambda_1 & -\mu_1 & \lambda_1 \\ \mu_2 & \lambda_2 & -\mu_2 & \lambda_2 \end{pmatrix}.$$

Let us compute the determinant of A. We first subtract the first column from the third, and add the second to the fourth:

$$\begin{vmatrix} \lambda_1 & -\mu_1 & \lambda_1 & \mu_1 \\ \lambda_2 & -\mu_2 & \lambda_2 & \mu_2 \\ \mu_1 & \lambda_1 & -\mu_1 & \lambda_1 \\ \mu_2 & \lambda_2 & -\mu_2 & \lambda_2 \end{vmatrix} = \begin{vmatrix} \lambda_1 & -\mu_1 & 0 & 0 \\ \lambda_2 & -\mu_2 & 0 & 0 \\ \mu_1 & \lambda_1 & -2\mu_1 & 2\lambda_1 \\ \mu_2 & \lambda_2 & -2\mu_2 & 2\lambda_2 \end{vmatrix} = -4 (\mu_1 \lambda_2 - \mu_2 \lambda_1)^2.$$

Since $\det A \neq 0$, A has rank 4. The system (5.15) therefore has no solution, and the lines l_1 and l_2 do not intersect. The lines cannot be parallel. To see this, we translate the planes determining l_1 and l_2 to the origin. We obtain the equations of the translated planes by setting the right-hand sides in (5.15) equal to zero:

$$\begin{cases} \lambda_1 \frac{x}{a} - \mu_1 \frac{y}{b} + \lambda_1 \frac{z}{c} = 0, \\ \lambda_2 \frac{x}{a} - \mu_2 \frac{y}{b} + \lambda_2 \frac{z}{c} = 0, \\ \mu_1 \frac{x}{a} + \lambda_1 \frac{y}{b} - \mu_1 \frac{z}{c} = 0, \\ \mu_2 \frac{x}{a} + \lambda_2 \frac{y}{b} - \mu_2 \frac{z}{c} = 0. \end{cases} \tag{5.16}$$

If the lines l_1 and l_2 were parallel, this system would have a nonzero solution. However, since A has rank 4, the rank of the matrix of coefficients of (5.16) must equal 3. The system therefore has only the zero solution, and the lines l_1 and l_2 cannot be parallel.

Property 5.70. *A line from one system is coplanar with every line from the other system.*

Proof. Consider a line l from the system (5.11) and a line m from the system (5.12):

$$l: \begin{cases} \lambda\left(\frac{x}{a} + \frac{z}{c}\right) = \mu\left(1 + \frac{y}{b}\right), \\ \mu\left(\frac{x}{a} - \frac{z}{c}\right) = \lambda\left(1 - \frac{y}{b}\right), \end{cases} \quad (5.17)$$

$$m: \begin{cases} \rho\left(\frac{x}{a} + \frac{z}{c}\right) = \sigma\left(1 - \frac{y}{b}\right), \\ \sigma\left(\frac{x}{a} - \frac{z}{c}\right) = \rho\left(1 + \frac{y}{b}\right). \end{cases} \quad (5.18)$$

The plane

$$\lambda\rho\left(\frac{x}{a} + \frac{z}{c}\right) + \mu\sigma\left(\frac{x}{a} - \frac{z}{c}\right) = \mu\rho\left(1 + \frac{y}{b}\right) + \lambda\sigma\left(1 - \frac{y}{b}\right)$$

is a linear combination both of the planes in (5.17) and of the planes in (5.18). Hence, if x, y, and z satisfy (5.17) or (5.18), they also satisfy the last equation. It follows that l and m both lie in this plane.

The existence of two systems of straight lines on the one-sheeted hyperboloid is used in the construction of cooling towers. The steel lattices in the reinforced concrete used in cooling towers is formed by systems of straight lines. The existence of these systems is also used in the construction of gears used to make a transfer between two axes making a sharp angle with each other.

Straight Lines on the Hyperbolic Paraboloid

There is a quadric that we have not discussed yet, the hyperbolic paraboloid, sometimes also called the *saddle*. The formula for this surface is

$$-\frac{x^2}{a^2} + \frac{y^2}{b^2} = 2z. \quad (5.19)$$

Figure 5.52 gives a sketch of a hyperbolic paraboloid. To understand this sur-

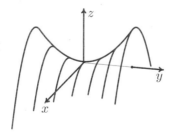

Fig. 5.52. The hyperbolic paraboloid

face, we intersect it with planes $y = k$ parallel to the xz-plane. The intersection of the hyperbolic paraboloid with the plane $y = k$ has equation

$$y = k, \quad z = -\frac{x^2}{2a^2} + \frac{k^2}{2b^2}.$$

The intersection is a parabola in the plane $y = k$. It is concave; that is, as a function of x, z has a maximum. The top of the parabola is at $(0, k, k^2/2b^2)$. As we let k vary, the tops of the intersection curves with the planes $y = k$ describe a convex parabola.

The following systems of straight lines lie on the one-sheeted paraboloid with equation (5.19):

$$\begin{cases} \lambda \left(\frac{y}{b} + \frac{x}{a} \right) = 2\mu z \,, \\ \mu \left(\frac{y}{b} - \frac{x}{a} \right) = \lambda \,; \end{cases} \quad \lambda \in \mathbb{R}, \; \mu \in \mathbb{R}, \; \mu \neq 0 \,, \qquad (5.20)$$

and

$$\begin{cases} \rho \left(\frac{y}{b} + \frac{x}{a} \right) = \sigma \,, \\ \sigma \left(\frac{y}{b} - \frac{x}{a} \right) = 2\rho z \,; \end{cases} \quad \rho \in \mathbb{R}, \; \sigma \in \mathbb{R}, \; \rho \neq 0 \,. \qquad (5.21)$$

In analogy to what we did for the one-sheeted hyperboloid, we can prove a number of properties.

Properties 5.71. *The following properties hold for systems (5.20) and (5.21) of straight lines and the hyperbolic paraboloid (5.19):*

1. *Every line in the systems lies entirely on the hyperbolic paraboloid.*
2. *There is exactly one line of each system passing through each point of the hyperbolic paraboloid.*
3. *Two distinct lines in one system are skew.*
4. *A line from one system intersects every line from the other system.*

Proof. The proofs are nearly the same as those for the one-sheeted hyperboloid. For (4): every line in the system (5.20) is parallel to the plane $x/a - y/b = 0$ and every line in the system (5.21) is parallel to the plane $x/a + y/b = 0$. If a line in one system were parallel to a line in the other system, both would be parallel to the intersection line of these planes, which is the z-axis. However, this line is excluded explicitly from the systems by the conditions $\mu \neq 0$ and $\rho \neq 0$.

Conics

What does the intersection curve of a quadric and a plane look like? The answer to this question is easy. The equation of the quadric has degree two. The equation of the plane has degree one in x, y, and z. Let us take one of the three variables, say x, write it as linear function in y and z, and substitute it in the equation of the quadric; this gives a quadratic equation in y and z. As we know from Sect. 4.4, this is the equation of a possibly degenerate conic in the yz-plane; see Fig. 4.1. The intersection curve of the quadric and the plane is equal to the intersection curve of the cylinder over this conic and the plane. It is therefore a conic of the same type.

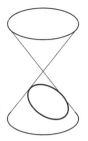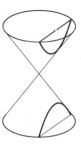

Fig. 5.53. Intersection of a conic and a plane

The conics are so called because they can be constructed as the intersection of a cone and a plane; see Fig. 5.53. This representation of the conics comes from Apollonius (ca. 262–ca. 190 BC), who also introduced the names *parabola*, *ellipse*, and *hyperbola*.

Consider the double cone

$$\frac{x^2}{a^2} + \frac{y^2}{b^2} - \frac{z^2}{c^2} = 0 .$$

If we intersect this cone with the plane $z = k$, the intersection curve is the ellipse

$$z = k, \quad \frac{x^2}{a^2} + \frac{y^2}{b^2} = \frac{k^2}{c^2} .$$

If we intersect the cone with the plane $y = k$, the intersection curve is the hyperbola

$$y = k, \quad \frac{x^2}{a^2} - \frac{z^2}{c^2} = -\frac{k^2}{b^2} .$$

The intersection of the cone with the plane $y = (b/c)z + k$ is the parabola

$$y = \tfrac{b}{c}z + k, \quad \frac{x^2}{a^2} + \frac{2k}{cb}z + \frac{k^2}{b^2} = 0 .$$

We see that there is no need to distinguish cones over ellipses, over parabolas, or over hyperbolas.

Different results concerning conics can be deduced directly from the representation of the conic as the intersection of a cone and a plane. We will discuss two properties here. The first is the characterization of the ellipse given at the end of Sect. 3.3: in the plane, the ellipse is the set of points whose distances to two fixed points, the foci, add up to a constant value. We recognize this property in Fig. 5.54. It shows a cone intersected with a plane β. The proof comes from Dandelin (1794–1847) and consists of three steps:

1. There are exactly two spheres that meet the cone in a circle and the plane β at a point; we call the tangent points of the two spheres on the plane F_1 and F_2. The tangent circles of the spheres cut off segments of equal length, which we call $2a$, from each *generatrix* of the cone; these are lines through the apex that lie entirely on the cone. In Fig. 5.54, $A_1 A_2 = 2a$.

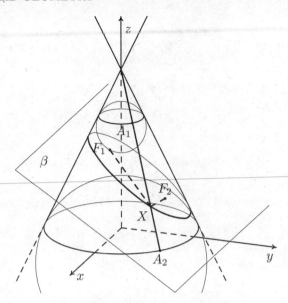

Fig. 5.54. The ellipse as a conic section

2. If we draw tangents from any fixed point to one of the spheres, we see that the tangent points are equidistant from the fixed point. In particular, a point X on the intersection curve of β and the cone satisfies $XF_1 = XA_1$ and $XF_2 = XA_2$. Since $A_1A_2 = 2a$, we have $XF_1 + XF_2 = 2a$. Hence every point of the intersection curve of the cone and the plane β lies on an ellipse K.

3. For a point X of the ellipse K, consider the plane α passing through X and the axis of the cone. The intersection of α and β is a straight line on which there are exactly two points X with $XF_1 + XF_2 = 2a$, namely the points on the intersection curve of the cone and β, which we found in (2). Consequently, every point of the ellipse K lies on the intersection curve.

The second property of the conics that can be deduced from their representation as the intersection of a cone and a plane is the definition of the conics using foci. We illustrate this with Fig. 5.55, which contains part of Fig. 5.54. In particular, β is the plane that intersects the cone. The apex of the cone is called T. The join of the point X on the intersection curve and the apex meets the tangent circle of the small sphere at A_1. The plane α through this tangent circle, which is not drawn, meets β in the line l. The plane through T and perpendicular to l meets α in the line CD. We choose the point D on this line in such a way that $TD \mathbin{/\mkern-5mu/} \beta$. The line A_1D meets l at B. We can now easily prove that l is the directrix of the intersection curve, F_1 is the focus, and that the eccentricity is equal to TA_1/TD. For this, we note that the lines TD and BX are parallel, and that $CD \perp l$. Consequently, $\triangle A_1XB \cong \triangle A_1TD$ and therefore

$$XF_1 : XB = XA_1 : XB = TA_1 : TD.$$

This implies the statement we want to prove.

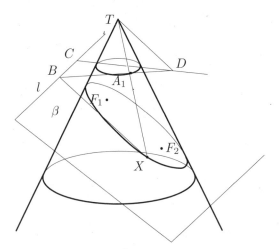

Fig. 5.55. Definition of the ellipse using foci

The Classification of Quadrics

In Sect. 4.4 we gave a classification of the quadratic curves, showing that the conics are the only ones. We now turn our attention to the classification of the quadrics, surfaces of degree two. We will show that we have already seen all nondegenerate quadrics. The classification of the quadrics is somewhat more difficult than that of the quadratic curves. While we were able to avoid the use of linear algebra for the quadratic curves, in particular the use of eigenvalues and eigenvectors, these will be essential for the quadrics. In fact, the theory of eigenvalues and eigenvectors is a particularly strong tool. Almost every book on linear algebra gives the application of that theory to the study of quadrics. In our discussion there is one spot where we use a property of eigenvectors, without further explanation.

The most general equation of a quadric $\mathbb{K}$ is

$$a_{11}x_1^2 + a_{22}x_2^2 + a_{33}x_3^2 + 2a_{12}x_1x_2 + 2a_{13}x_1x_3 + 2a_{23}x_2x_3 \quad (5.22)$$
$$+ 2a_{01}x_1 + 2a_{02}x_2 + 2a_{03}x_3 + a_{00} = 0.$$

Let us first concentrate on the quadratic part of this system. We may, and do, assume that not all coefficients of the quadratic terms are zero. In matrix notation, we write the quadratic part of system (5.22) as

326   5 SOLID GEOMETRY

$$\begin{pmatrix} x_1 & x_2 & x_3 \end{pmatrix} \begin{pmatrix} a_{11} & a_{12} & a_{13} \\ a_{21} & a_{22} & a_{23} \\ a_{31} & a_{32} & a_{33} \end{pmatrix} \begin{pmatrix} x_1 \\ x_2 \\ x_3 \end{pmatrix}.$$

The matrix (a_{ij}) is symmetric, that is, $a_{ij} = a_{ji}$ for all i and j. We will use the following property of eigenvalues and eigenvectors: since the characteristic equation of the matrix (a_{ij}) has degree three, the matrix has at least one real eigenvalue. We can now choose a new coordinate system such that the eigenvector corresponding to this eigenvalue lies along the first coordinate

Table 5.2. Classification of the quadrics

I	$b_{11}x_1^2 + b_{22}x_2^2 + b_{33}x_3^2 + b_{00} = 0,\qquad b_{11}b_{22}b_{33} \neq 0,$	
Ia	$b_{00} \neq 0$. After dividing by $-b_{00}$ we obtain $c_{11}x_1^2 + c_{22}x_2^2 + c_{33}x_3^2 = 1.$	
	$c_{11} > 0\ c_{22} > 0\ c_{33} > 0$	ellipsoid
	$c_{11} > 0\ c_{22} > 0\ c_{33} < 0$	one-sheeted hyperboloid
	$c_{11} > 0\ c_{22} < 0\ c_{33} < 0$	two-sheeted hyperboloid
	$c_{11} < 0\ c_{22} < 0\ c_{33} < 0$	imaginary ellipsoid
Ib	$b_{00} = 0$	
	$b_{11} > 0\ b_{22} > 0\ b_{33} > 0$	imaginary cone
	$b_{11} > 0\ b_{22} > 0\ b_{33} < 0$	cone
II	$b_{11}x_1^2 + b_{22}x_2^2 + 2b_{03}x_3 = 0,\qquad b_{03} \neq 0,\ b_{11}b_{22} \neq 0$	
	After dividing by $-b_{03}$ we find $c_{11}x_1^2 + c_{22}x_2^2 = 2x_3.$	
	$c_{11} > 0\ c_{22} > 0$	elliptic paraboloid
	$c_{11} > 0\ c_{22} < 0$	hyperbolic paraboloid
III	$b_{11}x_1^2 + b_{22}x_2^2 + b_{00} = 0,\qquad b_{11}b_{22} \neq 0$	
IIIa	$b_{00} \neq 0$. After dividing by $-b_{00}$ we find $c_{11}x_1^2 + c_{22}x_2^2 = 1.$	
	$c_{11} > 0\ c_{22} > 0$	elliptic cylinder
	$c_{11} > 0\ c_{22} < 0$	hyperbolic cylinder
	$c_{11} < 0\ c_{22} < 0$	imaginary cylinder
IIIb	$b_{00} = 0$	degenerate quadric
IV	$b_{11}x_1^2 + 2b_{02}x_2 = 0,\ b_{02} \neq 0,\ b_{11} \neq 0$	parabolic cylinder
V	$b_{11}x_1^2 + b_{00} = 0,\qquad b_{11} \neq 0$	degenerate quadric

axis. In the new coordinate system the equation of $\mathbb{K}$ looks as follows:

$$b_{11}x_1^2 + 2b_{01}x_1 + b_{22}x_2^2 + b_{33}x_3^2 + 2b_{23}x_2x_3 \qquad (5.23)$$
$$+ 2b_{02}x_2 + 2b_{03}x_3 + b_{00} = 0$$

with quadratic part

$$\begin{pmatrix} x_1 & x_2 & x_3 \end{pmatrix} \begin{pmatrix} b_{11} & 0 & 0 \\ 0 & b_{22} & b_{23} \\ 0 & b_{32} & b_{33} \end{pmatrix} \begin{pmatrix} x_1 \\ x_2 \\ x_3 \end{pmatrix}.$$

The matrix (b_{ij}) is again symmetric. We can now eliminate the mixed term $2b_{23}x_2x_3$ in the same way as we did when discussing the quadratic curves, by rotating the coordinate system around the x_1-axis. Then we can apply translations or rotations to eliminate as many linear terms as possible. After, if necessary, permuting the axes, this leads to the classification in Table 5.2.

Example 5.72 (Hessian). The Hessian, named after Hesse (1811–1874), is often used to study the extrema of functions in more than one variable in differential calculus. The method that is used has points in common with the classification of quadrics. Consider a function f in two variables defined on a subset D of $\mathbb{R}^2$. For simplicity of notation we work in a disk around $(0,0)$; we assume that this disk lies in D. We also assume that the function f is sufficiently differentiable. We write the first terms of the Taylor expansion of f around the point $(0,0)$. All partial derivatives are taken at $(0,0)$:

$$f(x,y) = f(0,0) + \frac{\partial f}{\partial x} x + \frac{\partial f}{\partial y} y$$
$$+ \frac{\partial^2 f}{\partial x^2} x^2 + 2\frac{\partial^2 f}{\partial x \partial y} xy + \frac{\partial^2 f}{\partial y^2} y^2 + \cdots. \qquad (5.24)$$

When studying the local extrema, we first look for the stationary points. These are the points with first partial derivatives equal to 0. We assume that $(0,0)$ is a stationary point, in which case the following is an approximation of f in a neighborhood of $(0,0)$:

$$f(x,y) \approx f(0,0) + \frac{\partial^2 f}{\partial x^2} x^2 + 2\frac{\partial^2 f}{\partial x \partial y} xy + \frac{\partial^2 f}{\partial y^2} y^2.$$

Locally, the graph of the function f is approximated by a quadric, whose form determines the type of stationary point; see Fig. 5.56. For brevity, we write

$$A = \frac{\partial^2 f}{\partial x^2}, \qquad B = \frac{\partial^2 f}{\partial x \partial y}, \qquad C = \frac{\partial^2 f}{\partial y^2}.$$

The approximation is then

$$f(x,y) \approx f(0,0) + Ax^2 + 2Bxy + Cy^2,$$

or in matrix notation,

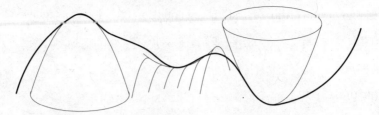

Fig. 5.56. Local approximation of a surface by quadratic equations

$$f(x,y) \approx f(0,0) + \begin{pmatrix} x & y \end{pmatrix} \begin{pmatrix} A & B \\ B & C \end{pmatrix} \begin{pmatrix} x \\ y \end{pmatrix}.$$

The *Hessian* of the function f at $(0,0)$ is

$$H = \begin{vmatrix} A & B \\ B & C \end{vmatrix} = AC - B^2 \, .$$

Theorem 5.73. *With the notation introduced above, if $(0,0)$ is a stationary point of f, we have:*

1. *If $H > 0$ and $A > 0$, f has a local minimum at $(0,0)$.*
2. *If $H > 0$ and $A < 0$, f has a local maximum at $(0,0)$.*
3. *If $H < 0$, f has neither a local maximum nor a local minimum at $(0,0)$.*

The Hessian determines the form of f in a neighborhood of $(0,0)$. In (1), f is close to an elliptic paraboloid. In (2) it is close to an upside-down elliptic paraboloid. In (3), f looks like a hyperbolic paraboloid; hence f can admit neither a maximum nor a minimum.

Proof. Let

$$z = f(0,0) + Ax^2 + 2Bxy + Cy^2 \, .$$

Let us first consider the case $H > 0$. Both A and C must be nonzero. By completing the squares we obtain

$$z = f(0,0) + A\left(x + \tfrac{B}{A}y\right)^2 + \tfrac{H}{A}y^2 \, . \tag{5.25}$$

If we also have $A > 0$, (5.25) is the equation of an elliptic paraboloid, and z has a minimum at $(0,0)$. If $A < 0$, (5.25) is again the equation of an elliptic paraboloid, and z has a maximum at $(0,0)$.

Now the case $H < 0$. If A and C are not both zero, we may, and do, assume that $A \neq 0$. We see that in (5.25), A and H/A have opposite signs; the formula is therefore the equation of a hyperbolic paraboloid. If A and C are both zero, $B \neq 0$. In this case the equation $z = f(0,0) + 2Bxy$ also represents a hyperbolic paraboloid.

Example 5.74 (Ruled surface). A *ruled surface* is a surface made up of straight lines. The cone, cylinder, one-sheeted hyperboloid, and hyperbolic paraboloid are examples of ruled surfaces. The straight lines making up the ruled surface are called *generatrices*. In general, a ruled surface is determined by three curves, the so-called *directrices*. Let us give an example.

We are given, in parametric form, the lines $l: \mathbf{x} = \lambda(1,1,1)$ and $m: \mathbf{x} = (0,1,0) + \mu(0,0,1)$ and the plane $z = 0$. What is the form of the ruled surface whose generatrices intersect the directrices l and m and are parallel to the directrix plane $z = 0$?

A "general point" on l is $\mathbf{p} = (\lambda, \lambda, \lambda)$, and a "general point" on m is $\mathbf{q} = (0, 1, \mu)$. The straight line through these points has parametric equation

$$\mathbf{x} = \nu\mathbf{p} + (1-\nu)\mathbf{q} = \nu(\lambda, \lambda, \lambda) + (1-\nu)(0, 1, \mu).$$

We obtain the following components for $\mathbf{x}$:

$$\begin{cases} x = \nu\lambda, \\ y = \nu\lambda + (1-\nu), \\ z = \nu\lambda + (1-\nu)\mu = \mu - \nu(\lambda - \mu). \end{cases} \quad (5.26)$$

When does this straight line lie on the ruled surface? If it does, it must be parallel to the directrix plane $z = 0$; this means that the z-coordinate of points on the line must be independent of ν. This is the case if and only if $\lambda = \mu$. We eliminate ν from (5.26) and express λ and μ as functions of x, y, and z:

$$\begin{cases} x = \nu\lambda, \\ y - x = 1 - \nu, \\ z - x = (1-\nu)\mu, \end{cases} \text{ and therefore } \begin{cases} \nu = 1 + x - y, \\ \lambda = x/(1 + x - y), \\ \mu = (z - x)/(y - x). \end{cases}$$

By setting $\lambda = \mu$, we obtain the equation of the ruled surface:

$$xz - yz + z - x = 0. \quad (5.27)$$

What type of surface is this? The matrix of the quadratic part is

$$\begin{pmatrix} 0 & 0 & \frac{1}{2} \\ 0 & 0 & -\frac{1}{2} \\ \frac{1}{2} & -\frac{1}{2} & 0 \end{pmatrix}.$$

We see that $(1, 1, 0)$ is an eigenvector with eigenvalue 0. Let us choose a new coordinate system whose axes are the spans of the eigenvectors, with standard basis vectors $(\sqrt{2}/2, \sqrt{2}/2, 0)$, $(-\sqrt{2}/2, \sqrt{2}/2, 0)$, and $(0, 0, 1)$. The coordinate transformation giving the old coordinates (x, y, z) in terms of the new coordinates (x', y', z') is

$$\begin{pmatrix} x \\ y \\ z \end{pmatrix} = \begin{pmatrix} \frac{1}{2}\sqrt{2} & -\frac{1}{2}\sqrt{2} & 0 \\ \frac{1}{2}\sqrt{2} & \frac{1}{2}\sqrt{2} & 0 \\ 0 & 0 & 1 \end{pmatrix} (x'\ y'\ z').$$

The eigenvector with eigenvalue 0 is in the first column. If we substitute this in (5.27), we obtain

$$-\sqrt{2}yz - \tfrac{1}{2}\sqrt{2}x - \tfrac{1}{2}\sqrt{2}y + z = 0,$$

where we leave out the primes for notational convenience. If we set $x = 0$, the discriminant is negative; hence the intersection with the plane $x = 0$ is a parabola. It follows from the classification that this ruled surface is a hyperbolic paraboloid.

Exercises

5.32. By rotating the line $x = 0$, $z = (c/a)\,y$, where $ac \neq 0$, around the z-axis, we obtain the elliptic cone

$$\frac{x^2}{a^2} + \frac{y^2}{a^2} - \frac{z^2}{c^2} = 0.$$

5.33. Let $0 < b < a$. The surface of revolution obtained by rotating the circle

$$x = 0, \quad (y - a)^2 + z^2 = b^2,$$

around the z-axis is called a *torus*. The equation of the torus is

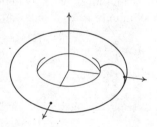

$$(x^2 + y^2 + z^2 + a^2 - b^2)^2 = 4a^2(x^2 + y^2).$$

5.34. Consider the paraboloid $x^2 + y^2 = 2z$.

(a) The intersection curve of the paraboloid with the plane $y = k$ is a convex parabola in the plane $y = k$.
(b) The vertices of the parabolas in (a) lie on a convex parabola.
(c) The intersection curves of the hyperbolic paraboloid $-x^2 + y^2 = 2z$ with planes $y = k$ are concave parabolas whose vertices lie on the convex parabola in (b).

5.35. Determine the type and position of the quadric

$$x^2 + 3y^2 + 3z^2 - 2yz - 2x - 2y + 6z + 3 = 0.$$

Hint/Answer: If we use local coordinates at $(1, 0, -1)$ with axes along the vectors $(1, 0, 0)$, $(0, \sqrt{2}/2, \sqrt{2}/2)$, and $(0, -\sqrt{2}/2, \sqrt{2}/2)$, the quadric has equation $x^2 + 2y^2 + 4z^2 = 1$.

5.36. Determine the equation of the ruled surface with directrices

$$l: \begin{cases} x = 0, \\ y + z = 1, \end{cases} \quad m: \begin{cases} y = 0, \\ x - z = 0, \end{cases} \quad n: \begin{cases} x = 1, \\ y - z = 0. \end{cases}$$

Answer: the one-sheeted hyperboloid $(x - 1/2)^2 + (y - 1/2)^2 - (z - 1/2)^2 = 1/4$.

5.37. Determine the equation of the ruled surface with directrices

$$l: \begin{cases} x + z = 0, \\ 1 - y = 0, \end{cases} \quad m: \begin{cases} 1 + y = 0, \\ x - z = 0, \end{cases} \quad C: \begin{cases} z = 0, \\ x^2 + y^2 = 1. \end{cases}$$

Answer: the one-sheeted hyperboloid $x^2 + y^2 - z^2 = 1$, the plane $x + z = 0$, and the plane $x - z = 0$.

A

Basic Assumptions

In this chapter, we repeat the basic assumptions for plane and solid geometry, plus a number of definitions.

Definition. A *metric space* is a set M with a *metric* d. We call the elements of M the *points* of the metric space. To every pair of points X and Y, the metric d associates a real number, the *distance* from X to Y. This distance satisfies the following properties: for every X, Y, Z in M,

1. $d(X,Y) \geq 0$;
2. $d(X,Y) = 0$ if and only if $X = Y$;
3. $d(X,Y) = d(Y,X)$;
4. $d(X,Z) \leq d(X,Y) + d(Y,Z)$.

Basic Assumption 1 *The Euclidean plane V and 3-space W are both metric spaces containing more than one point.*

Definition. For any pair of points A and B in a metric space M with metric d, the *line segment* $[AB]$ is given by

$$[AB] = \{\, X \in M : d(A,X) + d(X,B) = d(A,B) \,\} .$$

If A and B are distinct, the *line* AB is given by

$$AB = \{\, X : X \in [AB] \text{ or } B \in [AX] \text{ or } A \in [XB] \,\} .$$

Basic Assumption 2 *Every line l in the plane or in 3-space admits an isometric surjection φ from l to $\mathbb{R}$.*

Basic Assumption 3 *The Euclidean plane V and 3-space W each contain three noncollinear points.*

A Basic Assumptions

Definition. For lines in the plane V we define the notions parallel and intersecting. We say that lines m and n are *parallel* if $m = n$ or $m \cap n = \emptyset$. If m and n are not parallel, we say that m and n are *intersecting* lines. We call the point P the *intersection point* of the lines l and m if P lies on both l and m.

Basic Assumption 4 *For every point P and every line m of the plane V, there exists a unique line passing through P and parallel to m.*

Definition. Let l and m be intersecting lines of the plane V with common point C. We say that *l is perpendicular to m* if for every point A of l and every point B of m we have

$$d(C,A)^2 + d(C,B)^2 = d(A,B)^2 \ .$$

Basic Assumption 5 *For every point P and every line m of the plane V, there exists a line l through P that is perpendicular to m.*

In the plane V, two intersecting lines l and m define four *angles*. If P is the intersection point of the lines, any two points A and B other than P on l and m, respectively, determine exactly one of these angles. Every angle has an *interior*.

Definition. We say that an angle P is *congruent to* an angle Q if there is an isometric surjection from angle P onto angle Q.

Basic Assumption 6 *The intersection of any angle P in the plane V and the circle (P, r) with center P and radius r is an arc. If $r = 1$, we denote the length of this arc by $\angle P$; if $r = R$, the length of the arc is equal to $R \times \angle P$. Moreover, we have the following properties:*

1. $0 < \angle P < \pi$ for every angle P.
2. If angles P and Q are congruent, then $\angle P = \angle Q$.
3. For every real number c satisfying $0 < c < \angle APB$, there exists a point Q in the interior of angle APB such that $c = \angle APQ$.
4. If a point Q lies in the interior of angle APB, then

$$\angle APB = \angle APQ + \angle QPB \ .$$

Definition. Let A, B, and C be three noncollinear points in W. The *plane* α through A, B, and C is the subset of W with the following properties:

1. A, B, and C lie in α.
2. If D and E are distinct points of α, then the line DE lies completely inside α.
3. A point X lies in α if and only if there is a series of points $X_1, X_2, \ldots, X_n = X$ such that for $1 \leq i \leq n$ the point X_i is either equal to A, B, or C, or lies on the line $X_k X_l$ for some $1 \leq k < i$, $1 \leq l < i$.

Basic Assumption 7 *For every plane α in W there is an isometric surjection φ from α onto the Euclidean plane V.*

Definition. Two noncoplanar lines in W are called *skew*. Two coplanar lines l and m are called *parallel* if either $l = m$ or $l \cap m = \emptyset$; otherwise, they are called *intersecting lines*.

Definition. Two planes α and β in W are called *parallel* if either $\alpha = \beta$ or α and β have no point in common. If the planes α and β are not parallel, then we say that they *intersect each other*.

Basic Assumption 8 *Two intersecting planes α and β meet in a straight line.*

Definition. Let l and m be two lines, and let C' be an arbitrary point in 3-space. Draw the lines l' and m' passing through C' and parallel to l and m, respectively. We say that l and m are *perpendicular to each other* if l' and m' are perpendicular to each other.

Definition. Given a line l and a plane α in W, we say that l is *perpendicular* to α if it is perpendicular to every line in α.

Basic Assumption 9 *For every point P and every plane α in W, there is a line l through P that is perpendicular to α.*

References

1. BACHMANN, F., *Aufbau der Geometrie aus dem Spiegelungsbegriff*, Springer, Berlin, 1958; revised 2nd ed., 1973
2. BARNSLEY, M.F., *Fractals Everywhere*, Academic Press, Boston, 1988; revised 2nd ed., 1993
3. BERGER, M., *Géométrie*, vols. 1–5, CEDIC et Fernand Nathan, Paris, 1977; English translation *Geometry I & II*, Springer, Berlin, 1987
4. BOTTEMA, O., *Hoofdstukken uit de Elementaire Meetkunde*, Servire, Den Haag 1944; revised 2nd ed., Epsilon Uitgaven, Utrecht, 1997; to be published in English by Springer, New York
5. BRAUNER, H., *Lehrbuch der konstruktiven Geometrie*, Springer, Wien, 1986
6. BRIANCHON, C.J. AND J.V. PONCELET, *Géométrie des courves. Recherches sur la détermination d'une hyperbole équilatère, au moyen de quatres conditions données*, Annales de Gergonne, 11, 1820–1821, 205–220
7. BRIESKORN, E. AND H. KNÖRRER, *Plane Algebraic Curves*, Birkhäuser, Basel, 1986
8. CASSELS, J.W.S., *An Introduction to the Geometry of Numbers*, Springer, Berlin, 1997 (first published 1971)
9. COXETER, H.S.M., *Introduction to Geometry*, 2nd ed., Wiley, New York, 1969
10. COXETER, H.S.M., *Regular Polytopes*, 2nd edition, Macmillan, New York, 1963; revised edition, Dover, New York, 1973
11. COXETER, H.S.M. AND S.L. GREITZER, *Geometry Revisited*, New Mathematical Library 19, Mathematical Association of America, Washington, D.C., 1967
12. COXETER, H.S.M., M. EMMER, R. PENROSE, AND M.L. TEUBER, ed., *M.C. Escher, Arts and Science*, Elsevier, North-Holland, Amsterdam, 1986
13. CROMWELL, P.R, *Polyhedra*, Cambridge University Press, Cambridge, 1997
14. DESCARTES, *The Geometry of René Descartes, with a Facsimile of the First Edition*, translated by D. E. Smith and M. L. Latham, Open Court Publishing, 1925; reprinted Dover, New York, 1954
15. DESCHARNES, R. AND G. NÉRET, *Salvador Dalí*, Edita, Lausanne, 1993
16. DEVLIN, K., *Mathematics: The Science of Patterns*, Scientific Americain Library, HPHLP, New York, 1994
17. ERNST, B., *Avonturen met onmogelijke figuren*, Aramith, Amsterdam, 1985
18. EUCLID, *The Thirteen Books of the Elements*, Vols. I, II, III, translated with introduction and commentary by T. L. Heath, Cambridge University Press, Cambridge, 1908; revised 2nd ed., Dover, New York, 1956

19. FEUERBACH, K.W. *Eigenschaften einiger merkwürdigen Punkte des geradlinigen Dreiecks*, Haarlem; 2nd ed., P. Visser Azn., 1908
20. FIELD, J.V., *The Invention of Infinity, Mathematics and the Art in the Renaissance*, Oxford University Press, Oxford (USA), 1997
21. GOODSTEIN, D.L. AND J.R. GOODSTEIN, *Feynman's Lost Lecture: The Motion of the Planets Around the Sun*, Vintage, London, 1996
22. GREENBERG, M.J., *Euclidean and Non-Euclidean Geometries*, 3rd ed., Freeman, New York, 1993
23. GRÜNBAUM, B. AND G.C. SHEPHARD, *Tilings and Patterns*, Freeman, New York, 1987
24. HAHN, LIANG-SHIN, *Complex Numbers and Geometry*, Mathematical Association of America, Washington, D.C., 1994
25. HARDY, G.H. AND E.M. WRIGHT, *An Introduction to the Theory of Numbers*, Oxford University Press, Oxford; 5th ed., 1980
26. HARTSHORNE, ROBIN, *Geometry: Euclid and Beyond*, Springer, New York, 2000
27. HEILBRON, J.L., *Geometry Civilized; History, Culture, and Technique*, Clarendon Press, Oxford, 1998
28. HOFMANN, K.H. AND R. WILLE, ed., *Symmetry of Discrete Mathematical Structures and Their Symmetry Groups, A Collection of Essays*, Research and Exposition in Mathematics 15, Heldermann, Berlin, 1991
29. HOFSTEDE, G., H.J. MEEWIS, JAC. STOLK, AND C. KROS, *Machineonderdelen*, 18th edition, Morks, Dordrecht, 1970
30. HOLDEN, A, *Shapes, Space and Symmetry*, Columbia University Press, New York, 1971; reprinted Dover, New York, 1991
31. HONSBERGER, R., *Episodes in Nineteenth and Twentieth Century Euclidean Geometry*, New Mathematical Library 37, Mathematical Association of America, Washington, D.C., 1995
32. HUGHES, B.B. *Fibonacci's De Practica Geometrie*, Springer Verlag, New York, 2007
33. HUNTLEY, H.E., *The Divine Proportion, A Study in Mathematical Beauty*, Dover, New York, 1970
34. HUTCHINSON, J.E., *Fractals and Self Similarity*, Indiana University Mathematics Journal 30, 1981, 713–747
35. JACOBS, H.R., *Geometry*, Freeman, San Francisco, 1974
36. KINDT, M., *Lessen in projectieve meetkunde*, 2nd ed., Epsilon, Utrecht, 1996
37. KITTEL, C., *Introduction to Solid State Physics*, 5th ed., Wiley, New York, 1976
38. KOECHNER, M., *Lineare Algebra und analytische Geometrie*, 3rd ed., Springer, Berlin, 1992
39. KOECHNER, M. AND A. KRIEG, *Ebene Geometrie*, Springer, Berlin, 1993
40. LAUWERIER, H., *Symmetrie; regelmatige structuren in de kunst*, Aramith, Amsterdam, 1988
41. LAWRENCE, J.D., *A Catalog of Special Plane Curves*, Dover, New York, 1972
42. MARTIN, G.M., *Transformation Geometry; An Introduction to Symmetry*, Springer, New York, 1982
43. MARTIN, G.M., *Geometric Constructions*, Springer, New York, 1997
44. MEYER, W. FR. AND H. MOHRMANN, ed., *Encyklopädie der Mathematischen Wissenschaften mit Einschluss Ihrer Anwendungen*, Band III, Three parts, Leipzig, Teubner, 1902–1927
45. MEYERE, J. DE AND H. WEIJMA, *Anamorfosen: Kunst met een omweg*, Aramith, Bloemendaal, 1989

46. MILLER, W., *Symmetry Groups and Their Applications*, Academic Press, New York, 1972
47. MIYAZAKI, K., *An Adventure in Multidimensional Space; The Art and Geometry of Polygons, Polyhedra and Polytopes*, Wiley, New York, 1986
48. MOISE, E.E., *Elementary Geometry from an Advanced Standpoint*, Addison-Wesley, Reading (Mass), 1974
49. MOLEN, J.R. TER EN E. UITZINGER, red., *De Verboden Stad, Catalogus*, Museum Boymans–van Beuningen, Rotterdam, 1990
50. MOLENBROEK, P., *Leerboek der vlakke meetkunde*, Noordhoff, Groningen, 1943
51. MOLENBROEK, P., *Leerboek der stereometrie*, 12th ed., Noordhoff, Groningen, 1943
52. OGILVY, C.S., *Excursions in Geometry*, Oxford University Press, New York, 1969; reprinted Dover, New York, 1990
53. PEDOE, D. *Geometry, A Comprehensive Course*, Dover, New York, 1988
54. PEITGEN, H.-O., H. JÜRGENS, AND D. SAUPE, *Chaos and Fractals*, Springer, New York, 1992
55. PÓLYA, G., *Über die Analogie der Kristalsymmetrie in der Ebene*, Zeitschrift für Krystallographie und Mineralogie 60, 1924, 278–282
56. PRENOWITZ, W. AND M. JORDAN, *Basic Concepts of Geometry*, 2nd ed., Ardsley House Publishers, New York, 1989
57. REHBOCK, F., *Geometrische Perspective*, 2nd ed., Springer, Berlin, 1980
58. REHBOCK, F., *Darstellende Geometrie*, 2nd ed., Springer, Berlin, 1964.
59. REUTERSVÄRD, O., *Omöjliga Figurer*, Doxa, Bodafors, 1982
60. REUTERSVÄRD, O., *Omöjliga Figurer i färg*, Doxa, Lund, 1985
61. RUTTER, J.W., *Geometry of Curves*, Chapman & Hall/CRC, Boca Raton, 2000
62. SCHATTSCHNEIDER, D., *Visions of Symmetry; Notebooks, Periodic Drawings, and Related Work of M. C. Escher*, Freeman, New York, 1990
63. SCHOLZ, E., *Symmetrie, Gruppe, Dualität*, Science Networks, Historical Studies, Vol. 1, Birkhäuser, Basel, 1989
64. SCHROEDER, M., *Fractals, Chaos, Power Laws*, Freeman, New York, 1991
65. SENECHAL, M., *Quasicrystals and Geometry*, Cambridge University Press, Cambridge, 1995
66. SHIKIN, E.V., *Handbook and Atlas of Curves*, CRC Press, New York, 1995
67. STEIN, S.K. AND S. SZABO, *Algebra and Tiling; Homomorphisms in the Service of Geometry*, The Carus Mathematical Monographs 25, Mathematical Association of America, Washington, D.C., 1994
68. STEWART, J., *Cementing Relationships*, Mathematical Recreations, Scientific American, May 1998, 78–79
69. STILLWELL, J., *Geometry of Surfaces*, Springer, Berlin, 1992
70. TERQUEM, O., *Considérations sur le triangle rectiligne*, Nouvelles Annales de Mathématiques I, 1842, 196–200
71. VEEN, H.J. VAN, *Leerboek der beschrijvende meetkunde, Deel I, Projectiemethoden*, Noordhoff, Groningen, 1925
72. VEEN, H.J. VAN, *Leerboek der beschrijvende meetkunde, Deel II, Oppervlakken en Ruimtekrommen*, Noordhoff, Groningen, 1929
73. VEGT, A.K. VAN DER, *Regelmaat in de ruimte*, Delftse Uitgevers Maatschappij, Delft, 1991
74. VERWEIJ, A. AND M. KINDT, *Perspectief, hoe moet je dat zien?* Epsilon Uitgaven, Utrecht, 1999
75. VITRUVIUS AND H.M. MORGAN, *Ten Books on Architecture*; English translation, Dover, New York, 1960

Index

3-space, 256
C_n, 63
D_1, 98
D_n, 98, 303
G^+, 99
S_n, 303
V, 16
V_4, 63
$[AB]$, 16, 333
$\mathbb{R}$, 61
$\mathbb{R}^+$, 62
$\mathbb{R}^2$, 22
$\mathbb{R}^3$, 255
$\mathbb{Z}$, 66
$\angle$, 50
$\odot(M,r)$, 271
$/\!/$, 19, 258, 259
ex C, 146
$\mathcal{I}(D)^+$, 99
$\mathcal{I}(M)$, 63
in C, 146
$\langle,\rangle$, 31, 271
ex $\mathbb{S}$, 272
in $\mathbb{S}$, 272
$\odot(M,r)$, 146
$\perp$, 19, 261, 262
$\overset{\frown}{CD}$, 149
n, 96
3-4-5 method, 4

Aachen Cathedral, 100
acute angle, 58
addition
  modulo, 70
  of vectors, 266
admissible quadrilateral, 158
Alberti, 283
Alhambra, 95
altitude, 30, 275
angle, 46
  acute, 58
  central, 149
  complementary, 51
  dihedral, 264
  exterior, 57
  inscribed, 150
  interior, 57
  interior of, 47
  leg of, 48
  obtuse, 58
  right, 48
  straight, 48
  supplementary, 51
  top, 53
angle bisector, 59
  exterior, 59
  interior, 59
angle measure, 50, 149
  additivity, 50
  of an arc, 149
angles
  alternate exterior, 56
  alternate interior, 56
  congruent, 47
  corresponding, 56
  equal, 51

opposite, 158
supplementary, 46
vertical, 47
angular coordinate, 167
antiparallel, 84
apex, 298
Apollonius, 161, 182, 323
arc, 149
angle measure, 149
intercepted, 50, 149, 150
length, 244
Archimedean solids, 300
Archimedes, 188, 227
Archimedes triangle, 188
area, 91, 190
arithmetic mean, 163
associative, 14
astroid, 249
asymptote, 217
of a hyperbola, 201
asymptotic direction, 217
axiom, 15
axis
of a hyperbola, 194
of a parabola, 115, 187
radical, 153
reflection, 38, 72
rotation, 292
x_1-, 23, 266
x_2-, 23, 266
x_3-, 266

base plane, 298
base, of an isosceles triangle, 53
basic assumption
for plane geometry, 15
for solid geometry, 256, 257, 262
Bernoulli, 244
bijective, 11
bipyramid, 303
brachistochrone, 244
branch, of a hyperbola, 194
Brianchon, 80
bride's chair, 4
broken dimension, 91
bubble bath, 144
buckyball, 302

cabinet projection, 280

Cantor, 89
discontinuum, 90
set, 89
cardioid, 232
Cartesian coordinates, 166
Cauchy–Bunyakovskiĭ–Schwarz
inequality, 33, 34, 36, 147, 271
caustic curve, 251
cavalier projection, 279
center, 146, 271, 283
of a figure, 201
central angle, 149
central dilation, 77
center of a, 77
scale factor, 77
central projection, 277
centroid, 27, 269
Ceva, 157
chord, 147
circle, 48, 146
center of a, 146
chord, 147
circumscribed, 154
diameter, 147
escribed, 158
exterior, 146
inscribed, 156
interior, 146
nine-point, 80
of Apollonius, 161
point of contact, 148
radius, 146
squaring the, 169
tangent, 148
circumcircle, 154
cissoid, 173
coefficient matrix, 210
commutativity, 297
complementary, 51
complete 5-graph, 313
components, 23, 266
composition, 12
of maps, 14
of reflections, 13
of translations, 12
conchoid, 167
concur, lines, 19
concurrent lines, 19
cone, 317

elliptic, 317
generatrix, 323
congruence criterion
   AAS, 57
   ASA, 58
   SAS, 48, 51
   SSA, 58
   SSS, 43
congruent, 11, 43, 257
   angles, 47, 334
conic, 114, 323
conic section, 185
conjugate, 214
construction, 169
convex, 272, 298
convex subset, 109
coordinate
   3-space, 267
   plane, 24
   system, 25, 267
   transformation, 170
coordinates
   Cartesian, 166
   homogeneous, 204
   polar, 166
corkscrew rule, 287
cosine, 163
crystallography, 97, 127
cube, 299
cuboctahedron, 301
cuboid, 265
curve
   brachistochrone, 244
   isochrone, 248
   of Lissajous, 225
   tautochrone, 248
cusp, 168
cyclic group, 63, 97
cyclic quadrilateral, 158
cycloid, 228
   curtate, 232
   prolate, 232
cylinder, 317
   elliptic, 317
   hyperbolic, 317
   parabolic, 317

Dandelin, 323
degenerate triangle, 8

deltoid, 236
Descartes, 166, 250
determinant, 35, 208
   notation, 207, 274
diagonal, of a quadrilateral, 30
diameter, 147
difference formula
   cosine, 171
   sine, 171
difference of vectors, 23
dihedral angle, 264
dihedral group, 303
dilation, 76, 296
   central, 77
dimetric projection, 281
Diocles, 173
Diophantine equation, 123
Diophantus, 123
direct, 289
direction, 203, 277
directrix, 114, 185
   of a ruled surface, 329
   of an ellipse, 191
directrix circle, 115
directrix plane of a ruled surface, 329
Dirichlet, 123
discrete, 312
discrete metric, 9, 10
discriminant of a curve, 197
distance from a point to a line, 34
distance-preserving, 11
dodecahedron, 299
double point, 212
Droste effect, 87
duplication of a cube, 169
Dürer, 283

eccentricity, 185
edge, 269, 298
Einstein, 3
Elements, Euclid's, 15
ellipse, 115, 186, 323
   center of an, 191
   construction, 193
   directrix circle of, 115
   equation, 191
   focus, 115
   major axis, 191
   minor axis, 191

tangent, 192
vertex, 191
vertex equation, 193
ellipsoid, 316
elliptic paraboloid, 316
enantiomorphic pair, 312
envelope, 237
  computation of, 238
  convex, 111
epicycloid, 230
equivalent ordered triples, 204
Escher, 95, 102
Euclid, 15, 169
Euclidean metric, 9
Euclidean plane, 16, 256, 334
Euler, 79, 123, 299
excircle, 158
exterior angle, 57

face, 265, 269, 298
face diagonals, 265
Fedorov, 143
Fermat, 123
Feuerbach, 80
Feuerbach's theorem, 80, 181
finitely additive, 190
focal parameter, 217
focus, 185
  of a hyperbola, 115
  of a parabola, 114
  of an ellipse, 115, 191
folium of Descartes, 250
foot of a perpendicular, 20, 262
formula
  Euler, 299
  Heron's, 172
fractal, 86
fractal dimension, 93
frieze group, 104
frieze patterns, 102–108
  types of, 102
  essentially different, 102
  seven types of, 107
Fuller, 302
fullerenes, 302
fundamental parallelogram, 121

Gauss, 123
generatrix

of a ruled surface, 329
of cone, 323
geometric mean, 163
Gergonne, 157
Gergonne point, 157
glide reflection, 72, 293
golden ratio, 86
golden section, 86
group, 61
  C_n, 63
  D_1, 98
  D_n, 98
  S_n, 303
  V_4, 63
  $\mathbb{R}$, 61
  $\mathbb{R}^+$, 62
  $\mathbb{Z}$, 66
  $\mathbb{R}^2$, 62
  $\mathcal{I}(M)$, 63
  additive, abelian, 61
  additive, negative, 61
  additive, zero element, 61
  associativity, 61
  cyclic, 63, 97
  dihedral, 303
  dodecahedral, 307
  identity element, 61
  inverse, 61
  isotropy, 309
  octahedral, 306
  rotation, 298
  symmetry, 40
  transformation, 63
group isomorphism, 62

half-line, 109
half-plane
  closed, 46
  open, 45
harmonic oscillator, 246
Hausdorff
  dimension, 93
  measure, 92
Heron, 172
Hesse, 327
Hessian, 327, 328
homogeneous coordinates, 204
honeycomb, 311
hull, convex, 111

Hutchinson, 88
Huygens, 244
  -museum Hofwijck, 244
hyperbola, 115, 186, 323
  asymptote, 201
  axis, 194
  branch, 194
  center of a, 194
  construction, 195
  directrix circle of, 115
  equation, 194
  focus, 115
  tangent, 195
  vertex, 194
  vertex equation, 196
hyperbolic paraboloid, 321
hyperboloid
  one-sheeted circular, 315
  one-sheeted elliptic, 315
  two-sheeted, 315
hypocycloid, 230
hypotenuse, 2

icosahedron, 299
identity map, 11
image plane, 277
impossible figures, 284
improper rotation, 293
incident, point and line, 16
incircle, 156
indirect, 290
inequality
  C-B-S, 33, 34, 36, 147, 271
  triangle, 33
infinity
  line at, 206
  point at, 204
injective, 11
inner product, 31, 271
inscribed angle, 150
intercepted arc, 150
interior
  of a triangle, 57
  of an angle, 47
interior angle, 57
  remote, 57
intersecting
  lines, 19, 258, 334, 335
  planes, 259, 335

intersection, 146
  line, 259
  point, 18, 206, 259, 334
invariance under isometries, area, 190
invariant set, 88
inverse for inversion, 175
inverse map, 64
inversion, 175
  preserves angles, 178
  reverses orientation, 178
isochrone curve, 248
isometric
  projection, 281
  surjection, 17
isometry, 11
  direct, 73
  indirect, 73
  opposite, 73
isomorphic, 62
isosceles
  trapezoid, 60
  triangle, 53
isotropy group, 309
IFS, 88
iteration, 89

join, 18

Kepler, 299
knight's distance, 10
Koch curve, 87
Koch, von, 87

Lagrange, 123
lattice, 117, 126
  discrete, 118
  hexagonal, 122
  rectangular, 122
  rhomboid, 122
  spatial, 312
  square, 122
law of cosines, 166
law of sines, 165
leg
  isosceles triangle, 53
  of an angle, 48
  triangle, 2
length, 91
limaçon of Pascal, 233

Lindemann, 169
line, 16, 250
  equation of, 34
  Euler, 81
  in a projective plane, 205
  parametric equation, 25, 29
line at infinity, 206
line segment, 16, 256, 333
lines
  intersecting, 258, 335
  parallel, 258, 335
  skew, 258, 335
local coordinates, 25, 256, 267
logarithmic spiral, 228

Maclaurin, 218
Mandelbrot, 86
matrix
  2×2, 207
  3×3, 208
mean
  arithmetic, 163
  geometric, 163
median, 28, 269
Menger sponge, 88
metric, 8, 333
  discrete, 9, 10
  Euclidean, 9
metric space, 8, 333
mid-parallel, 30
midpoint, 22
military projection, 279
Miquel, 158
Miquel's theorem, 158
mirror, 285
mirror symmetry, 40
monotonic, 190
Monte, Castel del, 100
motion, 290, 298
Multatuli, 7

Nagel point, 162
Nicomedes, 167
nine-point circle, 80
node, 168
norm, 23, 266
normal, 252
normal plane, 264
normal subgroup, 297

obtuse angle, 58
octahedron, 299
opposite, 269
  angles, 158
  edges, 269
orbit of a point, 308
order
  infinite, 133
  of a rotation, 97
  of an element, 133
orientation
  preserving, 54, 73
  reversing, 52, 73, 288
orientation of a tripod
  negative, 288
  positive, 288
orientation, angle
  negative, 51
  positive, 51
origin, 23, 266
orthic triangle, 84
orthocenter, 30, 32, 80, 276
orthocentric system, 81
orthogonal circles, 7
outer product, 273

parabola, 114, 186, 323
  axis, 115, 187
  construction, 187
  equation, 187
  focus, 114
  tangent, 114, 117
  vertex, 117, 187
  vertex equation, 187
parallel, 19, 334, 335
  line and plane, 259
  lines, 258
  planes, 259, 265
  postulate, 206
  projection, 277
parallelepiped, 265
parallelogram, 21
  spanned, 28
parallelogram law, 33
parametric equation
  of a curve, 225
  of a line, 25, 28, 29, 268, 269
  of a plane, 270
Pascal, 233

Pasch, 46
pedal curve, 217
pendulum, mathematical, 246
Pentagon, 100
pericycloid, 234
permutation, 303
  even, 304
  odd, 304
perpendicular
  bisecting plane, 266
  bisector, 22
  line and plane, 262, 335
  lines, 19, 20, 261, 334, 335
  on a line, 20
  on a plane, 262
  position, 265
  skew lines, 261
perspective, 277, 284
plane, 16, 256, 334
Plato, 299
Platonic solids, 299
point, 333
  at infinity, 204, 205
  double, 212
  fixed, 38
  Gergonne, 157
  in a projective plane, 204
  intersection, 146, 334
  isolated, 168
  of a metric space, 8
  Torricelli, 156
  vanishing, 283
point of contact, 272
point-symmetric, 55
polar, 214
  angle, 167
  coordinates, 166
pole, 309
  k-tuple, 309
polyhedron, 298
  dual, 305
  regular, 298
  semiregular, 300
Poncelet, 80
position vector, 166
power, 152
primitive notion, 15
product
  inner, 31

  of maps, 64
  outer, 273
  scalar, 266
  vector, 273
projection, 36, 267
  cavalier, 279
  center of a, 283
  central, 283
  dimetric, 281
  isometric, 281
  military, 279
  oblique, 277
  orthogonal, 277
  parallel, 277
  trimetric, 281
projection line, 277
projective
  geometry, 220
  plane, 204
  transformation, 220
Ptolemy, 179, 243
pyramid, 298
Pythagoras, 2, 4
Pythagorean
  theorem, 2, 265
  triple, 4, 123
    primitive, 4

quadratic forms, 123
quadric, 314–330
quadrilateral, 21
  admissible, 158
  cyclic, 158
  of chords, 158
  tangential, 162
quadrilateral inequality, 10

$\mathbb{R}$, 8
radial coordinate, 167
radian, 50
radical axis, 153
radius, 146, 271
ratio, 204, 207
  golden, 86
real line, 8
rectangle, 21
reference point, 205, 220
reflection, 38
  in a line, 38

in a plane, 285
in a point, 12, 54, 292
reflection axis, 38, 72
reflexivity, 19
Reutersvärd, 284
rhombus, 36
right angle, 48
rotation, 67, 68
  angle, 68
  center of a, 68
  improper, 293
rotation group, 298
ruled surface, 329

saddle, 321
scalar
  multiplication, 266
scalar multiple, 23
scalar product, 266
scale factor
  of a central dilation, 77
  of a transformation, 82
scaling, 92
screw, 295
segment of a parabola, 189
self-similarity, 86
side, 21
sideline, 21
Sierpiński
  carpet, 88
  triangle, 86
similar, 82, 297
similarity, 82, 297
Simson, 159
Simson line, 159
sine, 163
skew lines, 258, 335
slope, 36
soccer ball, 302
solid geometry, 255
space diagonals, 265
span of a vector, 28, 269
sphere, 271
  center of a, 271
  exterior, 272
  interior, 272
  radius of a, 271
  tangent line, 272
  tangent plane, 272

spiral, 227
  Archimedean, 227
  logarithmic, 228
squaring the circle, 169
standard basis vector, 51
standard basis vectors, 274
star polygon, 97
Stewart, 172
straight
  angle, 48
  line, 16
strophoid, 184
subgroup, 66
sum
  of angles, 50
  of vectors, 23, 266
sum formula
  cosine, 171
  sine, 171
supplementary, 51
surface
  of revolution, 315
  ruled, 329
surjective, 11
symmetric, 8
symmetry, 19, 96
  mirror, 40
  point, 55
  rotational, 96
symmetry group, 40, 96, 298
  of periodic tiling, 142

tangent, 117, 164
  of a circle, 148
  of a hyperbola, 195
  of an ellipse, 192
  plane, 272
  vector, 249
tangent to
  a curve, 235
  a parabola, 114
tangential quadrilateral, 162
tautochrone curve, 248
technical drawings, 281
Terquem, 80
tetrahedron, 268, 299
Thales, 155
theorem, 15
  Apollonius, 203

Archimedes, 189
Ceva, 157
Pasch, 46
Pohlke, 281
Ptolemy, 179
Pythagorean, 2
Thales, 155
Thiessen polygon, 110
three-planes theorem, 259, 260
tiling, 125, 311, 314
    periodic, 125, 311
        of same type, 125, 141
        symmetry group of, 142
top, 53
top angle, 53
Torricelli, 156
Torricelli point, 156
torus, 330
transformation group, 63
transitivity, 19
translation, 11, 64, 290
    generating, 102, 117
translation vector, 65, 290
transposition, 304
trapezoid, 60
triangle inequality, 8, 33, 271
trimetric projection, 281
tripod, 288
trisection of an angle, 169
trisectrix, 218
trivial solution, 210

unit
    point, 220
    sphere, 309
utility graph, 313

vanishing point, 283
vector, 266
    addition, 23
    equality, 266
    substraction, 23
    translation, 65
vector product, 273
vector space, 22, 62, 266
vectors,equal, 23
vertex, 265
    of a hyperbola, 194
    of a parabola, 187
    of a polyhedron, 298
    of a quadrilateral, 21
    of a tetrahedron, 269
    of a triangle, 2
    of a Voronoi cell, 110
    of an ellipse, 191
vertex equation
    of a hyperbola, 196
    of a parabola, 187
    of an ellipse, 193
Vinci, Leonardo da, 100
Vitruvius, 4
vlak, 256
volume, 91
Voronoi, 123
    cell, 109
Voronoi cell, 119
Voronoi diagram, 109–116
    ellipse as, 115
    generalized, 113
    hyperbola as, 115
    parabola as, 114
    reconstruction, 112
    site, 110
Voronoi foam model, 144

Wallace, 159
wallpaper groups, 129
Wantzel, 169

x_1-axis, 23, 266
x_2-axis, 23, 266
x_3-axis, 266

zero vector, 23, 267

Printed in the United States